Mathematics

Phase 7

NOTION PRESS

NOTION PRESS

India. Singapore. Malaysia.

ISBN xxx-x-xxxxx-xx-x

Dedicated to our Beloved Prime Minister Shree Narendra Modi ji, whose life gives the Inspiration to every Individual

*If a Man **Decides** to achieve something in his life nothing is Impossible*

CHAPTERS

1. VECTORS

1. INTRODUCTION TO VECTOR ALGEBRA

1.1 Scalars and Vectors

Scalar: A scalar is a quantity that has only magnitude but no direction. Scalar quantity is expressed as a single number, followed by appropriate unit, e.g. length, area, mass, etc. In linear algebra, real numbers are called scalars.

Vector: A vector is a quantity that has both magnitude and direction, e.g. displacement, velocity, etc.

1.2 Representation of Vectors

(a) A vector is represented diagrammatically by a directed line segment or an arrow. A directed line segment has both magnitude (length) and direction. The length is denoted by $|V|$.

(b) If P and Q are the given two points, then the vector from P to Q is denoted by $\overrightarrow{PQ}$, where P is called the tail and Q is called the nose of the vector.

1.3 Vector Components

In a two-dimensional coordinate system, any vector can be resolved into x-component and y-component

$$\hat{v} = \langle v_x, v_y \rangle$$

Let us consider the figure shown (adjacent) here. In this figure, the components can be quickly read. The vector in the component form is $\hat{v} = \langle 4,5 \rangle$.

The relation between magnitude of the vector and the components of the vector can be calculated by using trigonometric ratios.

$$\cos\ \theta = \frac{\text{Adjacent side}}{\text{Hypotenuse}} = \frac{v_x}{v}$$

$$\sin\ \theta = \frac{\text{Opposite side}}{\text{Hypotenuse}} = \frac{v_y}{v}$$

$$v_x = v \cos \theta; \quad v_y = v \sin \theta$$

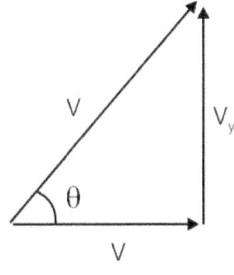

Figure 26.1

If v_x and v_y are the known lengths of a right triangle, then the length of the hypotenuse, V, is calculated by using the Pythagorean theorem

$$|v| = \sqrt{v_x^2 + v_y^2}$$

2. TYPE OF VECTORS

2.1 Null Vector/Zero Vector

A zero vector or null vector is a vector that has zero magnitude, i.e. initial and terminal points are coincident, so that its direction is in indeterminate form. It is denoted by ϕ.

2.2 Unit Vector

A unit vector is a vector of unit length. A unit vector is sometimes denoted by replacing the arrow on a vector with "^".

Unit vectors parallel to x-axis, y-axis and z-axis are denoted by $\hat{i}$, $\hat{j}$ and $\hat{k}$, respectively.

Unit vector $\hat{U}$ parallel to $\vec{V}$ can be obtained as $\hat{U} = \dfrac{\vec{V}}{|V|}$.

Illustration 1: Find unit vector of $\vec{i} - 2\vec{j} + 3\vec{k}$ (JEE MAIN)

Sol: Here unit vector of $\vec{a}$ is given by $\hat{a} = \dfrac{\vec{a}}{|\vec{a}|}$.

$\vec{a} = \vec{i} - 2\vec{j} + 3\vec{k}$

If $\vec{a} = a_x\hat{i} + a_y\hat{j} + a_z\hat{k}$ then it's magnitude $|\vec{a}| = \sqrt{a_x^2 + a_y^2 + a_z^2}$ $\Rightarrow |\vec{a}| = \sqrt{14} \Rightarrow \hat{a} = \dfrac{\vec{a}}{\sqrt{14}} = \dfrac{1}{\sqrt{14}}\left(\vec{i} - 2\vec{j} + 3\vec{k}\right)$

2.3 Collinear or Parallel Vectors

Two or more vectors are said to be collinear, when they are along the same lines or parallel lines irrespective of their magnitudes and directions.

2.4 Like and Unlike Vectors

Vectors having the same direction are called like vectors. Any two vectors parallel to one another, having unequal magnitudes and acting in opposite directions are called unlike vectors.

2.5 Co-Initial Vectors

All those vectors whose terminal points are same, are called co-terminal vectors.

2.6 Co-Terminal Vectors

Vectors that have the same initial points are called co-initial vectors.

Illustration 2: Which are co-initial and equal vectors in the given rectangle diagram?   (JEE MAIN)

Sol: By following above mentioned conditions we can obtain co-initial and equal vectors.

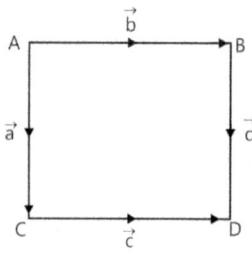

Figure 26.2

Here, $\vec{a}$ and $\dot{b}$ are co-initial vectors, $\dot{b}$ and $\ddot{c}$, $\ddot{a}$ and $\vec{d}$ are equal vectors.

2.7 Coplanar Vectors

Vectors lie on the same plane are called coplanar.

2.8 Negative vector

A vector that points to a direction opposite to that of the given vector is called a negative vector.

2.9 Reciprocal of a Vector

A vector having the same direction as that of a given vector $\vec{a}$, but magnitude equal to the reciprocal of the given vector is known as the reciprocal of $\vec{a}$ and is denoted by $\overline{a^{-1}}$.

2.10 Localized and Free vectors

A vector drawn parallel to a given vector through a specified point unlike free vector in space is called a localized vector. For example, the effect of force acting on a rigid body depends not only on the magnitude and direction but also on the line of action of the force. A vector that depends only on its length and direction and not on its position in the space is called a free vector, e.g. gravity. In this chapter, we will deal with free vectors, unless otherwise stated. Thus a free vector can be determined in space by choosing an arbitrary initial point.

Ilustration 3: Let $\vec{a} = \hat{i} + 2\hat{j}$ and $\hat{b} = 2\hat{i} + \hat{j}$. Is $|\vec{a}| = |\vec{b}|$? Are the vectors $\vec{a}$ and $\hat{b}$ equal? **(JEE MAIN)**

Sol: Two vectors are equal if their modulus and corresponding components both are equal.

We have $|\vec{a}| = \sqrt{1^2 + 2^2} = \sqrt{5}$ and $|\vec{b}| = \sqrt{2^2 + 1^2}$. So, $|\vec{a}| = |\vec{b}|$. But, the two vectors are not equal, since their corresponding components are distinct.

Illustration 4: Find a vector of magnitude 5 units which is parallel to the vector $2\hat{i} - \hat{j}$. **(JEE MAIN)**

Sol: As we know $\hat{a} = \dfrac{\vec{a}}{|\vec{a}|}$, therefore required vector will be $5\hat{a}$.

Let $\vec{a} = 2\hat{i} - \hat{j}$. Then, $\qquad |\vec{a}| = \sqrt{2^2 + (-1)^2} = \sqrt{5}$

$\therefore$ Unit vector parallel to $\vec{a} = \hat{a} = \dfrac{1}{|\vec{a}|} \cdot \vec{a} = \dfrac{1}{\sqrt{5}}\left(2\hat{i} - \hat{j}\right) = \dfrac{2}{\sqrt{5}}\hat{i} - \dfrac{1}{\sqrt{5}}\hat{j}$.

So, the required vector is $5\hat{a} = 5\left(\dfrac{2}{\sqrt{5}}\hat{i} - \dfrac{1}{\sqrt{5}}\hat{j}\right) = 2\sqrt{5}\hat{i} - \sqrt{5}\hat{j}$.

2.11 Position Vector

A vector that represents the position of a point P in space with respect to an arbitrary reference origin O is called a position vector (p.v.). It is also known as location vector or radius vector and usually denoted as x, r or s; it corresponds to the displacement from O to P.

$r = \overline{OP}$.

Illustration 5: Show that, the three points A(-2,3,5), B(1,2,3) and C(7,0,-1) are collinear. **(JEE MAIN)**

Sol: By obtaining $\overrightarrow{AB}$ and $\overrightarrow{BC}$, we can conclude that given points are collinear or not.

We have

$$\overrightarrow{AB} = \overrightarrow{OB} - \overrightarrow{OA} = \left(\hat{i} + 2\hat{j} + 3\hat{k} \right) - \left(-2\hat{i} + 3\hat{j} + 5\hat{k} \right) = 3\hat{i} - \hat{j} - 2\hat{k}$$

$$\overrightarrow{BC} = \overrightarrow{OC} - \overrightarrow{OB} = \left(7\hat{i} + 0\hat{j} - \hat{k} \right) - \left(\hat{i} + 2\hat{j} + 3\hat{k} \right) = 6\hat{i} - 2\hat{j} - 4\hat{k} = 2\left(3\hat{i} - \hat{j} - 2\hat{k} \right)$$

Therefore, $\overrightarrow{BC} = 2\overrightarrow{AB}$.

This shows that the vectors $\overrightarrow{AB}$ and $\overrightarrow{BC}$ are parallel. But, B is a common point. So, the given point A, B and C are collinear.

2.12 Equal Vectors

Two vectors having the same corresponding components and direction and represent the same physical quantity are called equal vectors.

Illustration 6: Find the values of x, y and z, so that the vectors $\vec{a} = x\hat{i} + 2\hat{j} + z\hat{k}$ and $\vec{b} = 2\hat{i} + y\hat{j} + \hat{k}$ are equal.
(JEE MAIN)

Sol: Two vectors are equal, if their corresponding components are equal.

Note that two vectors are equal, if their corresponding components are equal. Thus, the given vectors $\vec{a}$ and $\vec{b}$ will be equal, if and only if x = 2, y = 2, z = 1.

Illustration 7: Find the vector joining the point P (2, 3, 0) and Q (-1, -2, -4) directed from P to Q. **(JEE MAIN)**

Sol: By subtracting the component of P from Q we will get $\overrightarrow{PQ}$.

Since the vector is to be directed from P to Q. Clearly, P is the initial point and Q is the terminal point. So, the required vector joining P and Q is the vector PQ given by

$$\overrightarrow{PQ} = \overrightarrow{OQ} - \overrightarrow{OP} = (-1-2)\hat{i} + (-2-3)\hat{j} + (-4-0)\hat{K} \text{ i.e. } \overrightarrow{PQ} = -3\hat{i} \pm 5\hat{j} - 4\hat{k}$$

Illustration 8: Show that, the points $A\left(2\hat{i} - \hat{j} + \hat{k}\right)$, $B\left(\hat{i} - 3\hat{j} - 5\hat{k}\right)$, $C\left(3\hat{i} - 4\hat{j} - 4\hat{k}\right)$ are the vertices of a right-angled triangle.
(JEE MAIN)

Sol: Here if $\left|\overrightarrow{AB}\right|^2 = \left|\overrightarrow{BC}\right|^2 + \left|\overrightarrow{CA}\right|^2$ then only the given points are the vertices of right angled triangle. We have

$$\overrightarrow{AB} = (1-2)\hat{i} + (-3+1)\hat{j} + (-5-1)\hat{k} = -\hat{i} - 2\hat{j} - 6\hat{k}$$

$$\overrightarrow{BC} = (3-1)\hat{i} + (-4+3)\hat{j} + (-4+5)\hat{k} = 2\hat{i} - \hat{j} + \hat{k} \text{ and } \overrightarrow{CA} = (2-3)\hat{i} + (-1+4)\hat{j} + (1+4)\hat{k} = -\hat{i} + 3\hat{j} + 5\hat{k}$$

Moreover, $\left|\overrightarrow{AB}\right|^2 = 41 = 6 + 35 = \left|\overrightarrow{BC}\right|^2 + \left|\overrightarrow{CA}\right|^2$

Hence, it is proved that the points form a right-angled triangle.

3. RESULTANT OF VECTORS

When two or more vectors are added, they yield the resultant vector. If vectors A and B are added together, the result will be vector R, i.e. $\vec{R} = \vec{A} + \vec{B}$. Same technique can also be applied for multiple vectors.

4. VECTOR ADDITION

4.1 Triangular Law of Addition

It states that if two vectors can be represented in magnitude and direction by the two sides of a triangle taken in the same order, then their resultant is represented by the third side of the triangle, taken in the opposite direction of the sequence.

4.2 Parallelogram Law of Addition

It states that if two vectors can be represented in magnitude and direction by the two adjacent sides or a parallelogram, then their resultant is represented by the diagonal of the parallelogram.

4.3 Addition in Component Form

Consider two vectors A and B

$A = < a_1, b_1, c_1 >$

$B = < a_2, b_2, c_2 >$

Then, $A + B = < a_1 + a_2, b_1 + b_2, c_1 + c_2 >$

4.4 Properties of Vector Addition

The properties of vector addition are listed as follows:

(a)	$\pi / 2$	Commutative
(b)	$\pi / 3$	Associative
(c)	$\pi / 4$	Null vector is an additive identity
(d)	$\hat{A}$ and $\hat{B}$	Additive inverse
(e)	π	
(f)	$\left|\hat{A} - \hat{B}\right|$	
(g)	$\pi / 2$	

4.5 Vector Subtraction

Subtraction is taken as an inverse operation of addition. If $\vec{u}$ and $\vec{v}$ are two vectors, the difference $\vec{u} - \vec{v}$ of two vectors is defined to be the vector added to $\vec{v}$ to get $\vec{u}$. In order to obtain $\vec{u} - \vec{v}$, we put the tails of $\vec{u}$ and $\vec{v}$ together, the directed segment from the nose of $\vec{v}$ to the nose of $\vec{u}$ is a representative of $\vec{u} - \vec{v}$.

Illustration 9: If $\vec{a} = \hat{i} + 2\hat{j} + 3\hat{k}$ and $\vec{b} = 2\hat{i} + 4\hat{j} - 5\hat{k}$ represent two adjacent sides of a parallelogram, find the unit vectors parallel to the diagonals of the parallelogram. **(JEE MAIN)**

Sol: As mentioned above, if two vector quantities are represented by two adjacent sides or a parallelogram then the diagonal of parallelogram will be equal to the resultant of these two vectors.

Let ABCD be a parallelogram such that, $\overrightarrow{AB} // \vec{b}$ and $\overrightarrow{BC} // \vec{b}$.

Then,

$$\overrightarrow{AB} + \overrightarrow{BC} = \overrightarrow{AC} \Rightarrow \overrightarrow{AC} = \vec{a} + \vec{b} = 3\hat{i} + 6\hat{j} - 2\hat{k} \quad \text{and} \quad \overrightarrow{AB} + \overrightarrow{BD} = \overrightarrow{AD}$$

$$\Rightarrow \overrightarrow{BD} = \overrightarrow{AD} - \overrightarrow{AB} \Rightarrow \overrightarrow{BD} = \vec{b} - \vec{a} = \hat{i} + 2\hat{j} - 8\hat{k}$$

Now, $\overrightarrow{AC} = 3\hat{i} + 6\hat{j} - 2\hat{k} \Rightarrow \left|\overrightarrow{AC}\right| = \sqrt{9 + 36 + 4} = 7$

And $\overrightarrow{BD} = \hat{i} + 2\hat{j} - 8\hat{k}$.

$$\Rightarrow \left|\overrightarrow{BD}\right| = \sqrt{1 + 4 + 64} = \sqrt{69}$$

$\therefore$ Unit Vector along $\overrightarrow{AC} = \dfrac{\overrightarrow{AC}}{\left|\overrightarrow{AC}\right|} = \dfrac{1}{7}\left(3\hat{i} + 6\hat{j} - 2\hat{k}\right)$

$\therefore$ Unit vector along $\overrightarrow{BD} = \dfrac{\overrightarrow{BD}}{\left|\overrightarrow{BD}\right|} = \dfrac{1}{\sqrt{69}}\left(\hat{i} + 2\hat{j} - 8\hat{k}\right)$.

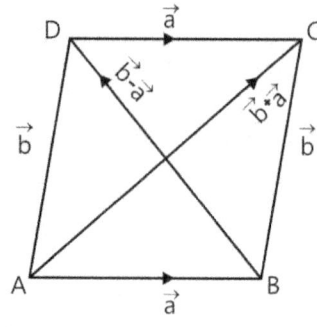

Figure 26.3

Illustration 10: ABCDE is a pentagon. Prove that the resultant of the forces $\overrightarrow{AB}$, $\overrightarrow{AE}$, $\overrightarrow{BC}$, $\overrightarrow{DC}$, $\overrightarrow{ED}$ and $\overrightarrow{AC}$ is $3\overrightarrow{AC}$.

(JEE MAIN)

Sol: By using method of finding resultant of vector we can prove required result.

Let R be the resultant force

$$\therefore R = \overrightarrow{AB} + \overrightarrow{AE} + \overrightarrow{BC} + \overrightarrow{DC} + \overrightarrow{ED} + \overrightarrow{AC}$$

$$\therefore R = \left(\overrightarrow{AB} + \overrightarrow{BC}\right) + \left(\overrightarrow{AE} + \overrightarrow{ED} + \overrightarrow{DC}\right) + \overrightarrow{AC}$$

$$= \overrightarrow{AC} + \overrightarrow{AC} + \overrightarrow{AC} = 3\overrightarrow{AC}. \text{ Hence proved.}$$

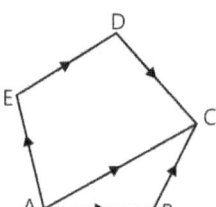

Figure 26.4

Illustration 11: ABCD is a parallelogram. If L and M are the middle points of BC and CD, respectively express $\overrightarrow{AL}$ and $\overrightarrow{AM}$ in terms of $\overrightarrow{AB}$ and $\overrightarrow{AD}$, also show that $\overrightarrow{AL} + \overrightarrow{AM} = \dfrac{3}{2}\overrightarrow{AC}$

(JEE MAIN)

Sol: By using mid – point formula and method of finding resultant of vector we can prove given relation.

Let $\vec{b}$ and $\vec{a}$ be the position vectors of points B and D, respectively be referred to A as the origin of reference.

Then $\overrightarrow{AC} = \overrightarrow{AD} + \overrightarrow{DC} = \overrightarrow{AD} + \overrightarrow{AB}$ $\quad \left[\therefore \overrightarrow{DC} = \overrightarrow{AB}\right]$

$= \vec{d} + \vec{b} \qquad \therefore \overrightarrow{AB} = \vec{b}, \quad \overrightarrow{AD} = \vec{d}$

i.e. the position vector of C referred to A is $\vec{d} + \vec{b}$

$\overrightarrow{AL} = $ p.v. of L, the midpoint of $\overrightarrow{BC}$. $\overrightarrow{AM} = \dfrac{1}{2}\left[\vec{a} + \vec{d} + \vec{b}\right] = \overrightarrow{AD} + \dfrac{1}{2}\overrightarrow{AB}$

$\therefore \overrightarrow{AL} + \overrightarrow{AM} = \vec{b} + \vec{d} + \dfrac{1}{2}\vec{b} = \dfrac{3}{2}\vec{b} + \dfrac{3}{2}\vec{d} = \dfrac{3}{2}(\vec{b} + \vec{d}) = \dfrac{3}{2}\overrightarrow{AC}$

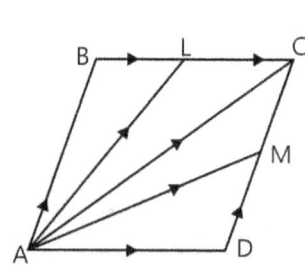

Figure 26.5

5. SCALAR MULTIPLE OF A VECTOR

If $\vec{a}$ is the given vector, then $k\vec{a}$ is a vector, whose magnitude is $|k|$ times the magnitude of $\vec{a}$ and whose direction is the same or opposite as that of $\vec{a}$ according to whether k is positive or negative.

6. SECTION FORMULA

(a) If $\vec{a}$ and $\vec{b}$ are the position vectors of two points A and B, then the position vector of a point which divides A and B in the ratio m:n is given by $r = \dfrac{\left(n\vec{a} + m\vec{b}\right)}{(m+n)}$.

(b) Position vector of the midpoint of $\overline{AB} = \dfrac{\left(\vec{a} + \vec{b}\right)}{2}$.

NOMORECLASS CONCEPTS

- If $\vec{a}, \vec{b}$ and $\vec{c}$ are the position vectors of the vertices of any $\triangle ABC$. Then the position vector of centroid G will be $\dfrac{\vec{a} + \vec{b} + \vec{c}}{3}$.

- The position vector of incenter of triangle with position vectors of triangle ABC, are A $(\vec{a})$, $B(\vec{b})$, $C(\vec{c})$ is $\vec{r} = \dfrac{a\vec{a} + b\vec{b} + c\vec{c}}{a+b+c}$.

Illustration 12: If ABCD is a quadrilateral and E and F are the mid points of AC and BD, respectively, prove that $\overline{AB} + \overline{AD} + \overline{CB} + \overline{CD} = 4\overline{EF}$.   **(JEE MAIN)**

Sol: By using mid-point theorem we can prove given relation.

Since F is the midpoint of BD. Applying the midpoint theorem in triangle ABD,

we have $\Rightarrow \overline{AB} + \overline{AD} = 2\overline{AF}$ ... (i)

Applying the midpoint theorem in triangle BCD, we have

$\Rightarrow \overline{CB} + \overline{CD} = 2\overline{CF}$ ... (ii)

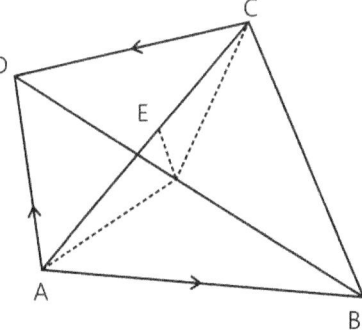

Adding equations (i) and (ii), we obtain

$\Rightarrow \overline{AB} + \overline{AD} + \overline{CB} + \overline{CD} = 2\left(\overline{AF} + \overline{CF}\right)$

Now applying the midpoint theorem in triangle CFA, we have $\overline{AF} + \overline{CF} = 2\overline{EF}$

$\Rightarrow \overline{AB} + \overline{AD} + \overline{CB} + \overline{CD} = 2\left(\overline{AF} + \overline{CF}\right) = 4\overline{EF}$ Hence proved.

Figure 26.6

Illustration 13: If G is the centroid of the triangle ABC, show that $\overrightarrow{GA} + \overrightarrow{GB} + \overrightarrow{GC} = 0$ and conversely $\overrightarrow{GA} + \overrightarrow{GB} + \overrightarrow{GC} = 0$, then G is the centroid of triangle ABC. **(JEE ADVANCED)**

Sol: As G is the centroid of triangle ABC, hence $G = \dfrac{\vec{a} + \vec{b} + \vec{c}}{3}$. Therefore

by obtaining $\overrightarrow{GA}$, $\overrightarrow{GB}$ and $\overrightarrow{GC}$ we can prove this problem.

Let the position vector of the vertices be $\vec{a}$, $\vec{b}$ and $\vec{c}$, respectively.

So, the position vector of centroid, G, is $\dfrac{\vec{a} + \vec{b} + \vec{c}}{3}$.

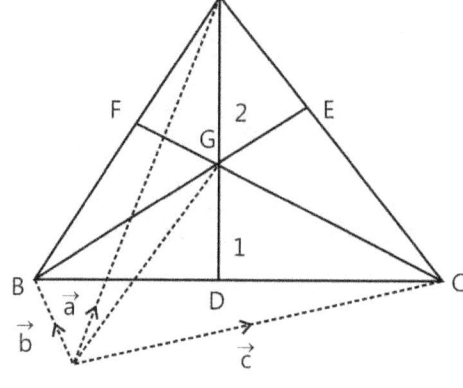

$\overrightarrow{GA} = \overrightarrow{OA} - \overrightarrow{OG} = \vec{a} - \dfrac{\vec{a} + \vec{b} + \vec{c}}{3} = \dfrac{2\vec{a} - \vec{b} - \vec{c}}{3}$

Similarly, $\overrightarrow{GB} = \dfrac{2\vec{b} - \vec{a} - \vec{c}}{3}$. $\overrightarrow{GC} = \dfrac{2\vec{c} - \vec{a} - \vec{b}}{3}$

$\Rightarrow \overrightarrow{GA} + \overrightarrow{GB} + \overrightarrow{GC} = \dfrac{1}{3}\left(2\vec{a} - 2\vec{a} + 2\vec{b} + 2\vec{b} + 2\vec{c} - 2\vec{c}\right) = 0$

Figure 26.7

Conversely if $\overrightarrow{GA} + \overrightarrow{GB} + \overrightarrow{GC} = 0$

$\Rightarrow \left(\overrightarrow{OA} - \overrightarrow{OG}\right) + \left(\overrightarrow{OB} - \overrightarrow{OG}\right) + \left(\overrightarrow{OC} - \overrightarrow{OG}\right) = 0 \Rightarrow \overrightarrow{OA} + \overrightarrow{OB} + \overrightarrow{OC} = 3\overrightarrow{OG} \Rightarrow \overrightarrow{OG} = \dfrac{\overrightarrow{OA} + \overrightarrow{OB} + \overrightarrow{OC}}{3}$

Hence, G is the centroid of the points A, B and C.

Illustration 14: Find the values of x and y, for which the vectors $\vec{a} = (x+2)\hat{i} - (x-y)\hat{j} + \hat{k}$, $\vec{b} = (x-1)\hat{i} + (2x+y)\hat{j} + 2\hat{k}$ are parallel **(JEE MAIN)**

Sol: Two vectors are parallel if ratio of there respective components are equal.

$\vec{a}$ and $\vec{b}$ are parallel if $\dfrac{x+2}{x-1} = \dfrac{y-x}{2x+y} = \dfrac{1}{2} \Rightarrow x = -5, y = \dfrac{-20}{3}$

Illustration 15: If ABCD is a parallelogram and E is the midpoint of AB, show by vector method, that DE trisects and is trisected by AC. **(JEE MAIN)**

Sol: By using section formula, we can solve this problem.

Let $\overrightarrow{AB} = \vec{a}$ and $\overrightarrow{BC} = \vec{b}$

Then $\overrightarrow{BC} = \overrightarrow{AD} = \vec{b}$ and $\overrightarrow{AC} = \overrightarrow{AB} + \overrightarrow{AD} = \vec{a} + \vec{b}$

Also, let K be a point on AC, such that AK:AC = 1:3

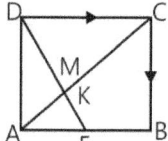

Figure 26.8

$\Rightarrow \overrightarrow{AK} = \dfrac{1}{3}\overrightarrow{AC} \Rightarrow \overrightarrow{AK} = \dfrac{1}{3}\left(\vec{a} + \vec{b}\right)$ ... (i)

Let E be the midpoint of AB, such that $\overrightarrow{AE} = \dfrac{1}{2}\vec{a}$

Let M be the point on $\overrightarrow{DE}$ such that DM: ME = 2:1

$\therefore \overrightarrow{AM} = \dfrac{\overrightarrow{AD} + 2\overrightarrow{AE}}{1+2} = \dfrac{\vec{b} + \vec{a}}{3}$ ... (ii)

Comparing equations (i) and (ii), we find that $\overrightarrow{AK} = \dfrac{\left(\vec{b} + \vec{a}\right)}{3} = \overrightarrow{AM}$, and thus we conclude that K and M coincide, i.e. $\overrightarrow{DE}$ trisects $\overrightarrow{AC}$ and is trisected by $\overrightarrow{AC}$. Hence proved.

7. LINEAR COMBINATION OF VECTORS

7.1 Collinear and Non-Collinear Vectors

Let $\vec{a}$ and $\vec{b}$ be non-zero vectors. These vectors are said to be collinear if there exists $\lambda \neq 0$ such that $\vec{a} + \lambda\vec{b} + \gamma\vec{c} = 0$.

Given a finite set of vectors $\vec{a}$, $\vec{b}$, $\vec{c}$....., then the vector $\vec{r} = x\vec{a} + y\vec{b} + z\vec{c} + ...$ is called a linear combination of $\vec{a}, \vec{b}, \vec{c}$....., for any scalar x, y, z ...... $\in R$.

7.2 Collinearity of Three Points

Let three points with position vectors (non-zero) $\vec{a}$, $\vec{b}$, and $\vec{c}$ be collinear. Then there exists λ, γ both not being 0 such that $\vec{a} + \lambda\vec{b} + \gamma\vec{c} = 0$

7.3 Coplanar Vectors

Let $\vec{a}$ and $\vec{b}$, be non-zero, non-collinear vectors. Then, any vector $\vec{r}$ coplanar with $\vec{a}, \vec{b}$ can be uniquely expressed as a linear combination of $\vec{a}, \vec{b}$, i.e. there exist some unique x, y $\in R$, such that.

NOMORECLASS CONCEPTS

- If $\vec{a}$, $\vec{b}$, $\vec{c}$ are non-zero, non-coplanar vectors, then $x\vec{a} + y\vec{b} + z\vec{c} = x'\vec{a} + y'\vec{b} + z'\vec{c} \Rightarrow x = x', y = y', z = z'$

- Let $\vec{a}$, $\vec{b}$, $\vec{c}$ be non-zero, non-coplanar vectors in space. Then any vector $\vec{r}$ can be uniquely expressed as a linear combination of $\vec{a}$, $\vec{b}$, $\vec{c}$ or there exists some unique x, y, z $\in R$, such that $x\vec{a} + y\vec{b} + z\vec{c} = \vec{r}$.

7.4 Linear Dependency of Vectors

A set of vectors $\{\vec{v}_1, \vec{v}_2, ... \vec{v}_p\}$ is said to be linearly independent if the vector equation $x_1 \vec{v}_1 + x_2 \vec{v}_2 + ... + x_p \vec{v}_p = 0$ has only a trivial solution. The set $\{\vec{v}_1, \vec{v}_2, ...\vec{v}_p\}$ is said to be linearly dependent if there exists weights $c_1, ..., c_p$, not all 0, such that $c_1\vec{v}_1 + c_2\vec{v}_2 + ... + c_p\vec{v}_p = 0$

NOMORECLASS CONCEPTS

- Two non-zero, non-collinear vectors are linearly independent.
- Any two collinear vectors are linearly dependent.
- Any three non-coplanar vectors are linearly independent.
- Any three coplanar vectors are linearly dependent.
- Any four vectors in three-dimensional space are linearly dependent.

Illustration 16: The position vectors of three points A = $\vec{a} - 2\vec{b} + 3\vec{c}$, B = $2\vec{a} + 3\vec{b} - 4\vec{c}$ and C = $-7\vec{b} + 10\vec{c}$. Prove that the vectors $\overrightarrow{AB}$ and $\overrightarrow{AC}$ are linearly dependent.   **(JEE MAIN)**

Sol: Here obtain $\overrightarrow{AB}$ and $\overrightarrow{AC}$ to check its linear dependency.

Let O be the point of reference, then, $\overrightarrow{OA} = \vec{a} - 2\vec{b} + 3\vec{c}$, $\overrightarrow{OB} = 2\vec{a} + 3\vec{b} - 4\vec{c}$, and $\overrightarrow{OC} = -7\vec{b} + 10\vec{c}$

$\Rightarrow \overrightarrow{AC} = \overrightarrow{OC} - \overrightarrow{OA} = \left(-7\vec{b} + 10\vec{c}\right) - \left(\vec{a} - 2\vec{b} + 3\vec{c}\right) = -\vec{a} - 5\vec{b} + 7\vec{c}$

$\overrightarrow{AB} = \overrightarrow{OB} - \overrightarrow{OA} = \left(2\vec{a} + 3\vec{b} - 4\vec{c}\right) - \left(\vec{a} - 2\vec{b} + 3\vec{c}\right) = \vec{a} + 5\vec{b} - 7\vec{c}$

$\therefore \overrightarrow{AC} = \lambda\, \overrightarrow{AB}$, where $\lambda = -1$.

Hence $\overrightarrow{AB}$ and $\overrightarrow{AC}$ are linearly dependent.

Illustration 17: Prove that the vectors $5\vec{a} + 6\vec{b} + 7\vec{c}$, $7\vec{a} - 8\vec{b} + 9\vec{c}$ and $3\vec{a} + 20\vec{b} + 5\vec{c}$ are linearly dependent and $\vec{a}, \vec{b}, \vec{c}$, being linearly independent vectors.   **(JEE MAIN)**

Sol: We know that if these vectors are linearly dependent, then we can express one of them as a linear combination of the other two.

Now, let us assume that the given vectors are coplanar, and then we can write

$5\vec{a} + 6\vec{b} + 7\vec{c} = \ell\left(7\vec{a} - 8\vec{b} + 9\vec{c}\right) + m\left(3\vec{a} + 20\vec{b} + 5\vec{c}\right)$, where ℓ and m are scalars.

Comparing the coefficients of $\vec{a}, \vec{b}$ and $\vec{c}$ on both sides of the equation

$5 = 7\ell + 3m$ ... (i)
$6 = -8\ell + 20m$ ... (ii)
$7 = 9\ell + 5m$ ... (iii)

From equations (i) and (iii), we get

$4 = 8\ell \Rightarrow \ell = \dfrac{1}{2} = m$, which evidently satisfies equation (ii) too.

Hence, the given vectors are linearly dependent.

Illustration 18: Prove that the four points $2\vec{a} + 3\vec{b} - \vec{c}, \vec{a} - 2\vec{b} + 3\vec{c}, 3\vec{a} + 4\vec{b} - 2\vec{c}$ and $\vec{a} - 6\vec{b} + 6\vec{c}$ are coplanar.   **(JEE MAIN)**

Sol: Let the given four points be P, Q, R and S respectively. These points are coplanar, if the vectors $\overrightarrow{PQ}$, $\overrightarrow{PR}$ and $\overrightarrow{PS}$ are coplanar. These vectors are coplanar if one of them can be expressed as a linear combination of other two.

So, let

$\overrightarrow{PQ} = x\overrightarrow{PR} + y\overrightarrow{PS}$

$\Rightarrow -\vec{a} - 5\vec{b} + 4\vec{c} = x\left(\vec{a} + \vec{b} - \vec{c}\right) + y\left(-\vec{a} - 9\vec{b} + 7\vec{c}\right)$

$\Rightarrow -\vec{a} - 5\vec{b} + 4\vec{c} = (x - y)\vec{a} + (x - 9y)\vec{b} + (-x + 7y)\vec{c}$

$\Rightarrow x - y = -1, x - 9y = -5, -x + 7y = 4$

Solving the first two of these three equations, we get $x = -\dfrac{1}{2}, y = \dfrac{1}{2}$

On substituting the values of x and y in the third equation, we find that the third equation is satisfied. Hence, the given four points are coplanar.

Illustration 19: Show that, the vectors $2\vec{a} - \vec{b} + 3\vec{c}$, $\vec{a} + \vec{b} - 2\vec{c}$ and $\vec{a} + \vec{b} - 3\vec{c}$ are non-coplanar vectors.

(JEE MAIN)

Sol: If vectors are coplanar then one of them can be expressed as a linear combination of other two otherwise they are non-coplanar. Assume the given vectors are coplanar.

Then one of the given vectors is expressible in terms of the other two.

Let $2\vec{a} - \vec{b} + 3\vec{c} = x(\vec{a} + \vec{b} - 2\vec{c}) + y(\vec{a} + \vec{b} - 3\vec{c})$ for some scalars x and y.

$\Rightarrow 2\vec{a} - \vec{b} + 3\vec{c} = (x + y)\vec{a} + (x + y)\vec{b} + (-2x - 3y)\vec{c} \Rightarrow 2 = x + y, -1 = x + y$ and $3 = 2x - 3y,$

Clearly, the first two equations contradict each other. Hence, it is proved that the given vectors are not coplanar.

8. SCALAR OR DOT PRODUCT

The scalar product of two vectors $\vec{a} = (a_1, a_2, a_3)$ and $\vec{b} = (b_1, b_2, b_3)$ is written using a dot as an operator (·) between the two vectors. The component form of the dot product is as follows:

$$\vec{a} \cdot \vec{b} = a_1 b_1 + a_2 b_2 + a_3 b_3 \qquad \text{... (i)}$$

And in geometrical form

$$\vec{a} \cdot \vec{b} = |\vec{a}||\vec{b}| \cos\theta \qquad \text{... (ii)}$$

where θ is the angle between the two vectors and $0 \le \theta \le \pi$.

From equation (i), it can also be written as

$$\cos\theta = \frac{\vec{a} \cdot \vec{b}}{|\vec{a}||\vec{b}|} = \frac{a_1 b_1 + a_2 b_2 + a_3 b_3}{|\vec{a}||\vec{b}|},$$

which can be used to find the angle between two vectors. If $\vec{a}$ and $\vec{b}$ are perpendicular then

$$\theta = 90° \Rightarrow \cos\theta = 0 \Rightarrow \vec{a} \cdot \vec{b} = 0$$

NOMORECLASS CONCEPTS

- $\vec{a} \cdot \vec{b} \le |\vec{a}||\vec{b}|$
- $\vec{a} \cdot \vec{b} > 0 \Rightarrow$ Angle between a and b is acute.
- $\vec{a} \cdot \vec{b} < 0 \Rightarrow$ Angle between a and b is obtuse.

Geometrical Interpretation of Dot Product

The scalar product is used to determine the projection of $\vec{r}$ vector along the given direction.

$\overrightarrow{ON}$ is the component of vector $\overrightarrow{OB} (= \vec{b})$ in the direction of vector $\overrightarrow{OA}(= \vec{a})$; $\overrightarrow{ON} = b\cos\theta$. Thus the projection of $\vec{b}$ along $\hat{a} = \dfrac{\vec{a} \cdot \vec{b}}{|\vec{a}|}\hat{a}$

$$\therefore \overrightarrow{ON} = \left(\frac{\vec{a} \cdot \vec{b}}{|\vec{a}|^2}\right)\vec{a}$$

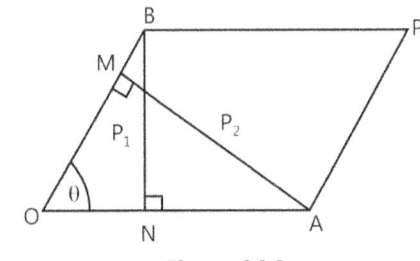

Figure 26.9

Projection of $\vec{a}$ along $\vec{b} = \left(\dfrac{\vec{a}.\vec{b}}{|\vec{b}|}\right)\hat{b}$ $\therefore \overline{OM} = \left(\dfrac{\vec{a}.\vec{b}}{|\vec{b}|^2}\right)\vec{b}$

8.1 Properties of Scalar Product

The properties of scalar product are listed as follows:

(a) $\vec{a}$, $\vec{b}$ are vectors and $\vec{a}.\vec{b}$ is a number   **(b)** $\vec{a}.\vec{a} = |\vec{a}|^2$ **(c)** $\vec{a}.\vec{b} = \vec{b}.\vec{a}$

(d) $\vec{a}.(\vec{b} + \vec{c}) = \vec{a}.\vec{b} + \vec{a}.\vec{c}$ **(e)** $(c\vec{a}).\vec{b} = c(\vec{a}.\vec{b})$ **(f)** $\vec{0}.\vec{a} = 0$

(g) $\vec{a}.\vec{b} = |\vec{a}||\vec{b}|\cos\theta$ **(h)** $\vec{a}.\vec{b} = 0 \Leftrightarrow \vec{a} = \vec{0}$ or $\vec{b} = \vec{0}$ or $\vec{a} \perp \vec{b}$

Illustration 20: Find the angle 'θ' between the vectors $\vec{a} = \hat{i} + \hat{j} - \hat{k}$ and vectors $\vec{b} = \hat{i} - \hat{j} + \hat{k}$ **(JEE MAIN)**

Sol: The angle θ between the two vectors $\vec{a}$ and $\vec{b}$ is given by $\cos\theta = \dfrac{\vec{a}.\vec{b}}{|\vec{a}||\vec{b}|}$

Now $\vec{a}.\vec{b} = \left(\hat{i} + \hat{j} - \hat{k}\right).\left(\hat{i} - \hat{j} + \hat{k}\right) = 1 - 1 - 1 = -1$

Therefore, we have $\cos\theta = \dfrac{-1}{3}$. Hence, the required angle is $\theta = \cos^{-1}\left(\dfrac{-1}{3}\right)$

Illustration 21: Find the length of the projection of vector $\vec{a} = 2\hat{i} + 3\hat{j} + 2\hat{k}$ on vector $\hat{b} = \hat{i} + 2\hat{j} + \hat{k}$.   **(JEE MAIN)**

Sol: The projection of vector $\vec{a}$ on the vector $\vec{b}$ is given by $\dfrac{1}{|\vec{b}|}(\vec{a}.\vec{b})$.

$\dfrac{1}{|\vec{b}|}(\vec{a}.\vec{b}) = \dfrac{(2\cdot1 + 3\cdot2 + 2\cdot1)}{\sqrt{(1)^2 + (2)^2 + (1)^2}} = \dfrac{10}{\sqrt{6}} = \dfrac{5}{3}\sqrt{6}$

Illustration 22: Let $\vec{a}$, $\vec{b}$, $\vec{c}$ be the vectors of lengths 3, 4 and 5, respectively. Let $\vec{a}$ be perpendicular to $(\vec{b} + \vec{c})$, $\vec{b}$ to $(\vec{c} + \vec{a})$ and $\vec{c}$ to $(\vec{a} + \vec{b})$. Then, find the length of the vector $(\vec{a} + \vec{b} + \vec{c})$.   **(JEE MAIN)**

Sol: By using property of scalar product of vector we can solve this illustration.

Given $|\vec{a}| = 3, |\vec{b}| = 4, |\vec{c}| = 5$

$\therefore |\vec{a} + \vec{b} + \vec{c}|^2 = (\vec{a} + \vec{b} + \vec{c}).(\vec{a} + \vec{b} + \vec{c}) = |\vec{a}|^2 + |\vec{b}|^2 + |\vec{c}|^2 + \vec{a}(\vec{b} + \vec{c}) + \vec{b}(\vec{c} + \vec{a}) + \vec{c}(\vec{a} + \vec{b}) = 9 + 16 + 25 + 0 + 0 + 0$

$\Rightarrow |\vec{a} + \vec{b} + \vec{c}| = 5\sqrt{2}$

Illustration 23: Let $\vec{a} = 4\hat{i} + 5\hat{j} - k$, $\vec{b} = \hat{i} - 4\hat{j} + 5\hat{k}$ and $\vec{c} = 3\hat{i} + \hat{j} - \hat{k}$. Find a vector $\vec{d}$, which is perpendicular to both $\vec{a}$ and $\vec{b}$, and satisfying $\vec{d}.\vec{c} = 21$.   **(JEE MAIN)**

Sol: If two vector are perpendicular then their product will be zero.

Let $\vec{d} = x\hat{i} + y\hat{j} + z\hat{k}$. Since $\vec{d}$ is perpendicular to both $\vec{a}$ and $\vec{b}$. Therefore,

$\vec{d}.\vec{a} = 0 \Rightarrow \left(x\hat{i} + y\hat{j} + z\hat{k}\right).\left(4\hat{i} + 5\hat{j} - \vec{k}\right) = 0 \Rightarrow 4x + 5y - z = 0$ ... (i)

$\vec{d}.\vec{b} = 0 \Rightarrow \left(x\hat{i} + y\hat{j} + z\hat{k}\right).\left(\hat{i} - 4\hat{j} + 5\hat{k}\right) = 0 \Rightarrow x - 4y + 5z = 0$ ... (ii)

$\vec{d}.\vec{c} = 21 \Rightarrow \left(x\hat{i} + y\hat{j} + z\hat{k}\right).\left(3\hat{i} + \hat{j} - \hat{k}\right) = 21 \Rightarrow 3x + y - z = 21$ ... (iii)

Solving equations (i), (ii) and (iii), we get x = 7, y = z = –7

Hence, $\vec{d} = 7\hat{i} - 7\hat{j} - 7\hat{k}$

Illustration 24: Three vectors $\vec{a}$, $\vec{b}$ and $\vec{c}$ satisfy the condition $\vec{a} + \vec{b} + \vec{c} = 0$. Evaluate the quantity $\mu = \vec{a}.\vec{b} + \vec{b}.\vec{c} + \vec{c}.\vec{a}$, if $|\vec{a}| = 1, |\vec{b}| = 4$ and $|\vec{c}| = 2$. **(JEE MAIN)**

Sol: Simply using property of scalar product we can calculate the value of μ.

Since $\vec{a} + \vec{b} + \vec{c} = \vec{0}$, we have $\vec{a}.(\vec{a} + \vec{b} + \vec{c}) = \vec{0} \Rightarrow \vec{a}.\vec{a} + \vec{a}.\vec{b} + \vec{a}.\vec{c} = 0$. Therefore, $\vec{a}.\vec{b} + \vec{a}.\vec{c} = -|\vec{a}|^2 = -1$

Similarly $\vec{a}.\vec{b} + \vec{b}.\vec{c} = -|\vec{b}|^2 = -16$, $\vec{a}.\vec{c} + \vec{b}.\vec{c} = -4$.

On adding these equations, we have $2(\vec{a}.\vec{b} + \vec{b}.\vec{c} + \vec{a}.\vec{c}) = -21$ or $2\mu = -21$, i.e., $\mu = \dfrac{-21}{2}$

Illustration 25: Prove, Cauchy–Schawarz inequality, $(\vec{a}.\vec{b})^2 \le |\vec{a}|^2 |\vec{b}|^2$, and hence show that

$$\left(a_1 b_1 + a_2 b_2 + a_3 b_3\right)^2 \le \left(a_1 + a_2 + a_3\right)^2 \left(b + b_2 + b_3\right)^2$$ **(JEE ADVANCED)**

Sol: As we know $\cos^2\theta \le 1$, solve it by multiplying both side by $|\vec{a}|^2 |\vec{b}|^2$.

We have, $\cos^2\theta \le 1$

$\Rightarrow |\vec{a}|^2 |\vec{b}|^2 \cos^2\theta \le |\vec{a}|^2 |\vec{b}|^2 \Rightarrow (\vec{a}.\vec{b})^2 \le |\vec{a}|^2 |\vec{b}|^2$

Let $\vec{a} = a_1\hat{i} + a_2\hat{j} + a_3\hat{k}$ and $\vec{b} = b_1\hat{i} + b_2\hat{j} + b_3\hat{k}$. Then,

$\vec{a}.\vec{b} = a_1 b_1 + a_2 b_2 + a_3 b_3, |\vec{a}|^2 = a_1^2 + a_2^2 + a_3^2$ and $|\vec{b}|^2 = b_1^2 + b_2^2 + b_3^2$.

$(\vec{a}.\vec{b})^2 \le |\vec{a}|^2 |\vec{b}|^2 \Rightarrow (a_1 b_1 + a_2 b_2 + a_3 b_3)^2 \le \left(a_1^2 + a_2^2 + a_3^2\right)\left(b_1^2 + b_2^2 + b_3^2\right)$

Illustration 26: If $\vec{a}$, $\vec{b}$, $\vec{c}$ are three mutually perpendicular vectors of equal magnitude, prove that $\vec{a} + \vec{b} + \vec{c}$ is equally inclined with vectors $\vec{a}$, $\vec{b}$ and $\vec{c}$. **(JEE ADVANCED)**

Sol: Here use formula of dot product to solve the problem. Let $|\vec{a}| = |\vec{b}| = |\vec{c}| = \lambda$ (say). Since $\vec{a}, \vec{b}, \vec{c}$ are mutually perpendicular vectors, We have $\vec{a}.\vec{b} = \vec{b}.\vec{c} = \vec{c}.\vec{a} = 0$

Now, $|\vec{a} + \vec{b} + \vec{c}|^2 = \vec{a}.\vec{a} + \vec{b}.\vec{b} + \vec{c}.\vec{c} + 2\vec{a}.\vec{b} + 2\vec{b}.\vec{c} + 2\vec{c}.\vec{a} = |\vec{a}|^2 + |\vec{b}|^2 + |\vec{c}|^2 = 3\lambda^2 \therefore |\vec{a} + \vec{b} + \vec{c}| = \sqrt{3}\lambda$

Let $\vec{a} + \vec{b} + \vec{c}$ makes angles $\theta_1, \theta_2, \theta_3$ with $\vec{a}$, $\vec{b}$ and $\vec{c}$, respectively. Then,

$\cos\theta_1 = \dfrac{\vec{a}.(\vec{a} + \vec{b} + \vec{c})}{|\vec{a}||\vec{a} + \vec{b} + \vec{c}|} = \dfrac{\vec{a}.\vec{a} + \vec{a}.\vec{b} + \vec{a}.\vec{c}}{|\vec{a}||\vec{a} + \vec{b} + \vec{c}|} = \dfrac{|\vec{a}|^2}{|\vec{a}||\vec{a} + \vec{b} + \vec{c}|} = \dfrac{\lambda}{\sqrt{3}} = \dfrac{1}{\sqrt{3}} \therefore \theta_1 = \cos^{-1}\left(\dfrac{1}{\sqrt{3}}\right)$

Similarly, $\theta_2 = \cos^{-1}\left(\dfrac{1}{\sqrt{3}}\right)$ and $\theta_3 = \cos^{-1}\left(\dfrac{1}{\sqrt{3}}\right)$ $\therefore \theta_1 = \theta_2 = \theta_3$

Hence, $\vec{a} + \vec{b} + \vec{c}$ is equally inclined with $\vec{a}$, $\vec{b}$ and $\vec{c}$.

Illustration 27: Using vectors, prove that cos(A+B) = cosA cosB – sinA sinB **(JEE ADVANCED)**

Sol: From figure, using vector method we can easily prove that cos(A+B)=cosA cosB – sinA sinB.

Let OX and OY be the coordinate axes and let $\hat{i}$ and $\hat{j}$ be unit vectors along OX and OY, respectively.

Let $\angle XOP = A$ and $\angle XOQ = B$. Draw $PL \perp OX$ and $QM \perp OX$

Therefore, the angle between $\overline{OP}$ and $\overline{OL}$ is A & $\overline{OQ}$ and $\overline{OL}$ is B.

In $\triangle OLP$, $OL = OP\cos A$ and $LP = OP\sin A$.

Therefore, $\overline{OL} = (OP\cos A)\hat{i}$ and $\overline{LP} = (OP\sin A)(-\hat{j})$

Now, $\overline{OL} + \overline{LP} = \overline{OP}$

$\Rightarrow \overline{OP} = OP\left[(\cos A)\hat{i} - (\sin A)\hat{j}\right]$ ... (i)

In $\triangle OMQ$, $\overline{OM} = OQ\cos B$ and $\overline{MQ} = OQ\sin B$.

Therefore, $|\overline{OM} = (OQ\cos B)\hat{i}$, $\overline{MQ} = (OQ\sin B)\hat{j}$

$\Rightarrow \overline{OQ} = OQ\left[(\cos\beta)\hat{i} + (\sin\beta)\hat{j}\right]$ ...(ii)

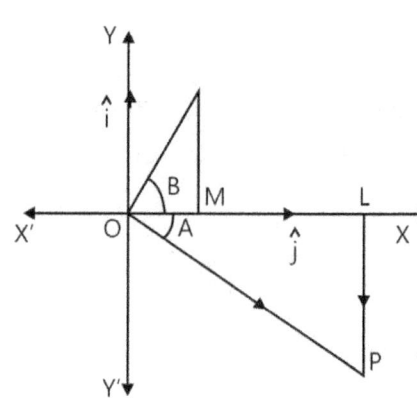

Figure 26.10

From (i) and (ii), we get

$\overline{OP}.\overline{OQ} = OP\left[(\cos A)\hat{i} - (\sin A)\hat{j}\right].OQ\left[(\cos B)\hat{i} + (\sin B)\hat{j}\right] = OP.OQ\left[\cos A\cos B - \sin A\sin B\right]$

But, $\overline{OP}.\overline{OQ} = |\overline{OP}||\overline{OQ}|\cos(A+B) = OP.OQ\cos(A+B)$

$\therefore OP.OQ\cos(A+B) = OP.OQ[\cos A\cos B - \sin A\sin B]$

$\Rightarrow \cos(A+B) = \cos A\cos B - \sin A\sin B$

Illustration 28: Find the values of c for which the vectors $\vec{a} = (c\log_2 x)\hat{i} - 6\hat{j} + 3\hat{k}$ and $\vec{b} = (\log_2 x)\hat{i} + 2\hat{j} + (2c\log_2 x)\hat{k}$ made an obtuse angle for any $x \in (0, \infty)$. **(JEE ADVANCED)**

Sol: For obtuse angle $\cos\theta < 0$, therefore by using formula $\cos\theta = \dfrac{\vec{a}.\vec{b}}{|\vec{a}||\vec{b}|}$, we can solve this problem.

Let θ be the angle between the vectors $\vec{a}$ and $\vec{b}$. Then,

$\cos\theta = \dfrac{\vec{a}.\vec{b}}{|\vec{a}||\vec{b}|}$

For θ to be an obtuse angle, we must have $\Rightarrow \cos\theta < 0$, for all $x \in (0, \infty) \Rightarrow \dfrac{\vec{a}.\vec{b}}{|\vec{a}||\vec{b}|} < 0$, for all $x \in (0, \infty)$

$\Rightarrow \vec{a}.\vec{b} < 0$, for all $x \in (0, \infty) \Rightarrow \vec{a}.\vec{b} < 0$, for all $x \in (0, \infty) \Rightarrow c(\log_2 x)^2 - 12 + 6c(\log_2 x) < 0$, for all $x \in (0, \infty)$

$\Rightarrow cy^2 + 6cy - 12 < 0$, for all $y \in R$, where $y = -\log_2 x$ $[\because x > 0 \Rightarrow y = \log_2 x \in R]$

$\Rightarrow c < 0$ and $36c^2 + 48c < 0$ $[\because ax^2 + bx^2 + c > 0$ for all $x \Rightarrow a < 0$ and Discriminant $< 0]$

$\Rightarrow c < 0$ and $c(3c+4) < 0$

$\Rightarrow c < 0$ and $-\dfrac{4}{3} < c < 0 \Rightarrow c \in \left(\dfrac{-4}{3}, 0\right)$

Illustration 29: D is the midpoint of the side $\overline{BC}$ of a triangle ABC, show that $AB^2 + AC^2 = 2\left(AD^2 + BD^2\right)$

(JEE MAIN)

Sol: By using the formula of resultant vector we will get the required result.

Given D is midpoint of BC $\Rightarrow \overline{BD} = \overline{DC}$

We have $\overline{AB} = \overline{AD} + \overline{DB} \Rightarrow AB^2 = \left(\overline{AD} + \overline{DB}\right)^2$

$AB^2 = AD^2 + DB^2 + 2\overline{AD} \cdot \overline{DB}$ ...(i)

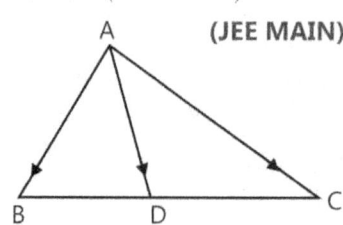

Figure 26.11

1.14

Also we have $\overrightarrow{AC} = \overrightarrow{AD} + \overrightarrow{DC} \Rightarrow AC^2 = \left(\overrightarrow{AD} + \overrightarrow{DC}\right)^2$

$AC^2 = AD^2 + DC^2 + 2\overrightarrow{AD}\cdot\overrightarrow{DC}$ ...(ii)

Adding equations (i) and (ii), we get

$AB^2 + AC^2 = 2AD^2 + 2BD^2 + 2\overrightarrow{AD}.\left(\overrightarrow{DB} + \overrightarrow{DC}\right) = 2\left(DA^2 + DB^2\right)$

9. VECTOR OR CROSS PRODUCT

Let $\vec{a}$ and $\vec{b}$ be two vectors. The vector product of these two vectors can be calculated as $\left(\vec{a} \times \vec{b}\right) = \left|\vec{a}\right|\left|\vec{b}\right|\sin\theta\,\hat{n}$, where θ is the angle between the vectors $\vec{a}$ and $\vec{b}$, $\left(0 \le \theta \le \pi\right)$ and $\hat{n}$ is the unit vector at right angles to both $\vec{a}$ and $\vec{b}$, i.e. $\hat{n}$ is vector normal to the plane that contains $\vec{a}$ and $\vec{b}$. $\vec{a}$, $\vec{b}$ and $\hat{n}$ are three vectors which form a right-handed set.

The convention is that we choose the direction specified by the right-hand screw rule. Imagine a screw in your right hand. If you turn a right-handed screw from $\vec{a}$ to $\vec{b}$, the screw advances along the unit vector $\hat{n}$. It is very important to realize that the result of a vector product is itself a vector.

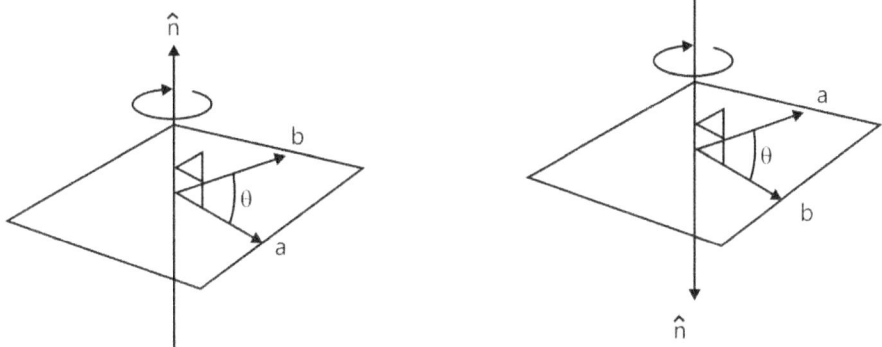

Figure 26.12

Let us see how the order of multiplication matters from the definition of the right-hand screw rule:

The vector given by $\left(\vec{b} \times \vec{a}\right)$ points in the opposite direction to $\left(\vec{a} \times \vec{b}\right)$. So, $\left(\vec{a} \times \vec{b}\right) = -\left(\vec{b} \times \vec{a}\right)$.

We can define vector product in terms of matrix notation as $\vec{a} \times \vec{b} = \begin{vmatrix} \hat{i} & \hat{j} & \hat{k} \\ a_1 & a_2 & a_3 \\ b_1 & b_2 & b_3 \end{vmatrix}$

and n terms of components as $\vec{a} = \left\langle a_1, a_2, a_3\right\rangle, \vec{b} = \left\langle b_1, b_2, b_3\right\rangle \Rightarrow \vec{a} \times \vec{b} = \left\langle a_2 b_3 - a_3 b_2, a_3 b_1 - a_1 b_3, a_1 b_2 - a_2 b_1\right\rangle$

From the definition, the angle can be calculated as $\sin\theta = \dfrac{\left|\vec{a} x \vec{b}\right|}{\left|\vec{a}\right|\left|\vec{b}\right|}$

If $\vec{a}$ and $\vec{b}$ are parallel then $\theta = 0° \Rightarrow \sin\theta = 0$ and $\vec{a} x \vec{b} = 0$

9.1 Properties of Vector Product

The properties of vector product are listed as follows:

$\vec{a}, \vec{b}$ and $\vec{a} \times \vec{b}$ are all vectors in three dimensions.

(a) $\vec{a} \times \vec{b} \perp \vec{a}$ and $\vec{b}$

(b) $\left|\vec{a} \times \vec{b}\right| = \left|\vec{a}\right|\left|\vec{b}\right| \sin\theta$

(c) $\hat{i} \times \hat{j} = \hat{k}, \hat{j} \times \hat{k} = \hat{i}, \hat{k} \times \hat{i} = \hat{j}$

(d) $\vec{a} \times \vec{b} = 0 \Leftrightarrow \vec{a} = \vec{0}$ or $\vec{b} = \vec{0}$ or $\vec{a} \parallel \vec{b}$

(e) $\vec{a} \times \vec{b} = -\vec{b} \times \vec{a}$

(f) $(c\vec{a}) \times \vec{b} = \vec{a} \times (c\vec{b}) = c(\vec{a} \times \vec{b})$

(g) $\vec{a} \times (\vec{b} + \vec{c}) = \vec{a} \times \vec{b} + \vec{a} \times \vec{c}$

(h) $\vec{a} \cdot (\vec{b} \times \vec{c}) = (\vec{a} \times \vec{b}) \cdot \vec{c}$

Geometrical interpretation of $\left|\vec{a} \times \vec{b}\right| = \left|\vec{a}\right|\left|\vec{b}\right| \sin\theta$, denotes the area of parallelogram, in which $\vec{a}$ and $\vec{b}$ are the two adjacent sides.

Vector area of the plane figure

Considering the boundaries of closed, bounded surface, which has been described in a specific manner and that do not cross, it is possible to associate a directed line segment $\vec{c}$, such that

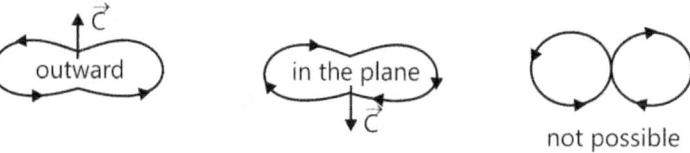

Figure 26.13

(a) $\left|\vec{c}\right|$ is the number of units of area enclosed by the plane figure.

(b) The support of $\vec{c}$ is perpendicular to the area and outside the surface.

(c) The sense of description of the boundaries and the direction of $\vec{c}$ is in accordance with the R.H.S. screw rule.

Vector area of a triangle

If $\vec{a}$ $\vec{b}$ are the position vectors, then the vector area of a triangle is given by the formula

$$\vec{\Delta} = \frac{1}{2}(\vec{a} \times \vec{b})$$

If $\vec{a}, \vec{b}, \vec{c}$ are the position vectors, then the vector area of $\triangle ABC$ is given by the formula

$$\vec{\Delta} = \frac{1}{2}\left[(\vec{c} - \vec{b}) \times (\vec{a} - \vec{b})\right]$$

$$\vec{\Delta} = \frac{1}{2}\left[(\vec{a} \times \vec{b}) + (\vec{b} \times \vec{c}) + (\vec{c} \times \vec{a})\right]$$

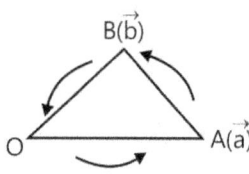

Figure 26.14

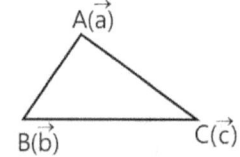

Figure 26.15

(i) If three points with position vectors $\vec{a}$, $\vec{b}$ and $\vec{c}$ are collinear, then $\vec{a} \times \vec{b} + \vec{b} \times \vec{c} + \vec{c} \times \vec{a} = 0$

(ii) Unit vector perpendicular to the plane of the $\triangle ABC$, when $\vec{a}, \vec{b}$ and $\vec{c}$ are the p.v. of its angular point is

$$\hat{n} = \pm \frac{\vec{a} \times \vec{b} + \vec{b} \times \vec{c} + \vec{c} \times \vec{a}}{2\Delta} \cdot$$

10. ANGULAR BISECTOR

As discussed earlier, the diagonal of a parallelogram is not necessarily the bisector of the angle formed by two adjacent sides. However, the diagonal of a rhombus bisects the angle formed between two adjacent sides. Consider vectors $\overrightarrow{AB} = \vec{a}$ and $\overrightarrow{AD} = \vec{b}$ forming a parallelogram ABCD as shown in the figure.

Consider the two unit vectors along the given vectors, forming a rhombus AB'C'D'.

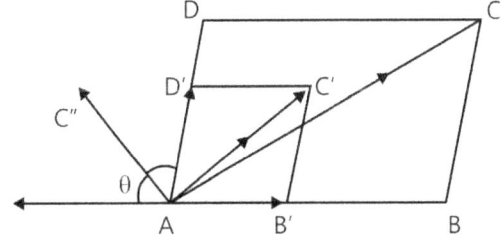

Figure 26.16

Now, $\overrightarrow{AB} = \dfrac{\vec{a}}{|\vec{a}|}$ and $\overrightarrow{AD} = \dfrac{\vec{b}}{|\vec{b}|}$. Therefore $\overrightarrow{AC'} = \dfrac{\vec{a}}{|\vec{a}|} + \dfrac{\vec{b}}{|\vec{b}|}$ Then, any vector along the internal bisector is $\lambda\left(\dfrac{\vec{a}}{|\vec{a}|} + \dfrac{\vec{b}}{|\vec{b}|}\right)$.

Similarly, any vector along the external bisector is $\lambda\left(\dfrac{\vec{a}}{|\vec{a}|} - \dfrac{\vec{b}}{|\vec{b}|}\right)$.

Illustration 30: Find a vector of magnitude 9, which is perpendicular to both the vectors $4\hat{i} - \hat{j} + 3\hat{k}$ and $-2\hat{i} + \hat{j} - 2\hat{k}$.

(JEE MAIN)

Sol: By using property of vector product, we can solve this problem. Let $\vec{a} = 4\hat{i} - \hat{j} + 3\hat{k}$ and $\hat{b} = -2\hat{i} + \hat{j} - 2\hat{k}$. Then,

$$\vec{a} \times \vec{b} = \begin{vmatrix} \hat{i} & \hat{j} & \hat{k} \\ 4 & -1 & 3 \\ -2 & 1 & -2 \end{vmatrix} = (2 - 3)\hat{i} - (-8 + 6)\hat{j} + (4 - 2)\hat{k} = -\hat{i} + 2\hat{j} + 2\hat{k} \Rightarrow |\vec{a} \times \vec{b}| = \sqrt{(-1)^2 + 2^2 + 2^2} = 3$$

$$\therefore \text{Required vector} = 9\left\{\frac{\vec{a} \times \vec{b}}{|\vec{a} \times \vec{b}|}\right\} = \frac{9}{3}(-\hat{i} + 2\hat{j} + 2\hat{k}) = -3\hat{i} + 6\hat{j} + 6\hat{k}$$

Illustration 31: Find the area of a parallelogram, whose adjacent sides are given by the vectors $\vec{a} = 3\hat{i} + \hat{j} + 4\hat{k}$ and $\hat{b} = \hat{i} - \hat{j} + \hat{k}$.

(JEE MAIN)

Sol: The area of a parallelogram with $\vec{a}$ and $\vec{b}$ as its adjacent sides is given by $|\vec{a} \times \vec{b}|$.

$$\text{Now, } \vec{a} \times \vec{b} = \begin{vmatrix} \hat{i} & \hat{j} & \hat{k} \\ 3 & 1 & 4 \\ 1 & -1 & 1 \end{vmatrix} = 5\hat{i} + \hat{j} - 4\hat{k}.$$

Therefore, $|\vec{a} \times \vec{b}| = \sqrt{25 + 1 + 16} = \sqrt{42}$; Hence, the required area is $\sqrt{42}$.

Illustration 32: Let $\vec{a}, \vec{b}, \vec{c}$ be the unit vectors such that $\vec{a} \cdot \vec{b} = \vec{a} \cdot \vec{c} = 0$ and the angle between $\vec{b}$ and $\vec{c}$ is $\dfrac{\pi}{6}$, Prove that $\vec{a} = \pm 2(\vec{b} \times \vec{c})$. **(JEE MAIN)**

Sol: Here $\vec{a} \cdot \vec{b} = \vec{a} \cdot \vec{c} = 0$, therefore $\vec{a}$ is perpendicular to the plane of $\vec{b}$ and $\vec{c}$ and it is parallel to $\vec{b} \times \vec{c}$.

We have $\vec{a} \cdot \vec{b} = \vec{a} \cdot \vec{c} = 0$

$\Rightarrow \vec{a} \perp \vec{b}$ and $\vec{a} \perp \vec{c}$ $\qquad \Rightarrow \vec{a}$ is perpendicular to the plane of $\vec{b}$ and $\vec{c}$.

$\Rightarrow \vec{a}$ is parallel to $\vec{b} \times \vec{c}$. $\Rightarrow \vec{a} = \lambda(\vec{b} \times \vec{c})$ for some scalar λ.

$\Rightarrow |\vec{a}| = |\lambda| |\vec{b}| |\vec{c}| \sin\dfrac{\pi}{6} \Rightarrow 1 = \dfrac{|\lambda|}{2} \qquad \left[\because |\vec{a}| = |\vec{b}| = |\vec{c}| \right]$

$\Rightarrow |\lambda| = 2 \qquad\qquad \Rightarrow \lambda = \pm 2$

$\therefore \vec{a} = \lambda(\vec{b} \times \vec{c}) \qquad \Rightarrow \vec{a} = \pm 2(\vec{b} \times \vec{c})$.

Illustration 33: If $\vec{a}, \vec{b}, \vec{c}$ are three non-zero vectors, such that $\vec{a} \times \vec{b} = \vec{c}$ and $\vec{b} \times \vec{c} = \vec{a}$, prove that $\vec{a}, \vec{b}, \vec{c}$ are mutually at right angles and $|\vec{b}| = 1$ and $|\vec{c}| = |\vec{a}|$. **(JEE MAIN)**

Sol: Use property of vector or cross product to prove this illustration.

We have, $\vec{a} \times \vec{b} = \vec{c}$ and $\vec{b} \times \vec{c} = \vec{a}$

$\Rightarrow \vec{c} \perp \vec{a}, \vec{c} \perp \vec{b}$ and $\vec{a} \perp \vec{b}, \vec{a} \perp \vec{c} \qquad \Rightarrow \vec{a} \perp \vec{b}, \vec{b} \perp \vec{c}$ and $\vec{c} \perp \vec{a}$.

$\Rightarrow \vec{a}, \vec{b}, \vec{c}$ are mutually perpendicular lines.

Again $\vec{a} \times \vec{b} = \vec{c}$ and $\vec{b} \times \vec{c} = \vec{a} \qquad \Rightarrow \left| \vec{a} \times \vec{b} \right| = |\vec{c}|$ and $\left| \vec{b} \times \vec{c} \right| = |\vec{a}|$

$\Rightarrow |\vec{a}| |\vec{b}| \sin\dfrac{\pi}{2} = |\vec{c}|$ and $|\vec{b}| |\vec{c}| \sin\dfrac{\pi}{2} = |\vec{a}| \qquad [\because \vec{a} \perp \vec{b}$ and $\vec{b} \perp \vec{c}]$

$\Rightarrow |\vec{a}| |\vec{b}| = |\vec{c}|$ and $|\vec{b}| |\vec{c}| = |\vec{a}|$

$\Rightarrow |\vec{b}|^2 |\vec{c}| = |\vec{c}| \qquad\qquad \left[\text{Putting } |\vec{a}| = |\vec{b}| |\vec{c}| \text{ in } |\vec{a}| |\vec{b}| = |\vec{c}| \right]$

$\Rightarrow |\vec{b}|^2 = 1 \qquad\qquad \left[\because |\vec{c}| \neq 0 \right]$

$\Rightarrow |\vec{b}| = 1$

Putting $|\vec{b}| = 1$ in $|\vec{a}| |\vec{b}| = |\vec{c}|$, we obtain $|\vec{a}| = |\vec{c}|$.

Illustration 34: Prove by vector method, that in a $\triangle ABC$, $\dfrac{a}{\sin A} = \dfrac{b}{\sin B} = \dfrac{c}{\sin C}$ **(JEE MAIN)**

Sol: As area of triangle ABC is equal to $\dfrac{1}{2} \left| \overrightarrow{AB} \times \overrightarrow{AC} \right| = \dfrac{1}{2} \left| \overrightarrow{BC} \times \overrightarrow{BA} \right| = \dfrac{1}{2} \left| \overrightarrow{CA} \times \overrightarrow{CB} \right|$, therefore by using cross product method we can prove this problem.

Let $\overrightarrow{BC} = \vec{a}, \overrightarrow{CA} = \vec{b}, \overrightarrow{AB} = \vec{c}$. Then

The area of $\triangle ABC = \dfrac{1}{2} \left| \overrightarrow{AB} \times \overrightarrow{AC} \right| = \dfrac{1}{2} \left| \overrightarrow{BC} \times \overrightarrow{BA} \right| = \dfrac{1}{2} \left| \overrightarrow{CA} \times \overrightarrow{CB} \right| \Rightarrow bc \sin A = ca \sin B = ab \sin C$

Dividing the above expression by abc, we get $\dfrac{\sin A}{a} = \dfrac{\sin B}{b} = \dfrac{\sin C}{c}$

Illustration 35: Given the vectors $\vec{a} = \hat{p} + 2\hat{q}$ and $\vec{b} = 2\hat{p} + \hat{q}$, where p and q are unit vectors forming an angle of $30°$. Find the area of the parallelogram constructed on these vectors. **(JEE MAIN)**

Sol: Simply by applying cross product between $\vec{a}$ and $\vec{b}$, we have $\vec{a} \times \vec{b} = (\hat{p} + 2\hat{q}) \times (2\hat{p} + \hat{q}) = -3(\hat{p} \times \hat{q})$.

$$\Rightarrow |\vec{a} \times \vec{b}| = 3|(\hat{p} + \hat{q})| = 3|\hat{p}||\hat{q}|\sin\frac{\pi}{6} = \frac{3}{2}$$

Illustration 36: Let $\overline{OA} = \vec{a}, \overline{OB} = 10\vec{a} + 2\vec{b}$ and $\overline{OC} = \vec{b}$, where O is the origin. Let p denote the area of the quadrilateral OABC and q denote the area of the parallelogram with $\overline{OA}$ and $\overline{OC}$ as adjacent sides. Prove that p = 6q. **(JEE MAIN)**

Sol: We have to obtain the area of quadrilateral and parallelogram using cross product method to get the required result.

We have, p = area of the quadrilateral OABC

$$= 1/2 |\overline{OB} \times \overline{AC}| = 1/2 |\overline{OB} \times (\overline{OC} - \overline{OA})| = 1/2 |(10\vec{a} + 2\vec{b}) \times (\vec{b} - \vec{a})|$$

$$= 1/2 |10(\vec{a} \times \vec{b}) - 10(\vec{a} \times \vec{a}) + 2(\vec{b} \times \vec{b}) - 2(\vec{b} \times \vec{a})|$$

$$= 1/2 |10(\vec{a} \times \vec{b}) - 0 + 0 + 2(\vec{a} \times \vec{b})| = 6(\vec{a} \times \vec{b}) \qquad \text{... (i)}$$

and q = area of the parallelogram with adjacent sides $\overline{OA}$ and $\overline{OC}$

$$= |\overline{OA} \times \overline{OC}| = (\vec{a} \times \vec{b}) \qquad \text{... (ii)}$$

From equations (i) and (ii), we get p = 6q.

Illustration 37: Given that D, E, F are the midpoints of the sides of a triangle ABC, using the vector method, prove that area of $\Delta DEF = \frac{1}{4}$ (area of ΔABC) **(JEE MAIN)**

Sol: Taking A as the origin, let the position vectors of B and C be $\vec{b}$ and $\vec{c}$ respectively.

Then, the position vector of D, E and F are $\frac{1}{2}(\vec{b} + \vec{c}), \frac{1}{2}\vec{c}$ and $\frac{1}{2}\vec{b}$ respectively. Therefore first obtain $\overline{DE}$ and $\overline{DF}$, and after that by applying formula of vector area of triangle DEF we can obtain the required result.

Now, $\overline{DE} = \frac{1}{2}\vec{c} - \frac{1}{2}(\vec{b} + \vec{c}) = \frac{-\vec{b}}{2}$

and $\overline{DF} = \frac{1}{2}\vec{b} - \frac{1}{2}(\vec{b} + \vec{c}) = \frac{-\vec{c}}{2}$

$\therefore$ Vector area of $\Delta DEF = \frac{1}{2}(\overline{DE} \times \overline{DE}) = \left(\frac{-\vec{b}}{2} \times \frac{-\vec{c}}{2}\right)$

$= \frac{1}{8}(\vec{b} \times \vec{c}) = \frac{1}{4}\left\{\frac{1}{2}(\overline{AB} \times \overline{AC})\right\} = \frac{1}{4}$(vector area of ΔABC)

Hence, area of $\Delta DEF = \frac{1}{4}$(area of ΔABC)

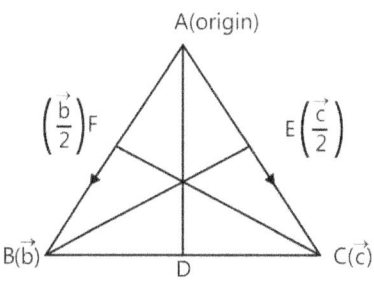

Figure 26.17

Illustration 38: Given that P, Q are the midpoints of the non-parallel sides BC and AD of a trapezium ABCD. Show that area of $\triangle APD = \triangle CQB$. **(JEE MAIN)**

Sol: Use formula of vector area of triangle to solve this problem. Let $\overrightarrow{AB} = \vec{b}$ and $\overrightarrow{AD} = \vec{d}$ Now DC is parallel to $\overrightarrow{AB} \Rightarrow$ there exists a scalar t, such that. $\overrightarrow{DC} = t\overrightarrow{AB} = t\vec{b}$

$\therefore \overrightarrow{AC} = \overrightarrow{AD} + \overrightarrow{DC} = \vec{d} + t\vec{b}$

From geometry we know that $QP = QP = \dfrac{AB + DC}{2}$

Now $\overrightarrow{AP}$ and $\overrightarrow{AQ}$ are $\dfrac{\vec{b} + \vec{d} + t\vec{b}}{2}$ and $\dfrac{\vec{d}}{2}$, respectively.

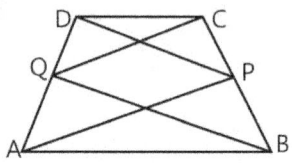

Figure 26.18

Now, $2\triangle APD = \overrightarrow{AP} \times \overrightarrow{AD} = \dfrac{1}{2}(\vec{b} + \vec{d} + t\vec{b}) \times \vec{d} = \dfrac{1}{2}(1+t)(\vec{b} \times \vec{d})$

Also $2\triangle CQB = \overrightarrow{BC} \times \overrightarrow{BQ} = \left[-\vec{b} + \vec{d} + t\vec{b}\right] \times \left[-\vec{b} + \dfrac{\vec{d}}{2}\right]$

$= -(\vec{d} \times \vec{b}) - \dfrac{\vec{b} \times \vec{d}}{2} + \dfrac{t\vec{b} \times \vec{d}}{2} = \dfrac{1}{2}(1+t)\vec{b} \times \vec{d} = 2\triangle APD \Rightarrow \triangle APD = \triangle CQB$

11. TRIPLE PRODUCT OF VECTORS

Two types of triple products are listed below:

Vector triple product $\Rightarrow (a \times b) \times c$

Scalar triple product $\Rightarrow (a \times b).c$

11.1 Scalar Triple Product

The scalar triple product has an interesting geometric interpretation:

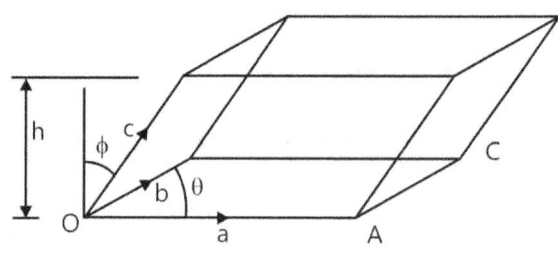

Figure 26.19

We know that $(\vec{a} \times \vec{b}) = |a||b|\sin\theta\hat{n}$ = (area of the parallelogram defined by a and b)

Thus, $(\vec{a} \times \vec{b}).\vec{c}$ = (area of the parallelogram) $\hat{n}.\vec{c}$ = (area of the parallelogram)$|\hat{n}||\vec{c}|\cos\phi$

But $|\vec{c}|\cos\theta = h = $ **height** of the parallelepiped normal to the plane containing $\vec{a}$ and $\vec{b}$. (ϕ is the angle between $\vec{c}$ and $\hat{n}$).

So, $(\vec{a} \times \vec{b}).\vec{c}$ = volume of the parallelepiped defined by $\vec{a}$, $\vec{b}$ and $\vec{c}$. Thus, the following conclusions are arrived:

(a) If any two vectors are parallel, then $(\vec{a} \times \vec{b}).\vec{c} = 0$ (zero volume)

(b) If the three vectors a co-planar, then $(\vec{a} \times \vec{b}).\vec{c} = 0$ (zero volume)

(c) If $(\vec{a} \times \vec{b}).\vec{c} = 0$, then either

    (i) $\vec{a} = 0$, or   (ii) $\vec{b} = 0$ or   (iii) $\vec{c} = 0$ or

    (iv) two of the vectors are parallel or   (v) the three vectors are co-planar

(d) $(\vec{a} \times \vec{b}).\vec{c} = \vec{a}.(\vec{b} \times \vec{c}) = \vec{b}.(\vec{c} \times \vec{a})$ = The same volume.

(e) $(\vec{a} \times \vec{b}).\vec{c}$ is also known as box product, which is represented as $[\vec{a}\vec{b}\vec{c}]$.

    Also $\left[\vec{a}+\vec{b} \quad \vec{c} \quad \vec{d}\right] = \left[\vec{a} \quad \vec{c} \quad \vec{d}\right] + \left[\vec{b} \quad \vec{c} \quad \vec{d}\right]$

(f) If $\vec{a}, \vec{b}, \vec{c}$ are non-coplanar, then $[\vec{a}, \vec{b}, \vec{c}] > 0$, for right-handed system and $[\vec{a}, \vec{b}, \vec{c}] < 0$, for left handed system.

(g) If O is the origin and $\vec{a}, \vec{b}, \vec{c}$ are the position vectors of A, B and C, respectively, of the tetrahedron OABC, then the volume is given by the formula $V = \dfrac{1}{6}\begin{bmatrix} \vec{a} & \vec{b} & \vec{c} \end{bmatrix}$.

Reciprocal system of vectors

(a) If $\vec{a}, \vec{b}, \vec{c}$ and $\vec{a'}, \vec{b'}, \vec{c'}$ are the two sets of non-coplanar vectors, such that $\vec{a}.\vec{a'} = \vec{b}.\vec{b'} = \vec{c}.\vec{c'} = 1$,

$\vec{a}.\vec{b'} = \vec{a}.\vec{c'} = 0$, $\vec{b}.\vec{a'} = \vec{b}.\vec{c'} = 0$ and $\vec{c}.\vec{a'} = \vec{c}.\vec{b'} = 0$,

Then $\vec{a}, \vec{b}, \vec{c}$ and $\vec{a'}, \vec{b'}, \vec{c'}$ constitute a reciprocal system of vectors.

(b) Reciprocal system of vectors exists only in the case of dot product.

(c) $\vec{a'}, \vec{b'}, \vec{c'}$ can be defined in terms of $\vec{a}, \vec{b}, \vec{c}$ as

$$\vec{a'} = \frac{\vec{b} \times \vec{c}}{\begin{bmatrix} \vec{a} & \vec{b} & \vec{c} \end{bmatrix}}; \vec{b'} = \frac{\vec{c} \times \vec{a}}{\begin{bmatrix} \vec{a} & \vec{b} & \vec{c} \end{bmatrix}}; \vec{c'} = \frac{\vec{a} \times \vec{b}}{\begin{bmatrix} \vec{a} & \vec{b} & \vec{c} \end{bmatrix}} \qquad \left(\begin{bmatrix} \vec{a} & \vec{b} & \vec{c} \end{bmatrix} \neq 0 \right)$$

Note:

(i) $\vec{a} \times \vec{a'} + \vec{b} \times \vec{b'} + \vec{c} \times \vec{c'} = 0 \Rightarrow \vec{a} \times (\vec{b} \times \vec{c}) + \vec{b} \times (\vec{c} \times \vec{a}) + \vec{c}(\vec{a} \times \vec{b}) = 0$

(ii) $\vec{a} \cdot \vec{a'} = \vec{b} \cdot \vec{b'} = \vec{c} \cdot \vec{c'} = 1$

(iii) $(\vec{a} + \vec{b} + \vec{c}).(\vec{a'} + \vec{b'} + \vec{c'}) = 3$

(iv) If $\begin{bmatrix} \vec{a} & \vec{b} & \vec{c} \end{bmatrix} = V$ then $\begin{bmatrix} \vec{a'} & \vec{b'} & \vec{c'} \end{bmatrix} = \dfrac{1}{V} \Rightarrow \begin{bmatrix} \vec{a} & \vec{b} & \vec{c} \end{bmatrix}\begin{bmatrix} \vec{a'} & \vec{b'} & \vec{c'} \end{bmatrix} = 1$

Illustration 39: If $\vec{l}, \vec{m}, \vec{n}$ three non-coplanar vectors, then prove that

$$\begin{bmatrix} \vec{l} & \vec{m} & \vec{n} \end{bmatrix}(\vec{a} \times \vec{b}) = \begin{vmatrix} \vec{l} \cdot \vec{a} & \vec{l} \cdot \vec{b} & \vec{l} \\ \vec{m} \cdot \vec{a} & \vec{m} \cdot \vec{b} & \vec{m} \\ \vec{n} \cdot \vec{a} & \vec{n} \cdot \vec{b} & \vec{n} \end{vmatrix}.$$ (JEE ADVANCED)

Sol: Use scalar triple product method as mentioned above to solve this problem.

Let, $\vec{l} = l_1\hat{i} + l_2\hat{j} + l_3\hat{k}$, $\vec{m} = m_1\hat{i} + m_2\hat{j} + m_3\hat{k}$, $\vec{n} = n_1\hat{i} + n_2\hat{j} + n_3\hat{k}$,

and $\vec{a} = a_1\hat{i} + a_2\hat{j} + a_3\hat{k}$, $\vec{b} = b_1\hat{i} + b_2\hat{j} + b_3\hat{k}$

Now $\begin{bmatrix} \vec{l} & \vec{m} & \vec{n} \end{bmatrix} = \begin{vmatrix} l_1 & l_2 & l_3 \\ m_1 & m_2 & m_3 \\ n_1 & n_2 & n_3 \end{vmatrix}$ and $= (\vec{a} \times \vec{b}) = \begin{vmatrix} \hat{i} & \hat{j} & \hat{k} \\ a_1 & a_2 & a_3 \\ b_1 & b_2 & b_3 \end{vmatrix}$

$$\begin{bmatrix} \vec{l} & \vec{m} & \vec{n} \end{bmatrix}(\vec{a} \times \vec{b}) = \begin{vmatrix} l_1 & l_2 & l_3 \\ m_1 & m_2 & m_3 \\ n_1 & n_2 & n_3 \end{vmatrix}\begin{vmatrix} \hat{i} & \hat{j} & \hat{k} \\ a_1 & a_2 & a_3 \\ b_1 & b_2 & b_3 \end{vmatrix} = \begin{vmatrix} l_1\hat{i} + l_2\hat{j} + l_3\hat{k} & \sum l_1 a_2 & \sum l_1 b_1 \\ m_1\hat{i} + m_2\hat{j} + m_3\hat{k} & \sum m_1 a_1 & \sum m_1 b_1 \\ n_1\hat{i} + n_2\hat{j} + n_3\hat{k} & \sum n_1 a_1 & \sum n_1 b_1 \end{vmatrix}$$

Now, $\vec{l}.\vec{a} = (l_1\hat{i} + l_2\hat{j} + l_3\hat{k}).(a_1\hat{i} + a_2\hat{j} + a_3\hat{k}) = \sum l_1 a_2$ etc.

1.21

$$\therefore \begin{bmatrix} \vec{\ell} & \vec{m} & \vec{n} \end{bmatrix}(\vec{a}\times\vec{b}) = \begin{vmatrix} \vec{\ell} & \vec{\ell}\cdot\vec{a} & \vec{\ell}\cdot\vec{b} \\ \vec{m} & \vec{m}\cdot\vec{a} & \vec{m}\cdot\vec{b} \\ \vec{n} & \vec{n}\cdot\vec{a} & \vec{n}\cdot\vec{b} \end{vmatrix} = \begin{vmatrix} \vec{\ell}\cdot\vec{a} & \vec{\ell}\cdot\vec{b} & \vec{\ell} \\ \vec{m}\cdot\vec{a} & \vec{m}\cdot\vec{b} & \vec{m} \\ \vec{n}\cdot\vec{a} & \vec{n}\cdot\vec{b} & \vec{n} \end{vmatrix},$$

Hence proved.

Illustration 40: Find the volume of a parallelepiped, whose sides are given by $-3\hat{i}+7\hat{j}+5\hat{k}, -5\hat{i}+7\hat{j}-3\hat{k}$ and $7\hat{i}-5\hat{j}-3\hat{k}$

(JEE MAIN)

Sol: We know that, the volume of a parallelepiped, whose three adjacent edges are $\vec{a},\vec{b},\vec{c}$ is $\begin{bmatrix} \vec{a} & \vec{b} & \vec{c} \end{bmatrix}$.

Let $\vec{a} = -3\hat{i}+7\hat{j}+5\hat{k}, \vec{b} = -5\hat{i}+7\hat{j}-3\hat{k}$ and $\vec{c} = 7\hat{i}-5\hat{j}-3\hat{k}$

We know that, the volume of a parallelepiped, whose three adjacent edges are $\vec{a},\vec{b},\vec{c}$ is $\begin{bmatrix} \vec{a} & \vec{b} & \vec{c} \end{bmatrix}$

Now, $\begin{bmatrix} \vec{a} & \vec{b} & \vec{c} \end{bmatrix} = \begin{vmatrix} -3 & 7 & 5 \\ -5 & 7 & -3 \\ 7 & -5 & -3 \end{vmatrix} = -3(-21-15)-7(15+21)+5(25-49) = 108-252-120 = -264$

So, the required volume of the parallelepiped $= \begin{bmatrix} \vec{a} & \vec{b} & \vec{c} \end{bmatrix} = |-264| = 264$ cubic units.

Illustration 41: Simplify $\begin{bmatrix} \vec{a}-\vec{b} & \vec{b}-\vec{c} & \vec{c}-\vec{a} \end{bmatrix}$

(JEE ADVANCED)

Sol: Here by using scalar triple product we can simplify this.

$\begin{bmatrix} \vec{a}-\vec{b} & \vec{b}-\vec{c} & \vec{c}-\vec{a} \end{bmatrix} = \left\{ (\vec{a}-\vec{b})\times(\vec{b}-\vec{c}) \right\}.(\vec{c}-\vec{a})$ [by def.]

$= (\vec{a}\times\vec{b} - \vec{a}\times\vec{c} - \vec{b}\times\vec{b} + \vec{b}\times\vec{c}).(\vec{c}-\vec{a})$ [by dist.law]

$= (\vec{a}\times\vec{b} + \vec{c}\times\vec{a} + \vec{b}\times\vec{c}).(\vec{c}-\vec{a})$ $\left[\because \vec{b}\times\vec{b} = 0 \right]$

$= (\vec{a}\times\vec{b}).\vec{c} - (\vec{a}\times\vec{b}).\vec{a} + (\vec{c}\times\vec{a}).\vec{c} - (\vec{c}\times\vec{a}).\vec{a} + (\vec{b}\times\vec{c}).\vec{c} - (\vec{b}\times\vec{c}).\vec{a}$ [by dist. law]

$= \begin{bmatrix} \vec{a} & \vec{b} & \vec{c} \end{bmatrix} - 0 + 0 - 0 + 0 - \begin{bmatrix} \vec{b} & \vec{c} & \vec{a} \end{bmatrix}$

$= \begin{bmatrix} \vec{a} & \vec{b} & \vec{c} \end{bmatrix} - \begin{bmatrix} \vec{b} & \vec{c} & \vec{a} \end{bmatrix} = \begin{bmatrix} \vec{a} & \vec{b} & \vec{c} \end{bmatrix} - \begin{bmatrix} \vec{a} & \vec{b} & \vec{c} \end{bmatrix} = 0$ $\left[\because \begin{bmatrix} \vec{b} & \vec{c} & \vec{a} \end{bmatrix} = \begin{bmatrix} \vec{a} & \vec{b} & \vec{c} \end{bmatrix} \right]$

Illustration 42: Find the volume of the tetrahedron, whose four vertices have position vectors $\vec{a},\vec{b},\vec{c}$ and $\vec{d}$, respectively.

(JEE MAIN)
... (i)

Sol: Here volume of tetrahedron is equal to $\frac{1}{6}\begin{bmatrix} \vec{a}-\vec{d} & \vec{b}-\vec{d} & \vec{c}-\vec{d} \end{bmatrix}$.

Let, four vertices be A, B, C, D with p.v. $\vec{a},\vec{b},\vec{c}$ and $\vec{d}$ respectively.

$\therefore \overline{DA} = (\vec{a}-\vec{d}), \overline{DB} = (\vec{b}-\vec{d}), \overline{DC} = (\vec{c}-\vec{d})$

Hence volume $= \frac{1}{6}\begin{bmatrix} \vec{a}-\vec{d} & \vec{b}-\vec{d} & \vec{c}-\vec{d} \end{bmatrix} = \frac{1}{6}(\vec{a}-\vec{d}).\left[(\vec{b}-\vec{d})\times(\vec{c}-\vec{d}) \right]$

$= \frac{1}{6}(\vec{a}-\vec{d}).\left[\vec{b}\times\vec{c} - \vec{b}\times\vec{d} + \vec{c}\times\vec{d} \right] = \frac{1}{6}\left\{ \begin{bmatrix} \vec{a} & \vec{b} & \vec{c} \end{bmatrix} - \begin{bmatrix} \vec{a} & \vec{b} & \vec{d} \end{bmatrix} + \begin{bmatrix} \vec{a} & \vec{c} & \vec{d} \end{bmatrix} - \begin{bmatrix} \vec{d} & \vec{b} & \vec{c} \end{bmatrix} \right\}$

$= \frac{1}{6}\left\{ \begin{bmatrix} \vec{a} & \vec{b} & \vec{c} \end{bmatrix} - \begin{bmatrix} \vec{d} & \vec{b} & \vec{c} \end{bmatrix} - \begin{bmatrix} \vec{a} & \vec{d} & \vec{c} \end{bmatrix} - \begin{bmatrix} \vec{a} & \vec{b} & \vec{d} \end{bmatrix} \right\}$.

Illustration 43: Let $\vec{u}$ and $\vec{v}$ be unit vectors and $\vec{w}$ is a vector, such that $\vec{u} \times \vec{v} + \vec{u} = \vec{w}$ and $\vec{w} \times \vec{u} = \vec{v}$, then find the value of $\begin{bmatrix} \vec{u} & \vec{v} & \vec{w} \end{bmatrix}$. **(JEE ADVANCED)**

Sol: Here as given $\vec{u} \times \vec{v} + \vec{u} = \vec{w}$ and $\vec{w} \times \vec{u} = \vec{v}$, solve it using scalar triple product.

Given, $\vec{u} \times \vec{v} + \vec{u} = \vec{w}$ and $\vec{w} \times \vec{u} = \vec{v}$

$\Rightarrow (\vec{u} \times \vec{v} + \vec{u}) \times \vec{u} = \vec{w} \times \vec{u}$

$\Rightarrow (\vec{u} \times \vec{v}) \times \vec{u} + \vec{u} \times \vec{u} = \vec{v}$ $\quad$ (as, $\vec{w} \times \vec{u} = \vec{v}$)

$\Rightarrow (\vec{u}.\vec{u})\vec{v} - (\vec{v}.\vec{u})\vec{u} + \vec{u} \times \vec{u} = \vec{v}$ $\quad$ (using $\vec{u}.\vec{u} = 1$ and $\vec{u} \times \vec{u} = 0$, since unit vector)

$\Rightarrow \vec{v} - (\vec{v}.\vec{u})\vec{u} = \vec{v} \Rightarrow (\vec{u}.\vec{v})\vec{u} = \vec{0}$

$\Rightarrow \vec{u}.\vec{v} = 0$ $\quad\quad$ (as; $\vec{u} \neq 0$) $\quad$ ....(i)

$\therefore \begin{bmatrix} \vec{u} & \vec{v} & \vec{w} \end{bmatrix} = \vec{u}.(\vec{v} \times \vec{w})$

$= \vec{u}.(\vec{v} \times (\vec{u} \times \vec{v} + \vec{u}))$ $\quad$ (given $\vec{w} = \vec{u} \times \vec{v} + \vec{u}$)

$= \vec{u}.(\vec{v} \times (\vec{u} \times \vec{v}) + \vec{v} \times \vec{u}) = \vec{u}.((\vec{v}.\vec{v})\vec{u} - (\vec{v}.\vec{u})\vec{v} + \vec{v} \times \vec{u})$

$= \vec{u}.(\vec{u} - 0 + \vec{v} \times \vec{u})$ (as $\vec{u}.\vec{v} = 0$ from (i))

$= (\vec{u}.\vec{u}) - \vec{u}.(\vec{v} \times \vec{u}) = 1 - 0 = 1$

$\therefore \begin{bmatrix} \vec{u} & \vec{v} & \vec{w} \end{bmatrix} = 1$

11.2 Vector Triple Product

Definition: $(\vec{a} \times \vec{b}) \times \vec{c}$ is a vector, which is coplanar to $\vec{a}$ and $\vec{b}$ and perpendicular to $\vec{c}$.

Hence $(\vec{a} \times \vec{b}) \times \vec{c} = x\vec{a} + y\vec{b}$ $\quad\quad$ [Linear Combination of $\vec{a}$ and $\vec{b}$] $\quad\quad\quad$ ... (i)

$\vec{c}.(\vec{a} \times \vec{b}) \times \vec{c} = x(\vec{a}.\vec{c}) + y(\vec{b}.\vec{c})$ $\quad\quad\quad\quad\quad\quad\quad\quad\quad\quad\quad\quad\quad$ ... (ii)

$0 = x(\vec{a}.\vec{c}) + y(\vec{b}.\vec{c})$

$\dfrac{x}{\vec{b}.\vec{c}} = -\dfrac{y}{\vec{a}.\vec{c}} = \lambda$

$\therefore x = \lambda(\vec{b}.\vec{c})$ and $y = -\lambda(\vec{a}.\vec{c})$

Substituting the values of x and y we get, $(\vec{a} \times \vec{b}) \times \vec{c} = \lambda(\vec{b}.\vec{c})\vec{a} - \lambda(\vec{a}.\vec{c})\vec{b}$

This identity must hold true for all values of $\vec{a}, \vec{b}, \vec{c}$

Substitute $\vec{a} = \hat{i}; \vec{b} = \hat{j}$ and $\vec{c} = \hat{k}$

$(\hat{i} \times \hat{j}) \times \hat{i} = \lambda(\hat{j}.\hat{i})\hat{i} - \lambda(\hat{i}.\hat{i})\hat{j}$

$\hat{j} = -\lambda\hat{j} \Rightarrow \lambda = -1$

$\Rightarrow (\vec{a} \times \vec{b}) \times \vec{c} = (\vec{a}.\vec{c})\vec{b} - (\vec{b}.\vec{c})\vec{a}$

Note: Unit vector coplanar with $\vec{a}$ and $\vec{b}$ perpendicular to $\vec{a}$ is $\pm\dfrac{(\vec{a} \times \vec{b}) \times \vec{a}}{\left|(\vec{a} \times \vec{b}) \times \vec{a}\right|}$

Illustration 44: Prove that $\vec{a} \times \left\{ \vec{b} \times \left(\vec{c} \times \vec{d} \right) \right\} = \left(\vec{b}.\vec{d} \right)\left(\vec{a} \times \vec{c} \right) - \left(\vec{b}.\vec{c} \right)\left(\vec{a} \times \vec{d} \right)$ **(JEE MAIN)**

Sol: By using vector triple product as mention above.

We have, $\vec{a} \times \left\{ \vec{b} \times \left(c \times d \right) \right\} = \vec{a} \times \left\{ \left(\vec{b}.\vec{d} \right)\vec{c} - \left(\vec{b}.\vec{c} \right)\vec{d} \right\}$

$= \vec{a} \times \left\{ \left(\vec{b}.\vec{d} \right)\vec{c} - \vec{a}\left(\vec{b}.\vec{c} \right)\vec{d} \right\}$ [by distributive law]

$= \left(\vec{b}.\vec{d} \right)\left(\vec{a} \times \vec{c} \right) - \left(\vec{b}.\vec{c} \right)\left(\vec{a} \times \vec{d} \right)$

Illustration 45: Let $\vec{a} = a\hat{i} + 2\hat{j} - 3\hat{k}, \vec{b} = \hat{i} + 2a\hat{j} - 2\hat{k}$, and $\vec{c} = 2\hat{i} + a\hat{j} - \hat{k}$. Find the value (s) of a, if any, such that $\left\{ \left(\vec{a} \times \vec{b} \right) \times \left(\vec{b} \times \vec{c} \right) \right\} \times \left(\vec{c} \times \vec{a} \right) = 0.$ **(JEE MAIN)**

Sol: Here use vector triple product to obtain the value of a.

$\left\{ \left(\vec{a} \times \vec{b} \right) \times \left(\vec{b} \times \vec{c} \right) \right\} \times \left(\vec{c} \times \vec{a} \right) = \begin{bmatrix} \vec{a} & \vec{b} & \vec{c} \end{bmatrix} \vec{b} \times \left(\vec{c} \times \vec{a} \right) = \begin{bmatrix} \vec{a} & \vec{b} & \vec{c} \end{bmatrix} \left\{ \left(\vec{a}.\vec{b} \right)\vec{c} - \left(\vec{b}.\vec{c} \right)\vec{a} \right\},$

Given $\left\{ \left(\vec{a} \times \vec{b} \right) \times \left(\vec{b} \times \vec{c} \right) \right\} \times \left(\vec{c} \times \vec{a} \right) = 0 \Rightarrow \left(\vec{a}.\vec{b} \right)\vec{c} = \left(\vec{b}.\vec{c} \right)\vec{a}$ or $\begin{bmatrix} \vec{a} & \vec{b} & \vec{c} \end{bmatrix} = 0$

$\left(\vec{a}.\vec{b} \right)\vec{c} = \left(\vec{b}.\vec{c} \right)\vec{a}$ leads to three different equations which do not have a common solution.

$\begin{bmatrix} \vec{a} & \vec{b} & \vec{c} \end{bmatrix} = 0 \Rightarrow \begin{vmatrix} a & 2 & -3 \\ 1 & 2a & -2 \\ 2 & a & -1 \end{vmatrix} = 0 \Rightarrow 9a - 6 = 0 \Rightarrow a = \frac{2}{3}$

Illustration 46: Solve for $\vec{r}$, from the simultaneous equations $\vec{r} \times \vec{b} = \vec{c} \times \vec{b}, \vec{r}.\vec{a} = 0$, provided $\vec{a}$ is not perpendicular to $\vec{b}$. **(JEE MAIN)**

Sol: As given $\vec{r} \times \vec{b} = \vec{c} \times \vec{b}$, solve this using vector triple product to get the result.

Given $\vec{r} \times \vec{b} = \vec{c} \times \vec{b}$

$\Rightarrow \left(\vec{r} - \vec{c} \right) \times \vec{b} = 0$ $\Rightarrow \left(\vec{r} - \vec{c} \right)$ and $\vec{b}$ are collinear

$\therefore \vec{r} - \vec{c} = k\vec{b}$ $\Rightarrow \vec{r} = \vec{c} + k\vec{b}$ ....(i)

$\vec{r}.\vec{a} = 0$ $\Rightarrow \left(\vec{c} + k\vec{b} \right).\vec{a} = 0$

$\Rightarrow k = -\dfrac{\vec{a}.\vec{c}}{\vec{a}.\vec{b}}$ putting in eq. (i) we get $\vec{r} = \vec{c} - \dfrac{\vec{a}.\vec{c}}{\vec{a}.\vec{b}}\vec{b} = \dfrac{\vec{a} \times \left(\vec{c} \times \vec{b} \right)}{\vec{a}.\vec{b}}$.

Illustration 47: If $\vec{x} \times \vec{a} + k\vec{x} = \vec{b}$, where k is a scalar and $\vec{a}, \vec{b}$ are any two vectors, then determine $\vec{x}$ in terms of $\vec{a}, \vec{b}$ and k. **(JEE MAIN)**

Sol: Here as given $\vec{x} \times \vec{a} + k\vec{x} = \vec{b}$, Apply cross product of $\vec{a}$ with both side and solve using vector triple product.

$\vec{x} \times \vec{a} + k\vec{x} = \vec{b}$ ... (i)

$\Rightarrow \vec{a} \times \left(\vec{x} \times \vec{a} \right) + k\left(\vec{a} \times \vec{x} \right) = \left(\vec{a} \times \vec{b} \right)$

$\Rightarrow \left(\vec{a}.\vec{a} \right)\vec{x} - \left(\vec{a}.\vec{x} \right)\vec{a} + k\left(\vec{a} \times \vec{x} \right) = \vec{a} \times \vec{b}$ ... (ii)

(i) $\Rightarrow \vec{a}.\left(\vec{x} \times \vec{a} \right) + k\left(\vec{a}.\vec{x} \right) = \vec{a}.\vec{b}$

$\Rightarrow \quad k\left(\vec{a}.\vec{x} \right) = \vec{a}.\vec{b}$ ... (iii)

Substituting the values from equations (i) and (iii) in equation (ii), we get,

$$\Rightarrow (\vec{a}.\vec{a})\vec{x} - \frac{1}{k}(\vec{a}.\vec{b})\vec{a} + k(k\vec{x} - \vec{b}) = \vec{a} \times \vec{b}$$

$$\Rightarrow (a^2 + k^2)\vec{x} = (\vec{a} \times \vec{b}) + \frac{1}{k}(\vec{a}.\vec{b})\vec{a} + k\vec{b} \quad \Rightarrow \vec{x} = \frac{1}{a^2 + k^2}\left[k\vec{b} + (\vec{a} \times \vec{b}) + \frac{\vec{a}.\vec{b}}{k}\vec{a}\right]$$

12. APPLICATION OF VECTORS IN 3D GEOMETRY

(a) Direction cosines of $\vec{r} = a\hat{i} + b\hat{j} + c\hat{k}$ are given by $\dfrac{a}{|\vec{r}|}, \dfrac{b}{|\vec{r}|}, \dfrac{c}{|\vec{r}|}$.

(b) Incentre formula: The position vector of the incentre of $\triangle ABC$ is $\dfrac{a\vec{a} + b\vec{b} + c\vec{c}}{a + b + c}$

(c) Orthocentre formula: The position vector of the orthocenter of

$\triangle ABC$ is $\dfrac{\vec{a}\tan A + \vec{b}\tan B + \vec{c}\tan C}{\tan A + \tan B + \tan C}$

(d) The vector equation of a straight line passing through a fixed point with position vector $\vec{a}$ and parallel to a given vector $\vec{b}$ is given by $\vec{r} = \vec{a} + \lambda\vec{b}$.

(e) The vector equation of a line passing through two points with position vectors $\vec{a}$ and $\vec{b}$ is given by $\vec{r} = \vec{a} + \lambda(\vec{b} - \vec{a})$.

(f) Perpendicular distance of a point from a line: Let L be the foot of perpendicular drawn $P(\vec{\alpha})$ on the line $\vec{r} = \vec{a} + \lambda\vec{b}$, where $\vec{r}$ is the position vector of any point on the give line. Therefore, let the position vector L be $a + \lambda b$.

$$PL = \frac{\left|(\vec{a} - \vec{\alpha}) \times \vec{b}\right|}{|\vec{b}|} \text{ and } \overline{PL} = \vec{a} - \vec{\alpha} + \lambda\vec{b} = (\vec{a} - \vec{\alpha}) - \left(\frac{(\vec{a} - \vec{\alpha}).\vec{b}}{|\vec{b}|^2}\right)\vec{b}$$

The length PL is the magnitude of $\overline{PL}$, and the required length of perpendicular.

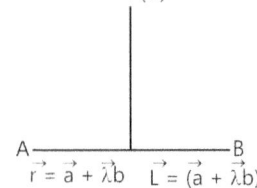

Figure 26.20

(g) Image of a point in a straight line: If $Q(\vec{\beta})$ is the image of P in $\vec{r} = \vec{a} + \lambda\vec{b}$, then

$$\vec{\beta} = 2\vec{a} - \left(\frac{2(\vec{a} - \vec{\alpha}).\vec{b}}{|\vec{b}|^2}\right)\vec{b} - \vec{\alpha}$$

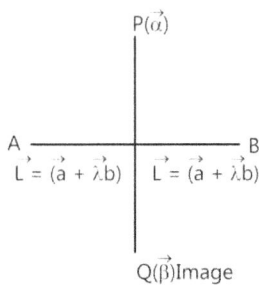

Figure 26.21

(h) **Shortest distance between two skew lines:** Let I_1 and I_2 be two lines whose equations are $I_1 : \vec{r} = \vec{a}_1 + \lambda\vec{b}_1$ and $I_2 : \vec{r} = \vec{a}_2 + \lambda\vec{b}_2$, respectively.

Then, shortest distance is given by $PQ = \left| \dfrac{\left(\vec{b}_1 \times \vec{b}_2\right).\left(\vec{a}_2 - \vec{a}_1\right)}{\left|\vec{b}_1 \times \vec{b}_2\right|} \right| = \left| \dfrac{\left[\vec{b}_1 \quad \vec{b}_2 \quad \vec{a}_2 - \vec{a}_1\right]}{\left|\vec{b}_1 \times \vec{b}_2\right|} \right|$

Shortest distance between two parallel lines: The shortest distance between the two given parallel lines

$\vec{r} = \vec{a}_1 + \lambda\vec{b}$ and $\vec{r} = \vec{a}_2 + \mu\vec{b}$ is given by $d = \dfrac{\left|\left(\vec{a}_2 - \vec{a}_1\right) \times \vec{b}\right|}{\left|\vec{b}\right|}$.

If the lines $\vec{r} = \vec{a}_1 + \lambda\vec{b}_1$ and $\vec{r} = \vec{a}_2 + \mu\vec{b}_2$ intersect, then the shortest distance between them is zero.

Therefore, $\left[\vec{b}_1 \quad \vec{b}_2 \quad \vec{a}_2 - \vec{a}_1\right] = 0$

(i) If the lines $\vec{r} = \vec{a}_1 + \lambda\vec{b}_1$ and $\vec{r} = \vec{a}_2 + \lambda\vec{b}_2$ are coplanar, then $\left[\vec{a}_1 \quad \vec{b}_1 \quad \vec{b}_2\right] = \left[\vec{a}_2 \quad \vec{b}_1 \quad \vec{b}_2\right]$ and the equation of the plane containing them is given by $\left[\vec{r} \quad \vec{b}_1 \quad \vec{b}_2\right] = \left[\vec{a}_1 \quad \vec{b}_1 \quad \vec{b}_2\right]$.

(j) The vector equation of a plane through the point A ($\vec{a}$) and perpendicular to the vector $\vec{a} \times \left(\vec{b} + \vec{c}\right) = \vec{a} \times \vec{b} + \vec{a} \times \vec{c}$ is given by $\left(\vec{r} - \vec{a}\right).\vec{n} = 0$.

(k) Vector The vector equation of a plane normal to unit vector $\hat{n}$ and at a distance d from the origin is given by $\vec{r}.\hat{n} = d$.

(l) The equation of the plane passing through a point having position vector $\vec{a}$ and parallel to $\vec{b}$ and $\vec{c}$ is given by $\vec{r} = \vec{a} + \lambda\vec{b} + \mu\vec{c} \Rightarrow \left[\vec{r} \quad \vec{b} \quad \vec{c}\right] = \left[\vec{a} \quad \vec{b} \quad \vec{c}\right]$, where λ and μ are scalars.

(m) The vector equation of plane passing through a point $\vec{a}, \vec{b}, \vec{c}$ is given by $\vec{r} = (1-s-t)\vec{a} + s\vec{b} + t\vec{c}$

or $\vec{r} \cdot \left(\vec{b} \times \vec{c} + \vec{c} \times \vec{a} + \vec{a} \times \vec{b}\right) = \left[\vec{a} \quad \vec{b} \quad \vec{c}\right]$

(n) The equation of any plane through the intersection of planes $\vec{r}.\vec{n}_1 = d_1$ and $\vec{r}.\vec{n}_2 = d_2$ is $\vec{r}.\left(\vec{n}_1 + \lambda\vec{n}_2\right) = d_1 + \lambda d_2$, where λ is an arbitrary constant.

(o) The perpendicular distance of a point having position vector $\vec{a}$ from the plane $\vec{r}.\vec{n} = d$ is given by $p = \dfrac{\left|\vec{a}.\vec{n} - d\right|}{\left|\vec{n}\right|}$.

(p) The angle θ between the planes $\vec{r}_1.\hat{n}_1 = d_1$ and $\vec{r}_2.\hat{n}_2 = d_2$ is given by $\cos\theta = \pm\dfrac{\vec{n}_1.\vec{n}_2}{\left|\vec{n}_1\right|\left|\vec{n}_2\right|}$

(q) The perpendicular distance of a point P($\vec{r}$) from a line passing through $\vec{a}$ and parallel to $\vec{b}$ is given by

$P = \dfrac{\left|\left(\vec{r} - \vec{a}\right) \times \vec{b}\right|}{\left|\vec{b}\right|} = \left[\left(\vec{r} - \vec{a}\right)^2 - \left\{\dfrac{\left(\vec{r} - \vec{a}\right).\vec{b}}{\left|\vec{b}\right|}\right\}^2\right]^{1/2}$

(r) The equation of the planes bisecting the angles between the planes $\vec{r}_1.\vec{n}_1 = d_1$ and $\vec{r}_2.\vec{n}_2 = d_2$ is

$\vec{r}.\left(\vec{n}_1 \pm \vec{n}_2\right) = \dfrac{d_1}{\left|\vec{n}_1\right|} \pm \dfrac{d_2}{\left|\vec{n}_2\right|}$

(s) The perpendicular distance of a point P($\vec{r}$) from a plane passing through a point $\vec{a}$ and parallel to points $\vec{b}$ and $\vec{c}$ is given by $PM = \dfrac{\left(\vec{r} - \vec{a}\right).\left(\vec{b} \times \vec{c}\right)}{\left|\vec{b} \times \vec{c}\right|}$

(t) The perpendicular distance of a point $P(\vec{r})$ from a plane passing through the points $\vec{a}$, $\vec{b}$ and $\vec{c}$ is given by

$$P = \frac{(\vec{r}-\vec{a}).(\vec{b}\times\vec{c}+\vec{c}\times\vec{a}+\vec{a}\times\vec{b})}{\left|\vec{b}\times\vec{c}+\vec{c}\times\vec{a}+\vec{a}\times\vec{b}\right|}$$

(u) **Angle between a line and the plane:** If θ is the angle between a line $\vec{r}=\left(\vec{a}+\lambda\vec{b}\right)$ and the plane $\vec{r}.\vec{n}=d$, then

$$\sin\theta = \frac{\vec{b}.\vec{n}}{\left|\vec{b}\right|\left|\vec{n}\right|}.$$

(v) The equation of sphere with center at $C(\vec{c})$ and radius 'a' is $\left|\vec{r}-\vec{c}\right|=a$. If center is the origin then $\left|\vec{r}\right|=a$.

(w) The plane $\vec{r}.\vec{n}=d$ touches the sphere $\left|\vec{r}-\vec{a}\right|=R$, if $\dfrac{\left|\vec{a}.\vec{n}-d\right|}{\left|\vec{n}\right|}=R$, i.e. the condition of tangency.

(x) If $\vec{a}$ and $\vec{b}$ are the position vectors of the extremities of a diameter of a sphere, then its equation is given by $\left(\vec{r}-\vec{a}\right).\left(\vec{r}-\vec{b}\right)=0$ or $\left|\vec{r}\right|^2-\vec{r}.\left(\vec{a}+\vec{b}\right)+\vec{a}.\vec{b}=0$ or $\left|\vec{r}-\vec{a}\right|^2+\left|\vec{r}-\vec{b}\right|^2=\left|\vec{a}-\vec{b}\right|^2$.

FORMULAE SHEET

(a) $\overrightarrow{OP}=x\,\hat{i}+y\,\hat{j}$

(b) $\left|\overrightarrow{OP}\right|=\sqrt{x^2+y^2}$ and direction is $\tan\theta=\dfrac{y}{x}$

(c) Unit vector $\hat{U}=\dfrac{\text{Vector}}{\text{Its modulus}}=\dfrac{\vec{a}}{\left|\vec{a}\right|}$

(d) Properties of vector addition:

i. $\vec{a}+\vec{b}=\vec{b}+\vec{a}$ commutative	(a) $\vec{a}+\left(\vec{b}+\vec{c}\right)=\left(\vec{a}+\vec{b}\right)+\vec{c}$ Associative
ii. $\vec{a}+\vec{0}=\vec{a}$ Null vector is an additive identity	(b) $\vec{a}+\left(-\vec{a}\right)=\vec{0}$ Additive inverse
iii. $c\left(\vec{a}+\vec{b}\right)=c\vec{a}+c\vec{b}$	(c) $(c+d)\vec{a}=c\vec{a}+d\vec{a}$
iv. $(cd)\vec{a}=c(d\vec{a})$	(d) $1\times\vec{a}=\vec{a}$

(e) **Section formula:**

(i) If $\vec{a}$ and $\vec{b}$ are the position vectors of two points A and B, then the position vector of a point which divides AB in the ratio m:n is given by $\vec{r}=\dfrac{\left(n\vec{a}+m\vec{b}\right)}{(m+n)}$.

(ii) Position vector of mid-point of $\overline{AB}=\dfrac{\left(\vec{a}+\vec{b}\right)}{2}$.

(f) Collinearity of three points: If $\vec{a}$, $\vec{b}$, and $\vec{c}$ are the position vectors (non-zero) of three points and given they are collinear then there exists λ, γ both not being 0 such that $\vec{a} + \lambda\vec{b} + \gamma\vec{c}$

(g) Coplanar vectors: Let $\vec{a}$, $\vec{b}$ be non-zero, non-collinear vectors. Then, any vector $\vec{r}$ coplanar with $\vec{a}$, $\vec{b}$ can be expressed uniquely as a linear combination of $\vec{a}$, $\vec{b}$ i.e. there exist some unique x, y $\in$ R, such that $x\vec{a} + y\vec{b} = \vec{r}$

(h) Product of two vectors:

(i) Scalar Product (dot product)

If $\vec{a}.\vec{b} = a_1b_1 + a_2b_2 + a_3b_3$

Note : • $\cos\theta = \dfrac{\vec{a}.\vec{b}}{|\vec{a}||\vec{b}|}$

• $\vec{a}$ and $\vec{b}$ are perpendicular if $\theta = 90°$

(ii) Properties of scalar product:

i. $\vec{a}.\vec{b} = \vec{b}.\vec{a}$	ii. $m\vec{a}.n\vec{b} = mn\vec{a}.\vec{b} = \vec{a}(.mn\vec{b})$
iii. $\vec{a}.(\vec{b}+\vec{c}) = \vec{a}.\vec{c} + \vec{a}.\vec{b}$	iv. $(\vec{a}+\vec{b})^2 = \vec{a}^2 + 2.\vec{a}.\vec{b} + \vec{b}^2$
v. If $\hat{i} = (1,0,0), \hat{j} = (0,1,0), \hat{k} = (0,0,1)$ then $\hat{i}.\hat{j} = \hat{j}.\hat{k} = \hat{k}.\hat{i} = 0$	

(iii) Vector (cross) Product of two vectors: Let $\vec{a} = (a_1, a_2, a_3), \vec{b} = (b_1, b_2, b_3)$ be two vectors then the cross product of $\vec{a}\times\vec{b}$ is devoted by $\vec{a}\times\vec{b}$ and defined by

$$\vec{a}\times\vec{b} = (a_1, a_2, a_3)\times(b_1, b_2, b_3) = \begin{pmatrix} a_2 & a_3 & a_1 & a_2 \\ b_2 & b_3 & b_1 & b_2 \end{pmatrix} = (a_2b_3 - a_3b_2, a_3b_1 - a_1b_3, a_1b_2 - a_2b_1)$$

OR

$$\vec{a}\times\vec{b} = \begin{vmatrix} \hat{i} & \hat{j} & \hat{k} \\ a_1 & a_2 & a_3 \\ b_1 & b_2 & b_3 \end{vmatrix} = (a_2b_3 - a_3b_2)\hat{i} + (a_3b_1 - a_1b_3)\hat{j} + (a_1b_2 - a_2b_1)\hat{k}$$

$$\left|\vec{a}\times\vec{b}\right| = |\vec{a}|\times|\vec{b}|\sin\theta\,\hat{n}$$

Note: (i) θ being angle between $\vec{a}$ & $\vec{b}$

(ii) If $\theta = 0$, The $\left|\vec{a}\times\vec{b}\right| = 0$ i.e. $\vec{a}\times\vec{b} = 0$ and $\vec{a}$ & $\vec{b}$ are parallel if $\vec{a}\times\vec{b} = 0$.

(iv) Properties of cross product

i. $\vec{a} \times \vec{b} = 0 \Rightarrow \vec{a} = 0$ or $\vec{b} = 0$ or $\vec{a} \parallel \vec{b}$	ii. $\vec{a} \times \vec{b} = -\vec{b} \times \vec{a}$
iii. $\vec{a} \times (\vec{b} + \vec{c}) = \vec{a} \times \vec{b} + \vec{a} \times \vec{c}$	iv. $(n\vec{a}) \times \vec{b} = n(\vec{a} \times \vec{b})$
v. $\vec{a} \times \vec{b}$ is perpendicular to both $\vec{a}$ and $\vec{b}$	vi. $\|\vec{a} \times \vec{b}\|$ is a Area of parallelogram with sides $\vec{a}$ and $\vec{b}$.

(v) Scalar Triple Product: If $\vec{a} = a_1\hat{i} + a_2\hat{j} + a_3\hat{k}, \vec{b} = b_1\hat{i} + b_2\hat{j} + b_3\hat{k}, \vec{c} = c_1\hat{i} + c_2\hat{j} + c_3\hat{k}$.

Then $\vec{a}.(\vec{b} \times \vec{c}) = \vec{b}.(\vec{c} \times \vec{a}) = \vec{c}.(\vec{a} \times \vec{b})$

$$\vec{a}.(\vec{b} \times \vec{c}) = \begin{vmatrix} a_1 & a_2 & a_3 \\ b_1 & b_2 & b_3 \\ c_1 & c_2 & c_3 \end{vmatrix}$$

$\vec{a}.(\vec{b} \times \vec{c})$ is also represented as $[\vec{a}\ \vec{b}\ \vec{c}]$

$$[\vec{a}\ \vec{b}\ \vec{c}] = [\vec{b}\ \vec{c}\ \vec{a}] = [\vec{c}\ \vec{a}\ \vec{b}]$$

$$[\vec{a}\ \vec{b}\ \vec{c}] = -[\vec{a}\ \vec{c}\ \vec{b}]$$

- If any of the two vectors are parallel, then $[\vec{a}\ \ \vec{b}\ \ \vec{c}] = 0$

- $[\vec{a}\ \ \vec{b}\ \ \vec{c}]$ is the volume of the parallelepiped whose coterminous edges are formed by $\vec{a}\ \vec{b}\ \vec{c}$

- If $\vec{a}\ \vec{b}\ \vec{c}$ are coplanar, $[\vec{a}\ \ \vec{b}\ \ \vec{c}] = 0$

- $\frac{1}{2}\|\vec{a} \times \vec{b} + \vec{b} \times \vec{c} + \vec{c} \times \vec{a}\|$ = area of triangle having $\vec{a}, \vec{b}, \vec{c}$ as position vectors of vertices of a triangle.

(vi) Vector Triple Product:

$$\vec{a} \times (\vec{b} \times \vec{c}) = (\vec{a}.\vec{c})\vec{b} - (\vec{a}.\vec{b})\vec{c}$$

$$(\vec{a} \times \vec{b}) \times \vec{c} = (\vec{a}.\vec{c})\vec{b} - (\vec{b}.\vec{c})\vec{a}$$

Unit vector coplanar with $\vec{a}$ and $\vec{b}$ perpendicular to $\vec{a}$ is $\pm \dfrac{(\vec{a} \times \vec{b}) \times \vec{a}}{\|(\vec{a} \times \vec{b}) \times \vec{a}\|}$.

JEE Main/Boards

Example 1: Show that the points A, B & C with position vector $2\hat{i} - \hat{j} + \hat{k}, \hat{i} - 3\hat{j} - 5\hat{k}$ and $3\hat{i} - 4\hat{j} - 4\hat{k}$ respectively, are the vertices of a right angled triangle. Also find the remaining angles of the triangle.

Sol: We have,

$$\overline{AB} = \left(\hat{i} - 3\hat{j} - 5\hat{k}\right) - \left(2\hat{i} - \hat{j} + \hat{k}\right) = -\hat{i} - 2\hat{j} - 6\hat{k}$$

$$\overline{BC} = \left(3\hat{i} - 4\hat{j} - 4\hat{k}\right) - \left(\hat{i} - 3\hat{j} - 5\hat{k}\right) = 2\hat{i} - \hat{j} + \hat{k} \text{ and}$$

$$\overline{CA} = \left(2\hat{i} - \hat{j} + \hat{k}\right) - \left(3\hat{i} - 4\hat{j} - 4\hat{k}\right) = -\hat{i} + 3\hat{j} + 5\hat{k}$$

Since $\overline{AB} + \overline{BC} + \overline{CA}$

$$= \left(-\hat{i} - 2\hat{j} - 6\hat{k}\right) + \left(2\hat{i} - \hat{j} + \hat{k}\right) + \left(-\hat{i} + 3\hat{j} + 5\hat{k}\right) = \vec{0}$$

So, A, B and C are the vertices of a triangle.

Now, $\overline{BC}.\overline{CA}$

$$= \left(2\hat{i} - \hat{j} + \hat{k}\right) + \left(-\hat{i} + 3\hat{j} + 5\hat{k}\right) = -2 - 3 + 5 = 0$$

$$\overline{BC} \perp \overline{CA} \Rightarrow \angle BCA = \frac{\pi}{2}$$

Hence, ABC is a right angled triangle. Since a is the angle between the vectors $\overline{AB}$ and $\overline{AC}$. Therefore,

$$\cos A = \frac{\overline{AB}.\overline{AC}}{|\overline{AB}||\overline{AC}|}$$

$$= \frac{\left(-\hat{i} - 2\vec{j} - 6\vec{k}\right).\left(\hat{i} - 3\hat{j} - 5\hat{k}\right)}{\sqrt{\left(-1\right)^2 + \left(-2\right)^2 + \left(-6\right)^2}\sqrt{1^2 + \left(-3\right)^2 + \left(-5\right)^2}}$$

$$= \frac{-1 + 6 + 30}{\sqrt{1+4+36}\sqrt{1+9+25}} = \frac{35}{\sqrt{41}\sqrt{35}} = \sqrt{\frac{35}{41}}$$

$$A = \cos^{-1}\sqrt{\frac{35}{41}}, \cos B = \frac{\overline{BA}}{|\overline{BA}||\overline{BC}|}$$

$$\frac{\left(\hat{i} + 2\hat{j} + 6\hat{k}\right).\left(2\hat{i} - \hat{j} + \hat{k}\right)}{\sqrt{1^2 + 2^2 + 6^2}\sqrt{2^2 + \left(-1\right)^2 + \left(1\right)^2}}$$

$$\Rightarrow \cos B = \frac{2-2+6}{\sqrt{41}\sqrt{6}} = \sqrt{\frac{6}{41}} \Rightarrow B = \cos^{-1}\sqrt{\frac{6}{41}}$$

Example 2: If ABCDE is a pentagon, prove that the resultant of $\overline{AB}, \overline{AE}, \overline{BC}, \overline{DC}, \overline{ED}$ and $\overline{AC}$ is $3\overline{AC}$

Sol: By using resultant vector formula, we can obtain required result.

If R be the resultant vector then

$$R = \overline{AB} + \overline{AE} + \overline{BC} + \overline{DC} + \overline{ED} + \overline{AC}$$

$$= \left(\overline{AB} + \overline{BC}\right) + \left(\overline{AE} + \overline{ED} + \overline{DC}\right) + \overline{AC}$$

$$= \overline{AC} + \overline{AC} + \overline{AC} = 3\overline{AC}$$

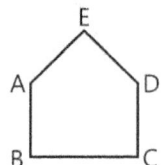

Example 3: Prove that the straight lines joining the mid points of the sides of a quadrilateral ABCD, taken in order, form a parallelogram.

Sol: Let the position vectors of A, B, C and D be $\vec{a}, \vec{b}, \vec{c}$ and $\vec{d}$. Hence position vectors of P, Q, R and S (the mid points of AB, BC, CD & DA respectively) are

$$\frac{\vec{a}+\vec{b}}{2}, \frac{\vec{b}+\vec{c}}{2}, \frac{\vec{c}+\vec{d}}{2} \text{ and } \frac{\vec{d}+\vec{a}}{2} \text{ respectively.}$$

$$\overrightarrow{PQ} = \frac{\vec{c}+\vec{b}}{2} - \frac{\vec{a}+\vec{b}}{2}$$

$$\overrightarrow{PQ} = PO + OQ = \frac{\vec{c}-\vec{a}}{2}$$

$$\overrightarrow{RS} = \frac{\vec{a}+\vec{d}}{2} - \frac{\vec{c}+\vec{d}}{2} = \frac{\vec{a}-\vec{c}}{2}$$

$$\Rightarrow \overrightarrow{SR} = \frac{\vec{c}-\vec{a}}{2}$$

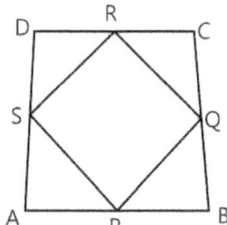

$\overrightarrow{PQ}$ is parallel and equal to $\overrightarrow{SR}$. Hence PQRS is a parallelogram.

Example 4: Write an equation for the plane that contains the points (2, 0, -3), (-4, -5, 2), and (0, 3, -4) in the form ax+by+cz = d.

Sol: Let $\vec{v} = \left(-4, -5, 2\right) - \left(2, 0, -3\right) = \left(-6, -5, 5\right)$ and $\vec{\omega} = \left(0, 3, -4\right) - \left(2, 0, -3\right) = \left(-2, 3, -1\right)$.

$$\vec{v} \times \vec{\omega} = \vec{i}\left(5 - 15\right) - \vec{j}\left(6 + 10\right) +$$

$$\hat{k}\left(-18 - 10\right) = \left(-10, -16, -28\right)$$

We can choose $\hat{n}$ to be any vector in the same direction as $\vec{v} \times \vec{\omega}$ so let $\hat{n} = (5, 8, 14)$. Then the plane has the form $5x + 8y + 14z = d$. Substituting the point $(2,0,-3)$ for (x,y,z) and solving for d gives $d = 10 + 0 + 0(-42) = -32$. So the plane has the equation $5x + 8y + 14z = -32$.

Example 5: Find a vector that is perpendicular to the vector $(1, 2, 3)$ with the same length. Also, find a plane perpendicular to $(1, 2, 3)$ that passes through the point $(3, 2, 1)$.

Sol: By using formula of perpendicular vector we can obtain the result.

A vector $\vec{v}$ is perpendicular to $(1, 2, 3)$ if $\vec{v}.(1,2,3) = 0$. There are infinite number of possibilities to choose from, but one possible choice for $\vec{v}$ is $(2,-1, 0)$. However, we want this vector to have length $\|(1,2,3)\| = \sqrt{14}$.

Since $\|(2,-1,0)\| = \sqrt{5}$, we need to rescale our vector to be $\dfrac{\sqrt{14}}{\sqrt{5}}(2,-1,0)$.

If $(1, 2, 3)$ is a normal vector to a plane then the plane will have the form $x+2y+3z=d$. since the plane passes through the point $(3, 2, 1)$, we substitute these values for x, y, and z to get $3+4+3=10=d$ so our plane equation is $x + 2y + 3x = 10$.

Example 6: Write an equation for the plane that contains the point $(1, 0, 3)$ and the line $(-3,-2,-2) + t(1, 2,-1)$ in the form $ax+by+cz=d$.

Sol: Since the plane contains the line $(-3,-2,-2)+ t(1,2,-1)$ we know that one tangent vector to the plane is $\vec{v} = (1,2,-1)$. We can get a second tangent vector by finding the vector between $(-3,-2,-2)$ and $(1, 0, 3)$. So let $\vec{\omega} = (4,2,5)$. Then

$$\vec{v} \times \vec{\omega} = \vec{i}(10 + 2) - \vec{j}(5 + 4) + \vec{k}(2 - 8) = (12,-9,-6)$$

So we can choose $\vec{n} = (4,-3,-2)$ and our plane has the form $4x-3y-2z=d$. Plugging in $(1, 0, 3)$ for (x,y,z) and solving for d yields $4x - 3y - 2z = -2$

Example 7: Find the minimum distance between the point $(3,-3,-3)$ and the plane $2x+y-z=3$.

Sol: The point in the plane closest to $(3,-3,-3)$ lies on a line that is perpendicular to the plane and passes through $(3,-3,-3)$. Since $(2, 1,-1)$ is a normal vector to the plane, we will use it as the direction of this line. Thus a parameterized form of the line is

$$c(t) = (3,-3,-3)+t(2,1,-1) = (3+2t,-3+t,,-3-t)$$

We substitute this into the plane equation to find its intersection with the plane and get:

$2(3+2t) + 1(-3 + t) - (-3 -t) = 6 + 6t = 3 \Rightarrow t = -\dfrac{1}{2}$.

So the point in the plane closest to $(3,-3,-3)$ is

$$c\left(-\frac{1}{2}\right) = \left(2,-\frac{7}{2},-\frac{5}{2}\right).$$

The distance between the point and the plane is thus

$$\sqrt{1^2 + \left(\frac{1}{2}\right)^2 + \left(-\frac{1}{2}\right)^2} = \sqrt{\frac{3}{2}}.$$

Example 8: Determine if the three vectors $\vec{a} = (1,4,-7)$ $\vec{b} = (2,-1,4)$ and $\vec{c} = (0,-9,18)$ lie in the same plane or not.

Sol: Three vectors lies in the same plane if volume of the parallelepiped formed by these three vectors is zero.

So, as we noted prior to this example all we need to do is compute the volume of the parallelepiped formed by these three vectors. If the volume is zero, then they lie in the same plane and if the volume isn't zero they don't lie in the same plane.

$$\vec{a} \cdot \left(\vec{b} \times \vec{c}\right) = \begin{vmatrix} 1 & 4 & -7 \\ 2 & -1 & 4 \\ 0 & -9 & 18 \end{vmatrix} = -18 + 126 - 144 + 36 = 0$$

So, the volume is zero and so they lie in the same plane.

JEE Advanced/Boards

Example 1: If O be the circumcenter; G, the centroid and H, the orthocenter of triangle ABC, prove that O, G, H are collinear and G divides OH in the ratio 1:2

Sol: Consider position vector of A, B, C be taken as $\vec{a}, \vec{b}, \vec{c}$. And then use geometry of triangle to solve this problem.

Let O, the circumcenter of the $\triangle ABC$ be chosen as origin and position vector of A, B, C be taken $\vec{a}, \vec{b}, \vec{c}$. Hence position vector of G the centroid is

$$\overline{OG} = \frac{\vec{a}+\vec{b}+\vec{c}}{3} \qquad \text{... (i)}$$

Since O is circumcenter

$\therefore \overline{OA} = \overline{OB} = \overline{OC} = \overline{OA}^2 = \overline{OB}^2 = \overline{OC}^2$ or $a^2 = b^2 = c^2$

$a^2 - b^2 = 0, \ b^2 - c^2 = 0, \ c^2 - a^2 = 0$

Or $(\vec{a}+\vec{b}).(\vec{a}-\vec{c})=0$

Or $\left(\vec{a}+\vec{b}+\vec{c}-\vec{c}\right).\left(\vec{a}-\vec{b}\right)=0$ ... (ii)

Let P be the point whose position vector is

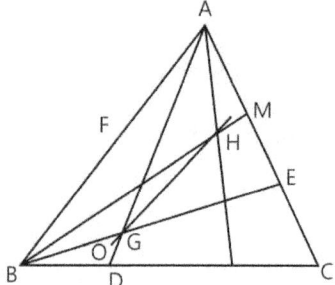

$\vec{a}+\vec{b}+\vec{c} \therefore \left(\overrightarrow{OP}-\overrightarrow{OC}\right).\left(\overrightarrow{OA}-\overrightarrow{OB}\right)=0$

Or $\overrightarrow{CP} \perp \overrightarrow{BA}$

In similar manner we can show that BP is perpendicular to AC and AP is perpendicular to CB.

Hence P is the orthocentre which is H.

$\overrightarrow{OP}=\overrightarrow{OH}=\vec{a}+\vec{b}+\vec{c}=3\overrightarrow{OG}$ ... (iii)

$\therefore \overrightarrow{OH}=3\overrightarrow{OG}$ or $\overrightarrow{GH}=2.\overrightarrow{OG}$ or $\dfrac{\overline{OG}}{\overline{GH}}=\dfrac{1}{2}$

Above show that O, G, H are collinear and G divides OH in the ratio 1:2

Example 2: Prove using vectors: If two medians of a triangle are equal, then it is isosceles.

Sol: Using mid – point formula of vector, we can solve this

Let ABC be a triangle and let BE and CF be two equal medians. Taking A as the origin, let the position vectors of B and C be $\vec{b}$ and $\vec{c}$ respectively. Then

Position vector of $E=\dfrac{1}{2}\vec{c}$ and

Position vector of $F=\dfrac{1}{2}\vec{b}$

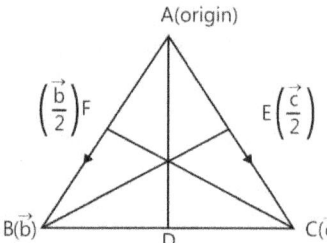

$\therefore \overrightarrow{BE}=\dfrac{1}{2}\left(\vec{c}-2\vec{b}\right),\overrightarrow{CF}=\dfrac{1}{2}\left(\vec{b}-2\vec{c}\right)$

Now, $BE=CF=\left|\overrightarrow{BE}\right|=\left|\overrightarrow{CF}\right|$

$\Rightarrow \left|\overrightarrow{BE}\right|^2=\left|\overrightarrow{CF}\right|^2 \Rightarrow \left|\dfrac{1}{2}\left(\vec{c}-2\vec{b}\right)\right|^2=\left|\dfrac{1}{2}\left(\vec{b}-2\vec{c}\right)\right|^2$

$\Rightarrow \dfrac{1}{4}\left|\vec{c}-2\vec{b}\right|^2=\dfrac{1}{4}\left|\vec{b}-2\vec{c}\right|^2 \Rightarrow \left|\left(\vec{c}-2\vec{b}\right)\right|^2=\left|\left(\vec{b}-2\vec{c}\right)\right|^2$

$\Rightarrow \left(\vec{c}-2\vec{b}\right)\bullet\left(\vec{c}-2\vec{b}\right)=\left(\vec{b}-2\vec{c}\right)\bullet\left(\vec{b}-2\vec{c}\right)$

$\Rightarrow \left|\vec{c}\right|^2-4\vec{b}.\vec{c}+4\left|\vec{b}\right|^2=\left|\vec{b}\right|^2-4\vec{b}.\vec{c}+4\left|\vec{c}\right|^2$

$\Rightarrow 3\left|\vec{b}\right|^2=3\left|\vec{c}\right|^2 \Rightarrow \left|\vec{b}\right|^2=\left|\vec{c}\right|^2$

$\Rightarrow AB=AC$

Hence, triangle ABC is an isosceles triangle.

Example 3: D, E, F are points dividing side $\overrightarrow{BC}$, $\overrightarrow{CA}$, $\overrightarrow{AB}$ of a triangle ABC in the ratio 2:3, 1:2 and 3:1 respectively. Show that the lines $\overrightarrow{AD}$, $\overrightarrow{BE}$, $\overrightarrow{CF}$ are concurrent and hence find the position vector of their point of intersection.

Sol: By using section formula we can obtain required result.

If $\vec{d}$, $\vec{e}$, $\vec{f}$ are position vector of points D, E & F respectively then, by section formula

$\vec{d}=\dfrac{2\vec{c}+3\vec{b}}{5}$ .... (i)

$\vec{e}=\dfrac{2\vec{c}+\vec{a}}{3}$ ....(ii)

$\vec{f}=\dfrac{3\vec{b}+\vec{a}}{4}$ ...(iii)

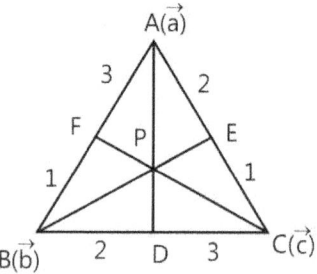

Equation of line AD is $\vec{r}=\vec{a}+t\left(\vec{d}-\vec{a}\right)$

Equation of line BE is $\vec{r}=\vec{b}+m\left(\vec{e}-\vec{b}\right)$

For intersection of $\overrightarrow{AD}$ and $\overrightarrow{BE}$ we need that

" $\vec{a}+t\left(\vec{d}-\vec{a}\right)=\vec{b}+m\left(\vec{e}-\vec{b}\right)$ " be true for some 0<t, m<1.

$\vec{a}+t\dfrac{\left(2\vec{c}+3\vec{b}-5\vec{a}\right)}{5}=\vec{b}+m\dfrac{\left(2\vec{c}+\vec{a}-3\vec{b}\right)}{3}$

$$\therefore\ 1-t=\frac{m}{3};\ \frac{3t}{5}=1-m;\ \frac{2t}{5}=\frac{2m}{3}$$

$$\therefore\ t=\frac{5}{6},\ m=\frac{1}{2}$$

The existence of t and m assures the intersection of $\overline{AD}$ and $\overline{BE}$.

The point of intersection is

$$\vec{r}=\vec{a}+\frac{5}{6}\left(\vec{d}-\vec{a}\right)=\frac{\left(\vec{a}+3\vec{b}+2\vec{c}\right)}{6}$$

Example 4: Find a parametric form for the line passing through the point (1,2) in the direction (3,4), which we will call $c_1(t)$. Set $c_1(t)$ equal to (x,y) and eliminate t to get the line into y = mx + b form. Now find a different parametrization $c_2(t)$ of the same line such that $c_2(0)$ =(-2,-2) and $c_2(2)$= (-5,-6).

Sol: $c_1(t)$ = (1,2) + t (3, 4) = (1 + 3t, 2 + 4t). Setting (x, y) = (1 + 3t, 2 + 4t) yields x = 1 + 3t and y = 2 + 4t.

Solving the former equation for t yields t = (x-1)/3. Substituting this into the second equation then gives us $y=\frac{4}{3}x+\frac{2}{3}$.

Let $c_2(t)=p+t\vec{v}$. c_2 will then be a parameterization of the same line given by c_1 if p is a point on the same line and $\vec{v}$ is in the same direction as (3,4) (i.e. some scalar multiple of (3,4)). Since $c_2(0)$ = (-2,-2) we will choose p= (-2,-2) (you can check that this point indeed lies on the line parameterized by c_1). Then

$c_2(2)=(-2,-2)+2\vec{v}=(-5,-6)$, so we get that

$\vec{v}=(-3/2,-2)$, which is indeed a scalar multiple of (3,4). So

$c_2(t)=(-2,-2)+t\left(\frac{-3}{2},-2\right)$ is a different

parameterization of the line parameterized by c_1.

Example 5: Find the vector projection of (3, 2) onto (-1,-1). Then find the area of the triangle with one side vector (3, 2) and another side the result of this projection.

Sol: Use projection method to obtain vector projection of (3, 2) and area of triangle will be half of the area of parallelogram.

$$proj_{(-1,-1)}(3,2)=\frac{-5}{2}(-1,-1)=\left(\frac{5}{2},\frac{5}{2}\right).$$

Then the area of the triangle with sides (3, 2) and $\left(\frac{5}{2},\frac{5}{2}\right)$ is one half the area of the parallelogram with sides (3, 2) and $\left(\frac{5}{2},\frac{5}{2}\right)$. So, the area of the triangle is

$$\frac{1}{2}\left\|(3,2,0)\times\left(\frac{5}{2},\frac{5}{2},0\right)\right\|=\frac{1}{2}\left\|\left(0,0,\frac{5}{2}\right)\right\|=\boxed{\frac{5}{4}}.$$

Example 6: Find the minimum distance between the point (4, 2,-3) and the line (1, 0, 2) + t (-1,-1, 2).

Sol: Let $\vec{v}(t)$ represents the vector from the point (4,2,-3) and line (1,0,2) + t(-1,-1,2) = (1-t,-t,2+2t) at any $t\in\mathbb{R}$.

So, $\overrightarrow{v(t)}$ = (4, 2,-3)-(1-t,-t, 2+2t) = (3+t, 2+t,-5-2t). We want to find the t such that $\overrightarrow{v(t)}$ is perpendicular to the line, which is when $\overrightarrow{v(t)}$. (-1,-1, 2) = 0.

(3 + t, 2 + t, -5 - 2t) . (-1, -1, 2) = -15 - 6t = 0

$\Rightarrow t=-\frac{5}{2}$. So the length of $\vec{v}\left(-\frac{5}{2}\right)$ should represent the minimum distance from (4, 2,-3) and the line.

$$\left\|\vec{v}\left(-5/2\right)\right\|=\left\|(1/2,-1/2,0)\right\|=\frac{1}{\sqrt{2}}.$$

Example 7: Prove that, in any triangle ABC

(i) $c^2=a^2+b^2-2ac\cos C$ (ii) $c=b\cos A+a\cos B$

Sol: By using simple scalar product method we can prove given relation.

(i) In $\triangle ABC$, $\overrightarrow{AB}+\overrightarrow{BC}+\overrightarrow{CA}=0$

or, $\overrightarrow{BC}+\overrightarrow{CA}=-\overrightarrow{AB}$ ....(i)

Squaring both sides

$$\left(\overrightarrow{BC}\right)^2+\left(\overrightarrow{CA}\right)^2+2\left(\overrightarrow{BC}\right).\left(\overrightarrow{CA}\right)=\left(\overrightarrow{AB}\right)^2$$

$\Rightarrow a^2+b^2+2\left(\overrightarrow{BC}.\overrightarrow{CA}\right)=c^2$

$\Rightarrow c^2=a^2+b^2=2ab\cos(\pi-C)$

$\Rightarrow c^2=a^2+b^2-2ab\cos C$

(ii) $\left(\overrightarrow{BC}+\overrightarrow{CA}.\right).\overrightarrow{AB}=-\overrightarrow{AB}.\overrightarrow{AB}\ \Rightarrow\overrightarrow{BC}.\overrightarrow{AB}+\overrightarrow{CA}.\overrightarrow{AB}$

$=-c^2-ac\cos B-bc\cos A$

$\Rightarrow a\cos B+b\cos A=c$

Example 8: In any triangle, show that the perpendicular bisectors of the sides are concurrent.

Sol: By using formula of Dot product and Mid – point we can solve this problem.

Let ABC be the triangle and D, E and F are respectively middle points of sides $\overline{BC}$ $\overline{CA}$ and $\overline{AB}$. Let the perpendicular through D and E meet of O join $\overrightarrow{OF}$. We are required to prove that $\overrightarrow{OF}$ is $\perp$ to $\overrightarrow{AB}$. Let the position vectors of A, B, C with O as origin of reference be $\vec{a}$, $\vec{b}$ and $\vec{c}$ respectively.

$$\therefore \overrightarrow{OD} = \frac{1}{2}\left(\vec{b} + \vec{c}\right),$$

$$\overrightarrow{OE} = \frac{1}{2}\left(\vec{c} + \vec{a}\right),$$

and $\overrightarrow{OF} = \frac{1}{2}\left(\vec{a} + \vec{b}\right)$

Also

$\overrightarrow{BC} = \vec{c} - \vec{b}$, $\overrightarrow{CA} = \vec{a} - \vec{c}$ and $\overrightarrow{AB} = \vec{b} - \vec{a}$

Since, $\overrightarrow{OD} \perp \overrightarrow{BC}$, $1/2(\vec{b} + \vec{c}).(\vec{c} - \vec{b}) = 0$

$\Rightarrow b^2 = c^2$ ... (i)

Similarly, $\overrightarrow{OE} \perp \overrightarrow{CA}$, $1/2(\vec{a} + \vec{c}).(\vec{a} - \vec{c}) = 0$

$\Rightarrow a^2 = c^2$ ... (ii)

From (i) and (ii) we have $b^2 - a^2 = 0$

$\left(\vec{a} + \vec{b}\right).\left(\vec{b} + \vec{a}\right) = 0 \Rightarrow 1/2\left(\vec{b} + \vec{a}\right).\left(\vec{b} - \vec{a}\right) = 0$

$\Rightarrow \overrightarrow{OF} \perp \overrightarrow{AB}$

Hence proved.

JEE Main/Boards

JEE Main/Boards

Exercise 1

Q.1 The line L_1 passes through the points (2,-3, 1) and (-1,-2,-4). The line L_2 passes through the point (3, 2,-9) and is parallel to the vector $4\hat{i} - 4\hat{j} + 5\hat{k}$.

(i) Find an equation for L_1 in the form $\vec{r} = \vec{a} + t\vec{b}$

(ii) Prove that L_1 and L_2 are skew.

Q.2 Two lines have vector equations

$$\vec{r} = \begin{pmatrix} 4 \\ 2 \\ -6 \end{pmatrix} + t \begin{pmatrix} -8 \\ 1 \\ -2 \end{pmatrix} \text{ and } \vec{r} = \begin{pmatrix} -2 \\ a \\ -2 \end{pmatrix} + s \begin{pmatrix} -9 \\ 2 \\ -5 \end{pmatrix},$$

Where 'a' is a constant.

(i) Calculate the acute angle between the lines.

(ii) Given that these two lines intersect, find the point of intersection.

Q.3 The points A and B have position vectors $\vec{a}$ and $\vec{b}$ relative to an origin O, where $\vec{a} = 4\hat{i} + 3\hat{j} - 2\hat{k}$ and $\vec{b} = -7\hat{i} + 5\hat{j} + 4\hat{k}$

(i) Find the length of AB.

(ii) Use a scalar product to find angle OAB.

Q.4 The position vectors of the points P and Q with respect to an origin O are $5\hat{i} + 2\hat{j} - 9\hat{k}$ and $4\hat{i} + 4\hat{j} - 6\hat{k}$ respectively.

(i) Find the vector equation for the line PQ

The position vector of the point T is $\hat{i} + 2\hat{j} - \hat{k}$

(ii) Write down a vector equation for the line OT and show that OT is perpendicular to PQ.

It is given that OT intersects PQ.

(iii) Find the position vector of the point of intersection of OT and PQ.

(iv) Hence find the perpendicular distance from O to PQ, giving your answer in an exact form.

Q.5 ABCD is a parallelogram. The position vectors of A, B and C are given respectively by

$\vec{a} = 2\hat{i} + \hat{j} + 3\hat{k}$, $\vec{b} = 3\hat{i} - 2\hat{j}$, $\vec{c} = \hat{i} - \hat{j} - 2\hat{k}$

(i) Find the position vector of D.

(ii) Determine, to the nearest degree, the angle ABC.

Q.6 The position vectors of three points A, B and C relative to an origin O are given respectively by

$\overrightarrow{OA} = 7\hat{i} + 3\hat{j} - 3\hat{k}$, $\overrightarrow{OB} = 4\hat{i} + 2\hat{j} - 4\hat{k}$, $\overrightarrow{OC} = 5\hat{i} + 4\hat{j} - 5\hat{k}$.

(i) Find the angle between AB and AC.

(ii) Find the area of triangle ABC.

Q.7 Two lines have vector equations

$\vec{r} = \hat{i} - 2\hat{j} + 4\hat{k} + \lambda(3\hat{i} + \hat{j} + a\hat{k})$ and

$\vec{r} = -8\hat{i} + 2\hat{j} + 3\hat{k} + \mu(\hat{i} - 2\hat{j} - \hat{k})$,

Where 'a' is a constant

(i) Given that the lines are skew, find the value that a cannot take.

(ii) Given instead that the lines intersect, find the point of intersection.

Q.8 Lines, L_1, L_2 and L_3 have vector equations

$L_1 : \vec{r} = (5\hat{i} - \hat{j} - 2\hat{k}) + s(-6\hat{i} + 8\hat{j} - 2\hat{k})$,

$L_2 : \vec{r} = (3\hat{i} - 8\hat{j}) + t(\hat{i} + 3\hat{j} + 2\hat{k})$

$L_3 : \vec{r} = (2\hat{i} + \hat{j} + 3\hat{k}) + u(3\hat{i} + c\hat{j} + \hat{k})$.

(i) Calculate the acute angle between L_1 and L_2.

(ii) Given that L_1 and L_3 are parallel, find the value of c.

(iii) Given that L_2 and L_3 intersect, find the value of c

Q.9 Given that $\vec{u} = \hat{i} - 2\hat{j} + 3\hat{k}$;

$\vec{v} = 2\hat{i} + \hat{j} + 4\hat{k}$; $\vec{w} = \hat{i} + 3\hat{j} + 3\hat{k}$ and

$(\vec{u} \cdot \vec{R} - 10)\hat{i} + (\vec{v}\vec{R} - 20)\hat{j} + (\vec{w} \cdot \vec{R} - 20)\hat{k} = 0$.

Find the unknown vector $\vec{R}$.

Q.10 The base vectors $\vec{a}_1$, $\vec{a}_2$, $\vec{a}_3$ are given in terms of base vectors $\vec{b}_1$, $\vec{b}_2$, $\vec{b}_3$ as,

$\vec{a}_1 = 2\vec{b}_1 + 3\vec{b}_2 - \vec{b}_3$;

$\vec{a}_2 = \vec{b}_1 - 2\vec{b}_2 + 2\vec{b}_3$ & $\vec{a}_3 = -2\vec{b}_1 + \vec{b}_2 - 2\vec{b}_3$.

If $\vec{F} = 3\vec{b}_1 - \vec{b}_2 + 2\vec{b}_3$. then express $\vec{F}$ in terms of $\vec{a}_1$, $\vec{a}_2$, $\vec{a}_3$.

Q.11 If $\vec{r}$ and $\vec{s}$ are non zero constant vectors and the scalar b is chosen such that $|\vec{r} + b\vec{s}|$ is minimum, then show that the value of $|b\vec{s}|^2 + |\vec{r} + b\vec{s}|$ is equal to $|r|^2$.

Exercise 2

Single Correct Choice Type

Q.1 Let $\vec{a}$, $\vec{b}$, $\vec{c}$ be three vectors such that $|\vec{a}| = |\vec{c}| = 1$; $|\vec{b}| = 4$ and $|\vec{b} \times \vec{c}| = \sqrt{15}$. If $\vec{b} - 2\vec{c} = \lambda\vec{a}$ then a value of λ is

(A) 1   (B) -1   (C) 2   (D) -4

Q.2 Vector $\vec{r}$ which is equally inclined to coordinate axes such that $|\vec{r}| = 15\sqrt{3}$ is

(A) $\hat{i} + \hat{j} + \hat{k}$ (B) $15(\hat{i} + \hat{j} + \hat{k})$

(C) $7(\hat{i} + \hat{j} + \hat{k})$ (D) None of these

Q.3 For 3 vectors $\vec{u}$, $\vec{v}$, $\vec{w}$, which of the following expressions is $\neq$ to any remaining three.

(A) $\vec{u} \cdot (\vec{v} \times \vec{w})$ (B) $(\vec{v} \times \vec{w}) \cdot \vec{u}$

(C) $\vec{v} \cdot (\vec{u} \times \vec{w})$ (D) $(\vec{w} \times \vec{u}) \cdot \vec{v}$

Q.4 If $\vec{a} + \vec{b} + \vec{c} = \vec{0}$, $|\vec{a}| = 3, |\vec{b}| = 5$ & $|\vec{c}| = 7$ then $\angle\theta$ between $\vec{a}$ and $\vec{b}$ is

(A) 40°   (B) 30°   (C) 150°   (D) None of these

Q.5 If 2 out of 3 vectors $\vec{a}, \vec{b}, \vec{c}$ are unit vectors, $\vec{a} + \vec{b} + \vec{c} = \vec{0}$ and $2(\vec{a}.\vec{b} + \vec{b}.\vec{c} + \vec{c}.\vec{a}) + 3 = 0$, then third vector is length-

(A) 3   (B) 1   (C) 2   (D) None of these

Q.6 Let $\vec{a} + \vec{b}$ is orthogonal to $\vec{b}$ and $\vec{a} + 2\vec{b}$ is orthogonal to $\vec{a}$, then

(A) $|\vec{a}| = \sqrt{2}|\vec{b}|$ (B) $|\vec{a}| = 2|\vec{b}|$

(C) $|\vec{a}| = |\vec{b}|$ (D) $2|\vec{a}| = |\vec{b}|$

Q.7 Magnitude of projection of vector $\hat{i}+2\hat{j}+\hat{k}$ on vector $4\hat{i}+4\hat{j}+7\hat{k}$ is

(A) 3 (B) $3\sqrt{6}$ (C) $\sqrt{6}/3$ (D) None of these

Q.8 Magnitude of moment of force $-2\hat{i}+6\hat{j}-8\hat{k}$ acting at point $2\hat{i}-\hat{j}+3\hat{k}$ about point $\hat{i}+2\hat{j}-\hat{k}$

(A) $\sqrt{211}$ (B) 0 (C) $\sqrt{54}$ (D) None of these

Q.9 If $\hat{a}$ & $\hat{b}$ are unit vectors represented by $\vec{O}A$ and $\vec{O}B$, then unit vector along bisector of $\angle AOB$ is scalar multiple of

(A) $\hat{a}-\hat{b}$ (B) $\hat{a}\times\hat{b}$ (C) $\hat{b}\times\hat{a}$ (D) None of these

Q.10 If $\left[2\vec{a}+4\vec{b}\;\;\vec{c}\;\;\vec{d}\right]=\lambda\left[\vec{a}\;\;\vec{c}\;\;\vec{d}\right]+\mu\left[\vec{b}\;\;\vec{c}\;\;\vec{d}\right]$ then $\lambda+\mu=$

(A) 6 (B) -6 (C) 10 (D) None of these

Previous Years' Questions

Q.1 The volume of the parallelepiped whose sides are given by $\vec{OA}=2\hat{i}-3\hat{j}$, $\vec{OB}=\hat{i}+\hat{j}-\hat{k}$, $\vec{OC}=3\hat{i}-\hat{k}$, is
(1983)

(A) $\dfrac{4}{13}$ (B) 4 (C) $\dfrac{2}{7}$ (D) None of these

Q.2 A vector $\vec{a}$ has components $2p$ and 1 with respect to a rectangular Cartesian system. This system is rotated through a certain angle about the origin in the counter clockwise sense. If, with respect to the new system, $\vec{a}$ has components $p+1$ and I, then **(1986)**

(A) $p=0$ (B) $p=1$ or $p=-\dfrac{1}{3}$

(C) $p=-1$ or $p=\dfrac{1}{3}$ (D) $p=1$ or $p=-1$

Q.3 Let a ,b, c be distinct non-negative numbers. If the vectors $a\hat{i}+a\hat{j}+c\hat{k}$, $\hat{i}+\hat{k}$ and $c\hat{i}+c\hat{j}+b\hat{k}$ lie in a plane, then c is **(1993)**

(A) The Arithmetic Mean of a and b.

(B) The Geometric Mean of a and b.

(C) The Harmonic Mean of a and b.

(D) Equal to zero.

Q.4 If $\vec{a},\vec{b},\vec{c}$ are non-coplanar unit vectors such that $\vec{a}\times(\vec{b}\times\vec{c})=\dfrac{(\vec{b}+\vec{c})}{\sqrt{2}}$, then the angle between $\vec{a}$ and $\vec{b}$ is **(1995)**

(A) $\dfrac{3\pi}{4}$ (B) $\dfrac{\pi}{4}$ (C) $\dfrac{\pi}{2}$ (D) π

Q.5 If $\vec{a}$ and $\vec{b}$ are two unit vectors such that $\vec{a}+2\vec{b}$ and $5\vec{a}-4\vec{b}$ are perpendicular to each other, then the angle between $\vec{a}$ and $\vec{b}$ is **(2002)**

(A) $45°$ (B) $60°$ (C) $\cos^{-1}\left(\dfrac{1}{3}\right)$ (D) $\cos^{-1}\left(\dfrac{2}{7}\right)$

Q.6 Let $\vec{V}=2\hat{i}+\hat{j}-\hat{k}$ and $\overline{W}=\hat{i}+3\hat{k}$. If $\vec{U}$ is a unit vector, then the maximum value of the scalar triple product $[\vec{U}\;\vec{V}\;\overline{W}]$ is **(2002)**

(A) -1 (B) $\sqrt{10}+\sqrt{6}$ (C) $\sqrt{59}$ (D) $\sqrt{60}$

Q.7 The unit vector which is orthogonal to the vector $3\hat{i}+2\hat{j}+6\hat{k}$ and is coplanar with the vectors $2\hat{i}+\hat{j}+\hat{k}$ and $\hat{i}-\hat{j}+\hat{k}$ is **(2004)**

(A) $\dfrac{2\hat{i}-6\hat{j}+\hat{k}}{\sqrt{41}}$ (B) $\dfrac{2\hat{i}-3\hat{j}}{\sqrt{13}}$

(C) $\dfrac{3\hat{j}-\hat{k}}{\sqrt{10}}$ (D) $\dfrac{4\hat{i}+3\hat{j}-3\hat{k}}{\sqrt{34}}$

Q.8 Two adjacent sides of a parallelogram ABCD are given by $\vec{AB}=2\hat{i}+10\hat{j}+11\hat{k}$ and $\vec{AD}=-\hat{i}+2\hat{j}+2\hat{k}$. The side AD is rotated by an acute angle α in the plane of the parallelogram so that AD becomes AD'. If AD' makes a right angle with the side AB, then the cosine of the angle α is given by **(2010)**

(A) $\dfrac{8}{9}$ (B) $\dfrac{\sqrt{17}}{9}$ (C) $\dfrac{1}{9}$ (D) $\dfrac{4\sqrt{5}}{9}$

Q.9 Let $\vec{a}=\hat{i}+\hat{j}+\hat{k}$, $\vec{b}=\hat{i}-\hat{j}+\hat{k}$ and $\vec{c}=\hat{i}-\hat{j}-\hat{k}$ be three vectors. A vector $\vec{v}$ in the plane of $\vec{a}$ and $\vec{b}$, whose projection on $\vec{c}$ is $\dfrac{1}{\sqrt{3}}$, is given by **(2011)**

Q.10 Let $\vec{a},\vec{b}$ and $\vec{c}$ be three unit vectors such that $\vec{a}\times\left(\vec{b}\times\vec{c}\right)-\dfrac{\sqrt{3}}{2}\left(\vec{b}+\vec{c}\right)$. If $\vec{b}$ is not parallel to $\vec{c}$ then the angle between $\vec{a}$ and $\vec{b}$ is: **(2016)**

(A) $\dfrac{\pi}{2}$ (B) $\dfrac{2\pi}{3}$ (C) $\dfrac{5\pi}{6}$ (D) $\dfrac{3\pi}{4}$

Q.11 Let $\vec{a}$, $\vec{b}$ and $\vec{c}$ be three non-zero vectors such that no two of them are collinear and $\left(\vec{a}\times\vec{b}\right)\times\vec{c} = \frac{1}{3}\left|\vec{b}\right|\left|\vec{c}\right|\vec{a}$. If θ is the angle between vectors $\vec{b}$ and $\left|\vec{c}\right|$ then a value of $\sin\theta$ is **(2015)**

(A) $\dfrac{-\sqrt{2}}{3}$ (B) $\dfrac{2}{3}$ (C) $-\dfrac{2\sqrt{3}}{3}$ (D) $\dfrac{2\sqrt{3}}{3}$

Q.12 If $\left[\vec{a}\times\vec{b}\ \vec{b}\times\vec{c}\ \vec{c}\times\vec{a}\right] = \lambda\left[\vec{a}\ \vec{b}\ \vec{c}\right]^2$ then λ is equal to **(2014)**

(A) 1   (B) 3   (C) 0   (D) 1

Q.13 If the vectors $\overrightarrow{AB} = 3\hat{j}+4\hat{k}$ and $\overrightarrow{AC} = 2\hat{j}+4\hat{k}$ are the sides of a triangle ABC, then the length of the median through A is **(2013)**

(A) $\sqrt{72}$ (B) $\sqrt{33}$ (C) $\sqrt{45}$ (D) $\sqrt{18}$

Q.14 Let $\hat{a}$ and $\hat{b}$ be two unit vectors. If the vectors $\vec{c} = \hat{a}+2\hat{b}$ and $\vec{d} = 5\hat{a}-4\hat{b}$ are perpendicular to each other then the angle between $\hat{a}$ and $\hat{b}$ is **(2012)**

(A) $\dfrac{\pi}{6}$ (B) $\dfrac{\pi}{2}$ (C) $\dfrac{\pi}{3}$ (D) $\dfrac{\pi}{4}$

Q.15 If the vectors $\vec{a} = \hat{i}-\hat{j}+2+4\hat{j}+\hat{k}$ and $\vec{c} = \lambda\hat{i}+\hat{j}+\mu\hat{k}$ are mutually orthogonal then $(\lambda, \mu) =$ **(2010)**

(A) $(2, -3)$ (B) $(-2, 3)$ (C) $(3, -2)$ (D) $(-3, 2)$

Q.16 Let $\vec{a} = \hat{j}-\hat{k}$. Then vector $\vec{b}$ satisfying $\vec{a}\times\vec{b}+\vec{c} = \vec{0}$ and $\vec{a}.\vec{b}$ is **(2010)**

(A) $2\hat{i}-\hat{j}+2\hat{k}$ (B) $\hat{i}-\hat{j}+2\hat{k}$

(C) $\hat{i}+\hat{j}-2\hat{k}$ (D) $-\hat{i}+\hat{j}-2\hat{k}$

Q.17 If $\vec{a} = \frac{1}{\sqrt{10}}\left(3\hat{i}+\hat{k}\right)$ and $\vec{b} = \frac{1}{7}\left(2\hat{i}+3\hat{j}-6\hat{k}\right)$, then the value of $\left(2\vec{a}-\vec{b}\right).\left[\left(\vec{a}\times\vec{b}\right)\times\left(\vec{a}+2\vec{b}\right)\right]$ is **(2011)**

(A) -3   (B) 5   (C) 3   (D) -5

Q.18 The vector $\vec{a}$ and $\vec{b}$ are not perpendicular and $\vec{a}$ and $\vec{b}$ are two vectors satisfying: $\vec{b}\times\vec{c} = \vec{b}\times\vec{d}$ and $\vec{a}.\vec{d} = 0$. Then the vector is equal to **(2011)**

(A) $\vec{c}+\left(\dfrac{\vec{a}.\vec{c}}{\vec{a}.\vec{b}}\right)\vec{b}$ (B) $\vec{b}+\left(\dfrac{\vec{b}.\vec{c}}{\vec{a}.\vec{b}}\right)\vec{c}$

(C) $\vec{c}-\left(\dfrac{\vec{a}.\vec{c}}{\vec{a}.\vec{b}}\right)\vec{b}$ (D) $\vec{b}+\left(\dfrac{\vec{b}.\vec{c}}{\vec{a}.\vec{b}}\right)\vec{c}$

JEE Advanced/Boards

Exercise 1

Q.1 What will be the value of $\dfrac{\left(\vec{a}\times\vec{b}\right)^2+\left(\vec{a}\cdot\vec{b}\right)^2}{\vec{a}^2\vec{b}^2}$?

Q.2 What will be the area of the triangle determined by the vectors 3i+4j and -5i+7j?

Q.3 What will be the value of a if points whose position vectors are $60\hat{i}+3\hat{j}$, $40\hat{i}-8\hat{j}$, $a\hat{i}-52\hat{j}$ are collinear?

Q.4 What will be the angle between diagonals which adjacent sides of *ll*gm are along $\vec{a} = \hat{i}+2\hat{j}$ & $\vec{b} = 2\hat{i}+\hat{j}$?

Q.5 What will be the angle between $\vec{a}$ and $\vec{b}$ if $\vec{a}$ & $\vec{b}$ are unit vectors such that $\vec{a} + 3\vec{b}$ is $\perp$ to $7\vec{a} - 5\vec{b}$?

Q. 6 If the unit vectors $\hat{A}$ and $\hat{B}$ are inclined at π then what will be the value of $\left|\hat{A} - \hat{B}\right|/2$?

Q.7 A particle acted upon by forces $3\hat{i}+2\hat{j}+5\hat{k}$ and $2\hat{i}+\hat{j}+3\hat{k}$ is displaced from a point P to a point Q whose respective position vectors are $2\hat{i}+\hat{j}+3\hat{k}$ and $4\hat{i}+3\hat{j}+7\hat{k}$. What will be the work done by the force?

Q.8 A force F = $6\hat{i}+\lambda\hat{j}+4\hat{k}$ acting on a particle displaces it from A (3,4,5) to B(1,1,1). If the work done is 2 units, then What will be the value of λ ?

Q.9 What will be the length of longer diagonal of *llgm* constructed on $5\vec{a} + 2\vec{b}$ & $\vec{a} - 3\vec{b}$. Given $\left|\vec{b}\right| = 3$ & $\left|\vec{a}\right| = 2\sqrt{2}$ and angle between $\vec{a}$ & $\vec{b}$ is $\pi/4$

Q.10 The vectors $\hat{i} + x\hat{j} + 3\hat{k}$ is rotated through an angle θ and doubled in magnitude, then it becomes $4\hat{i} + (4x - 2)\hat{j} + 2\hat{k}$. Find x ?

Exercise 2

Single Correct Choice Type

Q.1 Moment of couple formed by forces $5\hat{i} + \hat{j}$ & $-5\hat{i} + \hat{j}$ acting at [9,-1, 2] and [3,-2, 1]

(A) $-\hat{i} + 5\hat{j} + \hat{k}$ (B) $\hat{i} - \hat{j} - 5\hat{k}$

(C) $2\hat{i} - 2\hat{j} - 10\hat{k}$ (D) $-2\hat{i} - 2\hat{j} + 10\hat{k}$

Q.2 Let $\vec{a}, \vec{b}, \vec{c}$ be vectors such that

$\vec{a}.(\vec{b} + \vec{c}) + \vec{b}.(\vec{c} + \vec{a}) + \vec{c}.(\vec{a} + \vec{b}) = 0$ and

$\left|\vec{a}\right| = 1, \left|\vec{b}\right| = 4, \left|\vec{c}\right| = 8$ then $\left|\vec{a} + \vec{b} + \vec{c}\right|$ equals

(A) 13 (B) 81 (C) 9 (D) None of these

Q.3 Position vectors of A and B are $2\hat{i} + 2\hat{j} + \hat{k}$ and $2\hat{i} + 4\hat{j} + 4\hat{k}$. Length of internal bisector of $\angle BOA$ of $\triangle AOB$ is

(A) $\sqrt{\dfrac{136}{9}}$ (B) $\sqrt{\dfrac{136}{9}}$ (C) $\dfrac{20}{3}$ (D) None of these

Q.4 If $\vec{a} + 2\vec{b} + 3\vec{c} = 0$, then $\vec{a} \times \vec{b} + \vec{b} \times \vec{c} + \vec{c} \times \vec{a}$ is equal to

(A) $6(\vec{b} \times \vec{c})$ (B) $6(\vec{a} \times \vec{b})$

(C) $6(\vec{c} \times \vec{a})$ (D) None of these

Q.5 Value of $\left|\vec{a} - \vec{b}, \vec{b} - \vec{c}, \vec{c} - \vec{a}\right|$ where $\left|\vec{a}\right| = 1, \left|\vec{b}\right| = 2$, and $\left|\vec{c}\right| = 3$ is

(A) 1 (B) -6 (C) 0 (D) None of these

Q.6 If P and Q be two given points on the curve $y = x + 1/x$ such that OP.I = 1 and OQ.I = -1 where I is a unit vector along the x-axis, then the length of vector 2OP + 3OQ is

(A) $5\sqrt{5}$ (B) $3\sqrt{5}$ (C) $2\sqrt{5}$ (D) $\sqrt{5}$

Q.7 Let A, B, C be three vectors such that A (B + C) + B. C = 0 and IAI = 1, IBI = 4, ICI = 8, then IA+B+CI equals

(A) 13 (B) 81 (C) 9 (D) 5

Q.8 If the unit vectors $\hat{A}$ and $\hat{B}$ are inclined at an angle 2θ and $\left|\hat{A} - \hat{B}\right| \le 1$, then for $\theta \in [0, \pi], \theta$, may lie in the interval

(A) $[\pi/6, \pi/3]$ (B) $[\pi/6, \pi/2]$

(C) $[5\pi/6, \pi]$ (D) $[\pi/2, 5\pi/6]$

Q.9 If unit vectors $\hat{A}$ and $\hat{B}$ such that STP $\left|\hat{A} \ \hat{B} \ \hat{A} \times \hat{B}\right| = 1/4$ then $\hat{A}$ and $\hat{B}$ are inclined

(A) $\pi/6$ (B) $\pi/2$ (C) $\pi/3$ (D) $\pi/4$

Q.10 If $\hat{A}$ and $\hat{B}$ unit vectors then greatest value of $\left|\hat{A} - \hat{B}\right| + \left|\hat{A} - \hat{B}\right|$ is

(A) 2 (B) 4 (C) $2\sqrt{2}$ (D) $\sqrt{2}$

Previous Years' Questions

Q.1 (i) If C be a given non zero scalar and $\vec{A}$ and $\vec{B}$ be given non-zero vectors such that $\vec{A} \perp \vec{B}$, find the vector $\vec{X}$ which satisfies the equation $\vec{A} \cdot \vec{X} = c$ and $\vec{A} \times \vec{X} = \vec{B}$.

(ii) $\vec{A}$ vector A has components A_1, A_2, A_3 in a right-handed rectangular Cartesian coordinate system oxyz. The coordinate system is rotated about the x-axis through an angle $\dfrac{\pi}{2}$. Find the components of A in the new coordinate system, in terms of A_1, A_2, A_3. *(1983)*

Q.2 If vectors $\vec{a}, \vec{b}, \vec{c}$ are coplanar, show that

$$\begin{vmatrix} \vec{a} & \vec{b} & \vec{c} \\ \vec{a} \cdot \vec{a} & \vec{a} \cdot \vec{b} & \vec{a} \cdot \vec{c} \\ \vec{b} \cdot \vec{a} & \vec{b} \cdot \vec{b} & \vec{b} \cdot \vec{c} \end{vmatrix} = 0$$ *(1989)*

Q.3 Let $\vec{A} = 2\hat{i} + \hat{k}$, $\vec{B} = \hat{i} + \hat{j} + \hat{k}$, and $\vec{C} = 4\hat{i} - 3\hat{j} + 7\hat{k}$. Determine a vector $\vec{R}$ satisfying $\vec{R} \times \vec{B} = \vec{C} \times \vec{B}$ and $\vec{R} \cdot \vec{A} = 0$. *(1990)*

Q.4 If $\vec{a}, \vec{b}, \vec{c}, \vec{d}$ are four distinct vectors satisfying the conditions $\vec{a} \times \vec{b} = \vec{c} \times \vec{d}$ and $\vec{a} \times \vec{c} = \vec{b} \times \vec{d}$ then prove that $\vec{a} \cdot \vec{b} + \vec{c} \cdot \vec{d} \neq \vec{a} \cdot \vec{c} + \vec{b} \cdot \vec{d}$. *(2004)*

Q.5 Incident ray is along the unit vector $\hat{v}$ and the reflected ray is along the unit vector . The normal is along unit vector $\hat{a}$ outwards. Express $\hat{w}$ in terms of $\hat{a}$ and $\hat{v}$. *(2005)*

Q.6 Let $\vec{A}$ be vector parallel to line of intersection of planes P_1 and P_2 through origin. P_1 is parallel to the vectors $2\hat{j} + 3\hat{k}$ and $4\hat{j} - 3\hat{k}$ and P_2 is parallel to $\hat{j} - \hat{k}$ and $3\hat{i} + 3\hat{j}$, then the angle between vector $\vec{A}$ and $2\hat{i} + \hat{j} - 2\hat{k}$ is *(2006)*

(A) $\dfrac{\pi}{2}$ (B) $\dfrac{\pi}{4}$ (C) $\dfrac{\pi}{4}$ (D) $\dfrac{3\pi}{4}$

Q.7 The vector(s) which is /are coplanar with vectors $\hat{i} + \hat{j} + 2\hat{k}$ and $\hat{i} + 2\hat{j} + \hat{k}$, are perpendicular to the vector $\hat{i} + \hat{j} + \hat{k}$ is/are *(2011)*

(A) $\hat{j} - \hat{k}$ (B) $-\hat{i} + \hat{j}$ (C) $\hat{i} - \hat{j}$ (D) $-\hat{j} + \hat{k}$

Q.8 If $\vec{a}$ and $\vec{b}$ are vectors in space by $\vec{a} = \dfrac{\hat{i} - 2\hat{j}}{\sqrt{5}}$ and $\vec{b} = \dfrac{2\hat{i} + \hat{j} + 3\hat{k}}{\sqrt{14}}$, then the value of $\left(2\vec{a} + \vec{b}\right) \cdot \left[\left(\vec{a} \times \vec{b}\right) \times \left(\vec{a} - 2\vec{b}\right)\right]$ is *(2010)*

Q.9 Let $\vec{a} = -\hat{i} - \hat{k}$, $\vec{b} = -\hat{i} + \hat{j}$, and $\vec{c} = \hat{i} + 2\hat{j} + 3\hat{k}$ be three given vectors. If $\vec{r}$ is a vector such that $\vec{r} \times \vec{b} = \vec{c} \times \vec{b}$, $\vec{r} \cdot \vec{a} = 0$, then the value of $\vec{r} \cdot \vec{b}$ is...... *(2011)*

Q.10 Let $\hat{u} = u_1\hat{i} + u_2\hat{j} + u_3\hat{k}$ be a unit vector in R^2 and $\hat{w} = \dfrac{1}{\sqrt{6}}\left(\hat{i} + \hat{j} + 2\hat{k}\right)$. Given that there exists a vector $\vec{v}$ in R^2 such that $\left|\hat{u} \times \vec{v}\right| = 1$ and $\hat{w} = \left(\hat{u} \times \vec{v}\right) = 1$. Which of the following statements (s)is (are) correct? *(2016)*

(A) There is exactly one choice for such $\vec{v}$

(B) There are infinitely many choices for such $\vec{v}$

(C) If $\hat{u}$ lies in the xy-plane then $\left|u_1\right| = \left|u_2\right|$

(D) If $\hat{u}$ lies in the xz-plane then $2\left|u_1\right| = \left|u_3\right|$

Q.11 Let ΔPQR be a triangle. Let $\vec{a} = \overline{QR}$, $\vec{b} = \overline{RP}$ and $\vec{c} = \overline{PQ}$. If $|\vec{a}| = 12$, $|\vec{b}| = 4\sqrt{3}$ and $\vec{b} \cdot \vec{c} = 24$, then which of the following is (are) true *(2015)*

(A) $\dfrac{|\vec{c}|^2}{2} - |\vec{a}| = 12$

(B) $\dfrac{|\vec{c}|^2}{2} - |\vec{a}| = 30$

(C) $|\vec{a} \times \vec{b} + \vec{c} \times \vec{a}| = 48\sqrt{3}$

(D) $\vec{a} \cdot \vec{b} = -72$

Q.12 Let $\vec{a} = -\hat{i} - \hat{k}$, $\vec{b} = -\hat{i} + \hat{j}$ and $\vec{c} = \hat{i} + 2\hat{j} + 3\hat{k}$ be three given vectors. If $\vec{r}$ is a vector such that $\vec{r} \times \vec{b} = \vec{c} \times \vec{b}$ and $\vec{r} \cdot \vec{a} = 0$ then the value of $\vec{r} \cdot \vec{b}$ is *(2011)*

Q.13 If $\vec{a}$ and $\vec{b}$ are vectors in space given by $\vec{a} = \dfrac{\hat{i} - 2\hat{j}}{\sqrt{5}}$ and $\vec{b} = \dfrac{2\hat{i} + \hat{j} + 3\hat{k}}{\sqrt{14}}$ then the value of $\left(2\vec{a} + \vec{b}\right) \cdot \left[\left(\vec{a} \times \vec{b}\right) \times \left(\vec{a} - 2\vec{b}\right)\right]$ $\left(2\vec{a} + \vec{b}\right) \cdot \left[\left(\vec{a} \times \vec{b}\right) \times \left(\vec{a} - 2\vec{b}\right)\right]$ is *(2010)*

Q.14 Let $\vec{a}, \vec{b}$ and $\vec{c}$ be three non-coplanar unit vectors such that the angle between every pair of them is $\dfrac{\pi}{3}$. If $\vec{a} \times \vec{b} + \vec{b} \times \vec{c} = p\vec{a} + q\vec{b} + r\vec{c}$ where p, q and r are scalars, then the value of $\dfrac{p^2 + 2q^2 + r^2}{q^2}$ is *(2014)*

Q.15 Let $\overline{PR} = 3\hat{i} + \hat{j} - 2\hat{k}$ and $\overline{SQ} = \hat{i} - 3\hat{j} - 4\hat{k}$ determine diagonals of a parallelogram PQRS and $\overline{PT} = \hat{i} + 2\hat{j} - 3\hat{k}$ be another vector. Then the volume of the parallelepiped determined by the vectors $\overline{PT}, \overline{PQ}$ and $\overline{PS}$ is *(2013)*

(A) 5   (B) 20   (C) 10   (D) 30

Q.16 A line l passing through the origin is perpendicular to the lines

$$\ell_1 : (3+t)\hat{i} + (-1+2t)\hat{j} + (4+2t)\hat{k} \quad -\infty < t < \infty$$

$$\ell_2 : (3+2x)\hat{j} + (3+2s)\hat{j} + (2+s)\hat{k}, -\infty < s < \infty$$

Then, the coordinate (s) of the point(s) on ℓ_2 at a distance of $\sqrt{17}$ from the point of intersection of ℓ and ℓ_2 is (are) *(2013)*

(A) $\left(\dfrac{7}{3}, \dfrac{7}{3}, \dfrac{5}{3}\right)$

(B) $(-1, -1, 0)$

(C) $(1, 1, 1)$

(D) $\left(\dfrac{7}{9}, \dfrac{7}{9}, \dfrac{8}{9}\right)$

Q.17 Consider the set of eight vectors $V = \{a\hat{i} + b\hat{j} + c\hat{k} : a, b, c \in \{-1, 1\}\}$. Three non-coplanar vectors can be chosen from V in 2^n ways. Then p is *(2013)*

Q.18 Match list I with list II and select the correct answer using the code given below the lists : *(2013)*

List I	List II
(i) Volume of parallelepiped by vectors $\vec{a}, \vec{b}$ and $\vec{c}$ is 2. Then the volume of the parallelepiped determined by vectors $2(\vec{a} \times \vec{b}), 3(\vec{b} \times \vec{c})$ and $(\vec{c} \times \vec{a})$ is	(p) 100
(ii) Volume of parallelepiped determined by vectors $\vec{a}, \vec{b}$ and $\vec{c}$ is 5. Then the volume of the parallelepiped determined by vectors $3(\vec{a} + \vec{b}), (\vec{b} + \vec{c})$ and $2(\vec{c} + \vec{a})$ is	(q) 30
(iii) Area of a triangle with adjacent sides determined by vectors $\vec{a}$ and $\vec{b}$ is 20. Then the area of the triangle with adjacent sides determined by vectors $(2\vec{a} + 3\vec{b})$ and $(\vec{a} - \vec{b})$ is	(r) 24
(iv) Area of a parallelogram with adjacent sides determined by vectors $\vec{a}$ and $\vec{b}$ is 30. Then the area of the parallelogram with adjacent sides determined by vectors $(\vec{a} + \vec{b})$ and $\vec{a}$ is	(s) 60

(A) i → s, ii → q, iii → r, iv → p

(B) i → q, ii → r, iii → p, iv → s

(C) i → r, ii → s, iii → p, iv → q

(D) i → p, ii → s, iii → r, iv → q

Q.19 If $\vec{a}$, $\vec{b}$ and $\vec{c}$ are unit vectors satisfying $\left|\vec{a} - \vec{b}\right|^2 + \left|\vec{b} - \vec{c}\right|^2 + \left|\vec{c} - \vec{a}\right|^2 = 9$ then $\left|2\vec{a} + 5\vec{b} + 5\vec{c}\right|$ is *(2012)*

Q.20 If $\vec{a}$ and $\vec{b}$ are vectors such that $\left|\vec{a} + \vec{b}\right| = \sqrt{29}$ and $\vec{a} \times (2\hat{i} + 3\hat{j} + 4\hat{k}) = (2\hat{i} + 3\hat{j} + 4\hat{k}) \times \vec{b}$ then a possible value of $(\vec{a} + \vec{b}) \cdot (-7\hat{i} + 2\hat{j} + 3\hat{k})$ is *(2012)*

(A) 0   (B) 3   (C) 4   (D) 8

Q.21 Let $\vec{a} = \hat{i} + \hat{j} + \hat{k}, \vec{b} = \hat{i} - \hat{j} + \hat{k}$ and $\vec{c} = \hat{i} - \hat{j} + \hat{k}$ be three vectors. A vector $\vec{v}$ in the plane of $\vec{a}$ and $\vec{b}$ whose projection on $\vec{c}$ is $\dfrac{1}{\sqrt{3}}$, is given by *(2011)*

(A) $\hat{i} + 3\hat{j} + 3\hat{k}$

(B) $-3\hat{i} - 3\hat{j} + \hat{k}$

(C) $3\hat{i} - \hat{j} + 3\hat{k}$

(D) $\hat{i} - 3\hat{j} - 3\hat{k}$

Q.22 The vector(s) which is/are coplanar with vectors $\hat{i} + \hat{j} + 2\hat{k}$ and $\hat{i} + 2\hat{j} + \hat{k}$, and perpendicular to the vector $\hat{i} + \hat{j} + \hat{k}$ is/are to *(2011)*

(A) $\hat{j} - \hat{k}$ (B) $-\hat{i} + \hat{j}$ (C) $\hat{i} - \hat{j}$ (D) $-\hat{j} + \hat{k}$

Important Questions

JEE Main/Boards

Exercise 1

Q.2   Q.4   Q.7   Q.8
Q.10

Exercise 2

Q.3   Q.8   Q.10

Previous Years' Questions

Q.2   Q.8   Q.9

JEE Advanced/Boards

Exercise 1

Q.2   Q.9   Q.12   Q.15

Exercise 2

Q.1   Q.3

Previous Years' Questions

Q.1   Q.5   Q.6   Q.8
Q.10

Answer Key

JEE Main/Boards

Exercise 1

Q.1 (i) $r = \left(2\hat{i} - 3\hat{j} + \hat{k} \text{ or } -\hat{i} - 2\hat{j} - 4\hat{k}\right) + t\left(3\hat{i} - \hat{j} + 5\hat{k}\right)$

Q.2 (i) $15°$ $(15.38.....), 0.268$ rad   (ii) a = 1 and intersection is $(-20, 5, -12)$

Q.3 (i) $\sqrt{161}$ (ii) $43°$

Q.4 (i) r= (either point) + $t\left(\hat{i} - 2\hat{j} - 3\hat{k} \text{ or } -\hat{i} + 2\hat{j} + 3\hat{k}\right)$, (ii) s $(\hat{i} + 2\hat{j} - \hat{k})$ (iii) $3\hat{i} + 6\hat{j} - 3\hat{k}$ (iv) $\sqrt{54}$

Q.5 (i) $2\hat{j} + \hat{k}$ (ii) $86°$ **Q.6** (i) $45.3°$ (ii) 3.54 **Q.7** (i) A cannot be 2. (ii) $-5\hat{i} - 4\hat{j}$

Q.8 (i) $68.5°$ (ii) c = -4 (iii) c = -3   **Q.9** $-\hat{i} + 2\hat{j} + 5\hat{k}$ **Q.10** $\vec{F} = 2\vec{a}_1 + 5\vec{a}_2 + 3\vec{a}_3$

Exercise 2

Single Correct Choice Type

Q.1 D   **Q.2** B   **Q.3** C   **Q.4** B   **Q.5** B   **Q.6** A

Q.7 D   **Q.8** B   **Q.9** A   **Q.10** A

JEE Advanced/Boards

Exercise 1

Exercise 2

Single Correct Choice Type

Previous Years' Questions

Solutions

JEE Main/Boards

Exercise 1

Sol 1: (i) For (either point) + t(diff b/w vectors)

r = (2i – 3j + k or – i – 2j – 4k) + t(3i – j + 5k)

(ii) L(2)(r) = 3i + 2j – 9k + s(4i – 4j + 5k)

L(1) and L(2) must be of form r = a + tb

$2 + 3t = 3 + 4s, -3 - t = 2 - 4s, 1 + 5t = -9 + 5s$

(t, s) = (+ / –3,2) or (– / +1,1) or (– / +9, –7)

Or (+/–4,2) or (0, 1) or (–/+8, –7)

Sol 2: (i) Angle between the lines

$$\cos\theta = \frac{-8 \times 9 + 1 \times 2 + (-2) \times (-5)}{\sqrt{64 + 1 + 4}\sqrt{81 + 4 + 25}} = \frac{84}{\sqrt{69}\sqrt{110}} = 0.9641$$

$\Rightarrow \theta = \cos^{-1}(0.9641) = 15.38$ degree

(ii) Let P be the point of intersection

Equation of lines are $\dfrac{x-4}{-8} = \dfrac{y-2}{1} = \dfrac{z+6}{-2} = r_1$

Point P be $(-8r_1 + 4, r_1 + 2, -2r_1 - 6)$ ....(i)

Similarly for second line $\dfrac{x+2}{-9} = \dfrac{y-a}{2} = \dfrac{z+2}{-5} = r_2$

The point P be $(-9r_2 - 2, 2r_2 - a, -5r_2 - 2)$ ...(ii)

From (i) and (ii), we get

$-8r_1 + 4 = -9r_2 - 2$

$r_1 + 2 = 2r_2 + a$

$-2r_1 - 6 = -5r_2 - 2$

On solving, we get $r_1 = 3, r_2 = 2$ and $a = 1$

The points of intersection is (–20, 5, –12)

Sol 3: (i) Find $\vec{a} - \vec{b}$ or $\vec{b} - \vec{a}$ irrespective of label (expect $11\hat{i} - 2\hat{j} - 6\hat{k}$ or $-11\hat{i} + 2\hat{j} + 6\hat{k}$)

Magnitude of vector = $\sqrt{161}$

(ii) Using $(\overrightarrow{AO}$ or $\overrightarrow{OA})$ and $(\overrightarrow{AB}$ or $\overrightarrow{BA})$

$$\cos\theta = \frac{\text{Scalar product of any two vector}}{\text{Product of their moduli}}$$

$$\frac{(4\hat{i} + 3\hat{j} - 2\hat{k}) \cdot (11\hat{i} - 2\hat{j} - 6\hat{k})}{\sqrt{4^2 + 3^2 + 2^2}\sqrt{11^2 + 2^2 + 6^2}}$$

$$= \frac{44 - 6 + 12}{\sqrt{29}\sqrt{161}} = \frac{50}{\sqrt{29}\sqrt{161}} = 43°$$

Sol 4: (i) For (either point) + t(diff between position vectors)

(ii) r = s (i+2j-k) or (i + 2j – k) + s (i + 2j – k)

Evaluate scalar product of i + 2j – k and their dir vect in (i)

Show as $(1 \times 1$ or $1) + (2x - 2$ or $- 4) + (-1x - 3$ or $3)$

= 0 and

(iii) Obtain t = -2 or s = 3 (possibly – 3 or 2 or – 2)

Check if t = 2, 1 or -1

Subst. into eqn AB or OT and to produce 3i + 6j -3k

(iv) $\left|\overrightarrow{OC}\right|$ is to be found, where C is their point of intersection

$$\left|\overrightarrow{OC}\right| = \sqrt{54}$$

Sol 5: (i) OD = OA + AD or OB + BC + CD AEF

AD = BC or CD = BA

$\overrightarrow{OD} = 2\hat{i} + \hat{k}$

(ii) AB.CB = |AB||CB| cosθ $\Rightarrow$ cos θ = 86°

Sol 6: (i) Work out $\vec{b} - \vec{a}$ or $\vec{a} - \vec{b}$ or $\vec{c} - \vec{a}$ or $\vec{a} - \vec{c}$

$= \pm(-3i - j - k)$ or $\pm(-2i + j - 2k)$

Use cosine rule and find angle as $45.3°$

(ii) Use of $\dfrac{1}{2}\left|\overrightarrow{AB}\right| \times \left|\overrightarrow{AC}\right| \sin\theta$

$$= \frac{1}{2}(\sqrt{11})(3)\sin 45.3° = 3.54$$

Sol 7: (i) Produce at least 2 of the 3 relevant eqns in λ and μ

Solving we get

1st solution: $\lambda = -2$ or $\mu = 3$

2nd solution: $\mu = 3$ or $\lambda = -2$

Substitute their λ and μ into 3rd eqn and find 'a'

We get a=2 but a cannot be 2

(ii) Subst their λ or μ (& pass a) into either line eqn

Point of intersection is $-5\hat{i} - 4\hat{j}$

Sol 8: (i) $\cos\theta = \dfrac{-6 \times 1 + 8 \times 3 - 2 \times 2}{\sqrt{36 + 64 + 4}\sqrt{1 + 9 + 4}}$

$= \dfrac{14}{\sqrt{104}\sqrt{14}} = 68.47°$

(ii) Since, and are parallel

$\cos\phi = \dfrac{-6 \times 3 + 8 \times C - 2 \times 1}{\sqrt{104}\sqrt{9 + C^2 + 1}} = 1$

$\Rightarrow 8C - 20 = \sqrt{104}\sqrt{10 + C^2}$

$\Rightarrow (8C - 20)^2 = 104(10 + C^2)$

$\Rightarrow 64C^2 + 400 - 320C = 1040 + 104C^2$

$\Rightarrow 40C^2 + 320C + 640 = 0$

$\Rightarrow C^2 + 8C + 16 = 0$

$\Rightarrow (C + 4)^2 = 0 \Rightarrow C = -4$

(iii) $L_2 \equiv \dfrac{x - 3}{1} = \dfrac{y + 3}{3} = \dfrac{z - 0}{2} = m$

Any point $(m + 3, 3m - 8, 2m)$

$L_3 \equiv \dfrac{x - 2}{3} = \dfrac{y - 1}{C} = \dfrac{z - 3}{1} = n$

Any point $(3n + 2, Cn + 1, n + 3)$

If L_2 and L_3 intersect, then

$m + 3 = 3n + 2$ ...(i)

$3m - 8 = Cn + 1$ ...(ii)

$2m = n + 3$ ...(iii)

On solution, we get $C = -3$

Sol 9: Let $\vec{R} = R_1\hat{i} + R_2\hat{j} + R_3\hat{k}$

$\left(\vec{u}.\vec{R} - 10\right)\hat{i} + \left(\vec{V}.\vec{R} - 20\right)\hat{j} + \left(\vec{W}.\vec{R} - 20\right)\hat{k} = 0$

$\left(R_1 - 2R_2 + 3R_3 - 10\right)\hat{i} + \left(2R_1 + R_2 + 4R_3 - 20\right)\hat{j}$

$\quad + \left(R_1 + 3R_2 + 3R_3 - 20\right)\hat{k} = 0$

$\Rightarrow R_1 - 2R_2 + 3R_3 = 10$ ... (i)

$2R_1 + R_2 + 4R_3 = 20$ ... (ii)

$R_1 + 3R_2 + 3R_3 = 20$ ... (iii)

On solving, we get $R_1 = -1, R_2 = 2, R_3 = 5$

$\vec{R} = -\hat{i} + 2\hat{j} + 5\hat{k}$

Sol 10: $\vec{a}_1 = 2\vec{b}_1 + 3\vec{b}_2 - \vec{b}_3$

$\vec{a}_2 = \vec{b}_1 - 2\vec{b}_2 + 2\vec{b}_3$

$\vec{a}_3 = 3\vec{b}_1 - \vec{b}_2 + 2\vec{b}_3$

On solving, we get $\vec{F} = 2\vec{a}_1 + 5\vec{a}_2 + 3\vec{a}_3$

Sol 11: $\left|\vec{r} + b\vec{s}\right|^2 = \left|\vec{r}\right|^2 + b^2\left|\vec{s}\right|^2 + 2b\vec{r}.\vec{s}$

$\left|\vec{r} + b\vec{s}\right|$ is minimum when $\vec{r}.\vec{s} = -\left|\vec{r}\right|\left|\vec{s}\right|$ and $\vec{r} = -b\vec{s}$

Then, $\left|b\vec{s}\right|^2 + \left|\vec{r} + b\vec{s}\right|^2 = b^2\left|\vec{s}\right|^2 + \left|\vec{r}\right|^2 + b^2\left|\vec{s}\right|^2 - 2b\left|\vec{r}\right|\left|\vec{s}\right|$

$= 2b^2\left|\vec{s}\right|^2 + \left|\vec{r}\right|^2 - 2b^2\left|\vec{s}\right|^2 = \left|\vec{r}\right|^2$

Exercise 2

Single Correct Choice Type

Sol 1: (D) $(\vec{b} - 2\vec{c}) = \lambda\vec{a}$ or $\dfrac{1}{\lambda}|\vec{b} - 2\vec{c}| = 1$

or $(\vec{b} - 2\vec{c})^2 = (\lambda)^2$ or $(\vec{b} - 2\vec{c}).(\vec{b} - 2\vec{c}) = \lambda^2$

or $\vec{b}.\vec{b} - 2\vec{b}.\vec{c} - 2\vec{c}.\vec{b} + 4\vec{c}.\vec{c}$

$\Rightarrow 16 - 4.|\vec{b}|.|\vec{c}|.\cos\theta + 4 = \lambda^2$

$\sin\theta = \dfrac{\sqrt{15}}{4} \Rightarrow \cos\theta = \dfrac{1}{4}$

$\Rightarrow 16 - 4 \times \dfrac{1}{4} \times 4 \times 1 + 4 = \lambda^2$

$\Rightarrow \lambda^2 = 16 \Rightarrow \lambda = \pm 4$

Sol 2: (B) $l^2 + m^2 + n^2 = 1$

or $\cos^2\alpha + \cos^2\beta + \cos^2\gamma = 1$

or $3\cos^2\theta = 1$ or $\cos\theta = \dfrac{1}{\sqrt{3}}$

$\therefore$ The desired vector is

$15\sqrt{3}\left(\dfrac{1}{\sqrt{3}}\hat{i} + \dfrac{1}{\sqrt{3}}\hat{j} + \dfrac{1}{\sqrt{3}}\hat{k}\right) = 15(\hat{i} + \hat{j} + \hat{k})$

Sol 3: (C) Hint: Scalar Triple product

Sol 4: (B) $\vec{a} + \vec{b} + \vec{c} = 0$

$|\vec{a}| = 3, |\vec{b}| = 5, |\vec{c}| = 7$

$\cos\theta = \dfrac{(5)^2 + (3)^2 - (7)^2}{2 \times 5 \times 3} = \dfrac{25 + 9 - 49}{30}$

$= \dfrac{-15}{30} = -\dfrac{1}{2} \Rightarrow \theta = 150^\circ$ or -30°

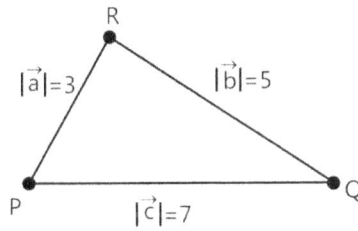

Sol 5: (B) Let $|\vec{a}| = 1 = |\vec{b}|$

$2(\vec{a}\,\vec{b} + \vec{b}.\vec{c} + \vec{c}.\vec{a}) + 3 = 0$

$(\vec{a} + \vec{b} + \vec{c}).(\vec{a} + \vec{b} + \vec{c}) = 0$

$\Rightarrow 1 + 1 + |\vec{c}|^2 + 2(\vec{a}\,\vec{b} + \vec{b}.\vec{c} + \vec{c}.\vec{a}) = 0$

$\Rightarrow |\vec{c}| + 2 - 3 = 0 \Rightarrow |\vec{c}| = 1$

Sol 6: (A) $(\vec{a} + \vec{b}).\vec{b} = 0 \Rightarrow 2\,\vec{a}.\vec{b} + 2\,\vec{b}.\vec{b} = 0$

Similarly $(\vec{a} + 2\vec{b}).\vec{a} = 0 \Rightarrow \vec{a}.\vec{a} + 2\,\vec{a}.\vec{b} = 0$

$\Rightarrow |\vec{a}|^2 = +2|\vec{b}|^2 \quad \Rightarrow |\vec{a}| = \sqrt{2}\,|\vec{b}|$

Sol 7: (D) Projection of $\vec{a}$ on $\vec{b} = \dfrac{\vec{a}.\vec{b}}{|\vec{b}|}$

Let $\vec{a} = \hat{i} + 2\hat{j} + \hat{k}$ and $\vec{b} = 4\hat{i} + 4\hat{j} + 7\hat{k}$

$\therefore$ Projection $= \dfrac{\vec{a}.\vec{b}}{|\vec{b}|}$

$= \dfrac{(\hat{i} + 2\hat{j} + \hat{k}).(4\hat{i} + 4\hat{j} + 7\hat{k})}{\sqrt{(4)^2 + (4)^2 + (7)^2}} = \dfrac{4 + 8 + 7}{\sqrt{16 + 16 + 49}} = \dfrac{19}{9}$

Hence, the correct option is d.

Sol 8: (B)

$\begin{vmatrix} \hat{i} & \hat{j} & \hat{k} \\ -2 & 6 & -8 \\ 1 & -3 & 4 \end{vmatrix} = \hat{i}(24 - 24) - \hat{j}(-8 + 8) + \hat{k}(6 - 6) = 0$

$\therefore$ Magnitude $= 0$

Sol 9: (A)

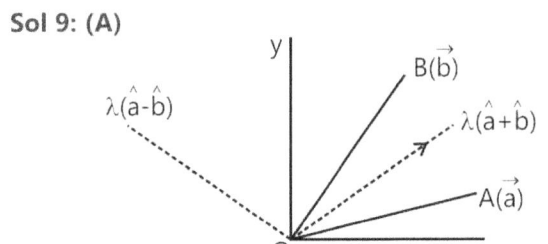

Sol 10: (A) $\left[2\vec{a} + 4\vec{b}\,\vec{c}\,\vec{d}\right] = \lambda\left[\vec{a}\,\vec{c}\,\vec{d}\right] + \mu\left[\vec{b}\,\vec{c}\,\vec{d}\right]$

$\left[2\vec{a} + 4\vec{b}.\vec{c}\,\vec{d}\right] = 2\vec{a} + 4\vec{b}.\left(\vec{c} \times \vec{d}\right) = 2\vec{a}.\left(\vec{c} \times \vec{d}\right) + 4\vec{b}.\left(\vec{c} \times \vec{d}\right)$

$= 2\left[\vec{a}\,\vec{c}\,\vec{d}\right] + 4\left[\vec{b}\,\vec{c}\,\vec{d}\right] \Rightarrow \lambda = 2, \mu = 4 \Rightarrow \lambda + \mu = 6$

Previous Years' Questions

Sol 1: (B) The volume of parallelopiped

$= \left[\vec{a}\,\vec{b}\,\vec{c}\right] = \begin{vmatrix} 2 & -3 & 0 \\ 1 & 1 & -1 \\ 3 & 0 & -1 \end{vmatrix}$

$= 2(-1) + 3(-1 + 3) = -2 + 6 = 4$

Sol 2: (B) $\vec{a} = 2p\hat{i} + \hat{j}$

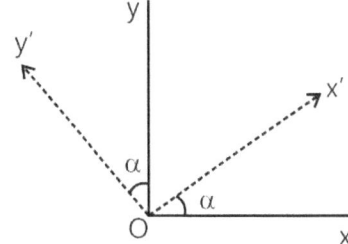

New vector

$\vec{a} = \sqrt{4P^2 + 1}\cos\alpha i + \sqrt{4P^2 + 1}\sin\alpha j$

$\Rightarrow \sqrt{4p^2 + 1}\cos\alpha = p + 1$

$\Rightarrow \cos\alpha = \dfrac{p + 1}{\sqrt{4p^2 + 1}}$

And $\sqrt{4p^2 + 1}\sin\alpha = 1$

$\Rightarrow \sin\alpha = \dfrac{1}{\sqrt{4p^2 + 1}}$

$\Rightarrow \sin^2\alpha + \cos^2\alpha = 1 = \dfrac{(p + 1)^2}{4p^2 + 1} + \dfrac{1}{4p^2 + 1}$

$\Rightarrow 4p^2 + 1 = p^2 + 1 + 2p + 1$

$\Rightarrow 3p^2 - 2p - 1 = 0 \Rightarrow 3p^2 - 3p + p - 1 = 0$

$\Rightarrow 3p(p-1)+1(p-1)=0$

$\Rightarrow (p-1)(3p+1)=0$

$\Rightarrow p=1,\dfrac{-1}{3}$

Sol 3: (B) Since, three vectors are coplanar.

$\begin{vmatrix} a & a & c \\ 1 & 0 & 1 \\ c & c & b \end{vmatrix}=0$

Applying $C_1 \rightarrow C_1 - C_2$

$\Rightarrow \begin{vmatrix} 0 & a & c \\ 1 & 0 & 1 \\ 0 & c & b \end{vmatrix}=0$

$\Rightarrow -1(ab-c^2)=0 \Rightarrow ab=c^2$

Sol 4: (A) Since, $\vec{a}\times(\vec{b}\times\vec{c})=\dfrac{\vec{b}+\vec{c}}{\sqrt{2}}$

$\Rightarrow (\vec{a}\cdot\vec{c})\vec{b}-(\vec{a}\cdot\vec{b})\vec{c}=\dfrac{1}{\sqrt{2}}\vec{b}+\dfrac{1}{\sqrt{2}}\vec{c}$

On equating the coefficient of $\vec{c}$, we get

$\vec{a}\cdot\vec{b}=-\dfrac{1}{\sqrt{2}}$

$\Rightarrow |\vec{a}||\vec{b}|\cos\theta=-\dfrac{1}{\sqrt{2}}$

$\therefore \cos\theta=-\dfrac{1}{\sqrt{2}} \Rightarrow \theta=\dfrac{3\pi}{4}$

Sol 5: (B) Since, $(\vec{a}+2\vec{b})\cdot(5\vec{a}-4\vec{b})=0$

$\Rightarrow 5|\vec{a}|^2+6\vec{a}\cdot\vec{b}-8|\vec{b}|^2=0$

$\Rightarrow 6\vec{a}\cdot\vec{b}=3 \; [\because |\vec{a}|=|\vec{b}|=1]$

$\Rightarrow \cos\theta=\dfrac{1}{2} \Rightarrow \theta=60°$

Sol 6: (C) Given, $\vec{v}=2\hat{i}+\hat{j}-\hat{k}$ and $\vec{w}=\hat{i}+3\hat{k}$

$\therefore [\vec{u}\ \vec{v}\ \vec{w}]=\vec{u}\cdot[(2\hat{i}+\hat{j}-\hat{k})\times(\hat{i}+3\hat{k})]$

$=\vec{u}\cdot(3\hat{i}-7\hat{j}-\hat{k})=|\vec{u}||3\hat{i}-7\hat{j}-\hat{k}|\cos\theta$

which is maximum, if angle between $\vec{u}$ and $3\hat{i}-7\hat{j}-\hat{k}$ is 0 and maximum value$=|3\hat{i}-7\hat{j}-\hat{k}|=\sqrt{59}$

Sol 7: (C) As we know, a vector coplanar to $\vec{a},\vec{b}$ and orthogonal to $\vec{c}$ is $\lambda\{(\vec{a}\times\vec{b})\times\vec{c}\}$

$\therefore$ A vector coplanar to $(2\hat{i}+\hat{j}+\hat{k}),(\hat{i}-\hat{j}+\hat{k})$ and

Orthogonal to $3\hat{i}+2\hat{j}+6\hat{k}$

$=\lambda[\{(2\hat{i}+\hat{j}+\hat{k})\times(\hat{i}-\hat{j}+\hat{k})\}\times(3\hat{i}+2\hat{j}+6\hat{k})]$

$=\lambda[(2\hat{i}-\hat{j}-3\hat{k})\times(3\hat{i}+2\hat{j}+6\hat{k})]$

$=\lambda(21\hat{j}-7\hat{k})$

$\therefore$ Unit vector $=\pm\dfrac{(21\hat{j}-7\hat{k})}{\sqrt{(21)^2+(7)^2}}=\pm\dfrac{(3\hat{j}-\hat{k})}{\sqrt{10}}$

Sol 8: (A) $\overrightarrow{AB}=2\hat{i}+10\hat{j}+11\hat{k}$

$\overrightarrow{AD}=-\hat{i}+2\hat{j}+2\hat{k}$

Angle 'θ' between $\overrightarrow{AB}$ and $\overrightarrow{AD}$ is

$\cos(\theta)=\left|\dfrac{\overrightarrow{AB}\cdot\overrightarrow{AD}}{|\overrightarrow{AB}||\overrightarrow{AD}|}\right|=\left|\dfrac{-2+20+22}{(15)(3)}\right|=\dfrac{8}{9}$

Sol 9: $\vec{V}=i+j+k+\lambda(i-j+k)$

$=(1+\lambda)i+(1-\lambda)j+(1+\lambda)k$

Projection on $\vec{C}$ is $\dfrac{1}{\sqrt{3}}$

$\dfrac{\vec{V}\cdot\vec{C}}{|\vec{C}|}=\dfrac{1}{\sqrt{3}}$

$\dfrac{(1+\lambda)-(1-\lambda)-(1+\lambda)}{\sqrt{3}}=\dfrac{1}{\sqrt{3}}$

$\Rightarrow 1+\lambda-1+\lambda-1-\lambda=1 \Rightarrow \lambda=2$

$\vec{V}=3\hat{i}-\hat{j}+3\hat{k}$

Sol 10: (C) $\vec{a}\times(\vec{b}\times\vec{c})-\dfrac{\sqrt{3}}{2}(\vec{b}+\vec{c})$

$\Rightarrow (\vec{a}\cdot\vec{c})\vec{b}-(\vec{a}\cdot\vec{b})\vec{c}=\dfrac{\sqrt{3}}{2}(\vec{b}+\vec{c})$

$\Rightarrow \vec{a}\cdot\vec{c}=\dfrac{\sqrt{3}}{2}$ and $\vec{a}\cdot\vec{b}=-\dfrac{\sqrt{3}}{2}$

$\Rightarrow |\vec{a}||\vec{b}|\cos\theta=-\dfrac{\sqrt{3}}{2}$

$\Rightarrow \cos\theta=\dfrac{-\sqrt{3}}{2}$

$\Rightarrow \theta=\dfrac{5\pi}{6}$

Sol 11: (D) $\left(\vec{a}\times\vec{b}\right)\times\vec{c}=\frac{1}{3}\left|\vec{b}\right|\left|\vec{c}\right|\vec{a}$

$\Rightarrow\left(\vec{c}.\vec{a}\right)\vec{b}-\left(\vec{c}.\vec{b}\right)\vec{a}=\frac{1}{3}\left|\vec{b}\right|\left|\vec{c}\right|\vec{a}$

$\Rightarrow\vec{c}.\vec{b}=-\frac{1}{3}\left|\vec{b}\right|\left|\vec{c}\right|\ \Rightarrow\cos\theta=-\frac{1}{3}$

$\Rightarrow\sin^2\theta=1-\cos^2\theta\ =1-\frac{1}{9}\ =\frac{8}{9}$

$\Rightarrow\sin\theta=\pm\frac{2\sqrt{3}}{3}$

But $\sin\theta=\frac{2\sqrt{3}}{3}$

Sol 12: (A) $\left[\vec{a}\times\vec{b}.\vec{b}\times\vec{c}\ \vec{c}\times\vec{a}\right]=\lambda\left[\vec{a}\ \vec{b}\ \vec{c}\right]^2$

We know that

$\left[\vec{a}\times\vec{b}\ \vec{b}\times\vec{c}\ \vec{c}\times\vec{a}\right]=\left[\vec{a}\ \vec{b}\ \vec{c}\right]^2$

$\Rightarrow\lambda=1$

Sol 13: (B) The length of median through A

$=\dfrac{\overrightarrow{AB}+\overrightarrow{AC}}{2}\ =\dfrac{3\hat{i}+4\hat{k}+5\hat{i}-2\hat{j}+4\hat{k}}{2}=\dfrac{8\hat{i}-2\hat{j}+8\hat{k}}{2}$

$=4\hat{i}-\hat{j}+4\hat{k}$

Length $=\sqrt{16+1+16}=\sqrt{33}$

Sol 14: (C) $\vec{c}=\vec{a}+2\vec{b}$ and $\vec{d}=5\vec{a}-4\vec{b}$

$\vec{c}\perp\vec{d}$

$\vec{c}.\vec{d}=0$

$\Rightarrow5\left|\vec{a}\right|^2-4\vec{a}.\vec{b}+10\vec{a}.\vec{b}-8\left|\vec{b}\right|^2=0$

$\Rightarrow5+6\vec{a}.\vec{b}-8=0$

$\Rightarrow6\,\vec{a}.\vec{b}=3$

$\Rightarrow\vec{a}.\vec{b}=\frac{1}{2}$

$\Rightarrow\cos\phi=\frac{1}{2}$

$\Rightarrow\phi=\frac{\pi}{3}$

Sol 15: (D) $\vec{a}.\vec{b}=0\qquad\vec{b}.\vec{c}=0\qquad\vec{c}.\vec{a}=0$

$\Rightarrow2\lambda+4+\mu=0\qquad\lambda-1+2\mu=0$

Solving we get : $\lambda=-3,\ \mu=2$

Sol 16: (D) $\vec{a}=\hat{j}-\hat{k}$ and $\vec{c}=\hat{i}-\hat{j}-\hat{k}$

Let $\vec{b}=b_1\hat{i}+b_2\hat{j}+b_3\hat{k}$

$\because\ \vec{a}\times\vec{b}+\vec{c}=0,\ \vec{a}\times\vec{b}=-\vec{c}$

$\begin{vmatrix}\hat{i}&\hat{j}&\hat{k}\\0&1&-1\\b_1&b_2&b_3\end{vmatrix}=-\hat{i}+\hat{j}+\hat{k}$

$\hat{i}(b_3+b_2)-\hat{j}(b_1)+k(-b_1)=-\hat{i}+\hat{j}+\hat{k}$

$b_3+b_2=-1$ ... (i)

$b_1=-1$ ... (ii)

$\vec{a}\cdot\vec{b}=3$

$b_2-b_3=3$ ... (iii)

Solve (i) and (iii)

$2b_2=2\qquad\qquad b_1=2\qquad\qquad b_3=-2$

$\therefore\ b_1=-1\qquad\quad b_2=1\qquad\qquad b_3=-2$

Hence $\vec{b}=-\hat{i}+\hat{j}-2\hat{k}$

Sol 17: (D) $\left(2\vec{a}-\vec{b}\right)\cdot\left\{\left(\vec{a}\times\vec{b}\right)\times\left(\vec{a}+2\vec{b}\right)\right\}$

$=\left(2\vec{a}-\vec{b}\right)\cdot\left\{\left[\vec{a}\cdot\left(\vec{a}+2\vec{b}\right)\right]\vec{b}-\left[\vec{b}\cdot\left(\vec{a}+2\vec{b}\right)\vec{a}\right]\right\}$

$=-5\left(\vec{a}\right)^2\left(\vec{b}\right)^2+5\left(\vec{a}\cdot\vec{b}\right)^2=-5$

Sol 18: (C) $\vec{b}\times\vec{c}=\vec{b}\times\vec{d}$

$\Rightarrow\vec{a}\times\left(\vec{b}\times\vec{c}\right)=\vec{a}\times\left(\vec{b}\times\vec{d}\right)$

$\Rightarrow\left(\vec{a}.\vec{c}\right)\vec{b}-\left(\vec{a}.\vec{b}\right)\vec{c}=\left(\vec{a}.\vec{d}\right)\vec{b}-\left(\vec{a}.\vec{b}\right)\vec{d}$

$\Rightarrow\left(\vec{a}.\vec{c}\right)\vec{b}-\left(\vec{a}.\vec{b}\right)\vec{c}=-\left(\vec{a}.\vec{b}\right)\vec{d}$

$\therefore\ \vec{d}=\vec{c}-\left(\dfrac{\vec{a}.\vec{c}}{\vec{a}.\vec{b}}\right)\vec{b}$

JEE Advanced/Boards

Exercise 1

Sol 1: $\dfrac{\left(\vec{a}\times\vec{b}\right)^2+\left(\vec{a}.\vec{b}\right)^2}{2\vec{a}^2\vec{b}^2}$

$$= \frac{(\vec{a} \times \vec{b}).(\vec{a} \times \vec{b}) + (\vec{a}.\vec{b}).(\vec{a}.\vec{b})}{2|\vec{a}|^2.|\vec{b}|^2} = \frac{|\vec{a} \times \vec{b}|^2 + |\vec{a}.\vec{b}|^2}{2|\vec{a}|^2.|\vec{b}|^2}$$

$$= \frac{|\vec{a}|^2.|\vec{b}|^2.\sin^2\theta + |\vec{a}|^2.|\vec{b}|^2.\cos^2\theta}{2|\vec{a}|^2.|\vec{b}|^2}$$

Sol 2: Area $= \frac{1}{2}\left|\left(3\hat{i} + 4\hat{j}\right) \times \left(-5\hat{i} + 7\hat{j}\right)\right| = \frac{1}{2}\left|21\hat{K} + 20\,\hat{k}\right| = \frac{41}{2}$

Sol 3: $20\hat{i} + 11\hat{j} = \lambda\left\{(40 - a)\hat{i} + 44\hat{j}\right\}$

$\therefore \lambda(40 - a) = 20$ and $\lambda(4) = 11$

$\Rightarrow a = -40$

Sol 4: Note that since $|\vec{a}| = |\vec{b}|$ hence the parallelogram will be a rhombus.

Sol 5: $\because (\vec{a} + 3\vec{b})$ is perpendicular to $(7\vec{a} - 5\vec{b})$

$\therefore (\vec{a} + 3\vec{b}).(7\vec{a} - 5\vec{b}) = 0$

$\Rightarrow 7.|\vec{a}|^2 - 5\,\vec{a}.\vec{b} + 21\,\vec{a}.\vec{b} - 15|\vec{b}|^2 = 0 \Rightarrow 16\,\vec{a}.\vec{b} = 8$

$\therefore \vec{a}.\vec{b} = \frac{1}{2} \Rightarrow |\vec{a}|.|\vec{b}|.\cos\theta = \frac{1}{2} \Rightarrow \cos\theta = \frac{1}{2} \Rightarrow \theta = \frac{\pi}{3}$

Sol 6: $\left|\hat{A} - \hat{B}\right|^2 = (\hat{A} - \hat{B})(\hat{A} - \hat{B})$

$= |\hat{A}|^2 + |\hat{B}|^2 - 2\hat{A}.\hat{B} = 1 + 1 - 2\,|1||1|\cos\pi = 2 + 2 = 4$

$\Rightarrow |\hat{A} - \hat{B}| = 2 \Rightarrow \frac{|\hat{A} - \hat{B}|}{2} = 1$

Sol 7: $\vec{F_1} = 3\hat{i} + 2\hat{j} + 5\hat{k}$ and $\vec{F_2} = 2\hat{j} + \hat{j} + 3\hat{k}$

$\therefore \vec{F} = \vec{F_1} + \vec{F_2} = 5\hat{i} + 3\hat{j} + 8\hat{k}$

$\Delta\vec{x} = (4\hat{i} + 3\hat{j} + 7\hat{k}) - (2\hat{i} + \hat{j} + 3\hat{k}) = 2\hat{i} + 2\hat{j} + 4\hat{k}$

$\therefore$ Work done $= \vec{F}.\Delta\vec{x} = 10 + 6 + 32 = 48$ units

Sol 8: $\Delta\vec{x} = -2\hat{i} - 3\hat{j} - 4\hat{k}$

Work done $= \vec{F}.\Delta\vec{x}$

$\Rightarrow 2 = -12 - 3\lambda - 16 \Rightarrow 30 = -3\lambda \Rightarrow \lambda = -10$

Sol 9: Let $\vec{b} = 3\hat{i}$ and $\vec{a} = 2(\hat{i} + \hat{j})$

The two diagonals will be $6\vec{a} - \vec{b}$ and $4\vec{a} + 5\vec{b}$

Length of $6\vec{a} - \vec{b} = |9\hat{i} + 12\hat{j}| = 15$

Length of $4\vec{a} + 5\vec{b}$

$= |8\hat{j} + 23\hat{i}| = \sqrt{(23)^2 + (8)^2} = \sqrt{593}$

Sol 10: $2 \times |\hat{i} + x\hat{j} + 3\hat{k}| = |4\hat{i} + (4x - 2)\hat{j} + 2\hat{k}|$

or, $2.\sqrt{1^2 + x^2 + 9} = \sqrt{16 + (4x - 2)^2 + 4}$

$\therefore 4(10 + x^2) = 20 + 16x^2 + 4 - 16x$

or, $40 + 4x^2 = 24 + 16x^2 - 16x$

or, $12x^2 - 16x - 16 = 0$ or, $3x^2 - 4x - 4 = 0$

$\therefore x = \frac{4 \pm \sqrt{16 + 4.4.3}}{6} = \frac{4 \pm 8}{6} = 2$ or $\frac{-2}{3}$

Exercise 2

Sol 1: (A) $\vec{r} = (9\hat{i} - \hat{j} + 2\hat{k}) - (3\hat{i} - 2\hat{j} + \hat{k}) = (6\hat{i} + \hat{j} + \hat{k})$

$\therefore$ Moment of couple $= \vec{r} \times \vec{F} = (6\hat{i} + \hat{j} + \hat{k}) \times (5\hat{i} + \hat{j})$

$= \begin{vmatrix} \hat{i} & \hat{j} & \hat{k} \\ 6 & 1 & 1 \\ 5 & 1 & 0 \end{vmatrix} = \hat{i}(0 - 1) - \hat{j}(0 - 5) + \hat{k}(6 - 5)$

$= -\hat{i} + 5\hat{j} + \hat{k}$

Sol 2: (C) $(\vec{a} + \vec{b} + \vec{c}).(\vec{a} + \vec{b} + \vec{c})$

$= |\vec{a}|^2 + |\vec{b}|^2 + |\vec{c}|^2 + \vec{a}(\vec{b} + \vec{c}) + \vec{b}(\vec{a} + \vec{c}) + \vec{c}(\vec{a} + \vec{b})$

$= (1)^2 + (4)^2 + (8)^2 + 0 = 81$

$\therefore |\vec{a} + \vec{b} + \vec{c}| = 9$

Sol 3: (A) Let M be the point of intersection of internal bisector with AB.

$\therefore \frac{AM}{MB} = \frac{1}{2}$

$\therefore \overline{OM} = \frac{1(2\hat{i} + 4\hat{j} + 4\hat{k}) + 2(2\hat{i} + 2\hat{j} + \hat{k})}{3} = \frac{6\hat{i} + 8\hat{j} + 6\hat{k}}{3}$

$\therefore |\overline{OM}| = \sqrt{4 + \left(\frac{8}{3}\right)^2 + 4} = \sqrt{\frac{72 + 64}{9}} = \sqrt{\frac{136}{9}}$

Sol 4: (A) $\vec{a} + 2\vec{b} + 3\vec{c} = 0$

$\Rightarrow \vec{a} \times \vec{b} + 2\vec{b} \times \vec{b} + 3\vec{c} \times \vec{b} = 0$

$\Rightarrow \vec{a} \times \vec{b} + 3\vec{c} \times \vec{b} = 0 \Rightarrow \vec{a} \times \vec{b} = 3\vec{b} \times \vec{c}$ ... (i)

Similarly, $\vec{a} + 2\vec{b} + 3\vec{c} = 0$

$\vec{a} \times \vec{a} + 2\vec{b} \times \vec{a} + 3\vec{c} \times \vec{a} = 0 \Rightarrow \vec{c} \times \vec{a} = \dfrac{2}{3}\vec{a} \times \vec{b}$

$= \dfrac{2}{3} \times 3(\vec{b} \times \vec{c}) = 2(\vec{b} \times \vec{c})$ ... (ii)

$\Rightarrow \vec{a} \times \vec{b} + \vec{b} \times \vec{c} + \vec{c} \times \vec{a} = 6(\vec{b} \times \vec{c})$

Sol 5: (C) $\left[\vec{a} - \vec{b} \quad \vec{b} - \vec{c} \quad \vec{c} - \vec{a}\right] = (\vec{a} - \vec{b}) \cdot (\vec{b} - \vec{c}) \times (\vec{c} - \vec{a})$

$= (\vec{a} - \vec{b}) \cdot \{\vec{b} \times \vec{c} - \vec{b} \times \vec{a} + \vec{c} \times \vec{a}\} = \vec{a} \cdot (\vec{b} \times \vec{c}) - \vec{b} \cdot (\vec{c} \times \vec{a}) = 0$

Sol 6: (D) A general point on the curve will have vector

$x\hat{i} + y\hat{j} = x\hat{i} + \left(x + \dfrac{1}{x}\right)\hat{j}$

$\because \overrightarrow{OP} \cdot \vec{I} = 1$

$\therefore \left\{x i + \left(+\dfrac{1}{x}\right)\hat{j}\right\} \cdot \hat{i} = 1 \Rightarrow x = 1$

$\therefore \overrightarrow{OP} = \hat{i} + 2\hat{j}$

Again, $\overrightarrow{OQ} \cdot \vec{I} = -1 \Rightarrow x = -1$

$\therefore \overrightarrow{OQ} = -\vec{i} - 2\vec{j}$

$\therefore 2\overrightarrow{OP} + 3\overrightarrow{OQ} = -\hat{i} - 2\hat{j}$

$\therefore \left|2\overrightarrow{OP} + 3\overrightarrow{OQ}\right| = \sqrt{(-1)^2 + (2)^2} = \sqrt{5}$

Sol 7: (C) $\vec{A} \cdot (\vec{B} \cdot \vec{C}) + \vec{B}\vec{C}(\vec{A} + \vec{B}) = 0$

$\Rightarrow A.B + A.C + B.C = 0$

$|A + B + C|^2 = |A|^2 + |B|^2 + |C|^2 + 2(A.B + B.C + C.A)$

$= 1 + 16 + 64 + 0 = 81$

$\Rightarrow |A + B + C| = 9$

Sol 8: (C) Given

$|A - B| \le 1$

$\therefore \sqrt{1^2 + 1^2 - 2\cos 2\theta} \le 1$

$\Rightarrow 2 - 2\cos 2\theta \le 1 \Rightarrow 2\cos 2\theta \ge 1 \Rightarrow \cos 2\theta \ge \dfrac{1}{2}$

$\therefore \dfrac{5\pi}{3} \le 2\theta \le 2\pi \Rightarrow \dfrac{5\pi}{6} \le \theta \le \pi$

Sol 9: (C) $\left|\vec{A} \ \vec{B} \ \vec{A} \times \vec{B}\right| = \dfrac{1}{4}$

$\Rightarrow \vec{A} \cdot (\vec{B} \times \vec{A} \times \vec{B}) = \dfrac{1}{4} \Rightarrow \vec{A}\left[(\vec{B} \cdot \vec{B})\vec{A} - (\vec{A} \cdot \vec{B})\vec{B}\right] = \dfrac{1}{4}$

$\Rightarrow \vec{A} \cdot \left[\vec{A} - (\vec{A} \cdot \vec{B})\vec{B}\right] = \dfrac{1}{4} \Rightarrow \vec{A} \cdot \vec{A} - (\vec{A} \cdot \vec{B})^2 = \dfrac{1}{4}$

$\Rightarrow (\vec{A} \cdot \vec{B})^2 = 1 - \dfrac{1}{4} = \dfrac{3}{4} \Rightarrow \vec{A} \cdot \vec{B} = \dfrac{\sqrt{3}}{2}$

$\Rightarrow \cos\theta = \dfrac{\sqrt{3}}{2} \Rightarrow \theta = \dfrac{\pi}{3}$

Sol10:(C) $|\vec{A} - \vec{B}| + |\vec{A} + \vec{B}| = \sqrt{2 - 2\cos\theta} + \sqrt{2 + 2\cos\theta}$

$= \sqrt{2}\left(\sqrt{1 - \cos\theta} + \sqrt{1 + \cos\theta}\right)$

$= \sqrt{2} \cdot \left\{\sqrt{2}\sin\dfrac{\theta}{2} + 2\cos\dfrac{\theta}{2}\right\} = 2\left(\sin\dfrac{\theta}{2} + \cos\dfrac{\theta}{2}\right)$

Greatest value is $2\sqrt{2}$

Previous Years' Questions

Sol 1: $\vec{A} \cdot \vec{X} = C$ and $\vec{A} \times \vec{X} = \vec{B}$

Let $\vec{A} = A_1\hat{i} + A_2\hat{j} + A_3\hat{k}$

(i) $\vec{A} \times \vec{X} = \vec{B}$

$\vec{A} \times (\vec{A} \times \vec{X}) = \vec{A} \times \vec{B} \Rightarrow (\vec{A} \cdot \vec{X})\vec{A} - (\vec{A} \cdot \vec{A})\vec{X} = \vec{A} \times \vec{B}$

$\Rightarrow |\vec{A}|^2 \vec{X} = C\vec{A} - \vec{A} \times \vec{B} \Rightarrow \vec{X} = \dfrac{C\vec{A}}{|\vec{A}|^2} - \dfrac{\vec{A} \times \vec{B}}{|\vec{A}|^2}$

(ii) If coordinate system is rotated about the x-axis through an angle $\dfrac{\pi}{2}$, then

x- component = A_2

y – component = A_1

z – component = A_3

$\vec{A} = A_2\hat{i} - A_1\hat{j} + A_3\hat{k}$

[New coordinates system]

Sol 2: $\begin{vmatrix} \vec{a} & \vec{b} & \vec{c} \\ \vec{a}.\vec{a} & \vec{a}.\vec{b} & \vec{a}.\vec{c} \\ \vec{b}.\vec{a} & \vec{b}.\vec{b} & \vec{b}.\vec{c} \end{vmatrix} = \begin{vmatrix} \vec{a}+\vec{b}+\vec{c} & \vec{b} & \vec{c} \\ \vec{a}.(\vec{a}+\vec{b}+\vec{c}) & \vec{a}.\vec{b} & \vec{a}.\vec{c} \\ \vec{b}.(\vec{a}+\vec{b}+\vec{c}) & \vec{b}.\vec{b} & \vec{b}.\vec{c} \end{vmatrix}$

$$= \left(\vec{a}+\vec{b}+\vec{c}\right).\begin{vmatrix} 1 & \vec{b} & \vec{c} \\ \vec{a} & \vec{a}.\vec{b} & \vec{a}.\vec{c} \\ \vec{b} & \vec{b}.\vec{b} & \vec{b}.\vec{c} \end{vmatrix}$$

$$= \left(\vec{a}+\vec{b}+\vec{c}\right).\begin{bmatrix} \left(\vec{a}.\vec{b}\right)\left(\vec{b}.\vec{c}\right)-\left(\vec{b}.\vec{b}\right)\left(\vec{a}.\vec{c}\right)-\left(\vec{a}.\vec{b}\right)\left(\vec{b}.\vec{c}\right)+ \\ \left(\vec{b}.\vec{b}\right)\left(\vec{a}.\vec{c}\right)+\left(\vec{a}.\vec{c}\right)\left(\vec{b}.\vec{b}\right)-\left(\vec{b}.\vec{c}\right)\left(\vec{a}.\vec{b}\right) \end{bmatrix}$$

$$= \left(\vec{a}+\vec{b}+\vec{c}\right)\left[\left(\vec{b}.\vec{b}\right)\left(\vec{a}.\vec{c}\right)-\left(\vec{a}.\vec{b}\right)\left(\vec{b}.\vec{c}\right)\right]=0$$

$\Rightarrow$ if $\vec{a}+\vec{b}+\vec{c}=0$ [coplanar condition]

Sol 3: $\vec{R}\times\vec{B}=\vec{C}\times\vec{B} \Rightarrow \vec{A}\times\left(\vec{R}\times\vec{B}\right)=\vec{A}\times\left(\vec{C}\times\vec{B}\right)$

$\Rightarrow \left(\vec{A}.\vec{B}\right)\vec{R}-\left(\vec{A}.\vec{R}\right)\vec{B}=\left(\vec{A}.\vec{B}\right)\vec{C}-\left(\vec{A}.\vec{C}\right)\vec{B}$

$\left(2+0+1\right)\vec{R}-0=\left(2+0+1\right)\vec{C}-\left(8+0+7\right)\vec{B}$

$\Rightarrow 3\vec{R}=3\vec{C}-15\vec{B} \Rightarrow \vec{R}=\vec{C}-5\vec{B}$

$= 4\hat{i}-3\hat{j}+7\hat{k}-5\hat{i}-5\hat{j}-5\hat{k} = -\hat{i}-8\hat{j}+2\hat{k}$

Sol 4: Given, $\vec{a}\times\vec{b}=\vec{c}\times\vec{d}$

and $\vec{a}\times\vec{c}=\vec{b}\times\vec{d}$

$\Rightarrow \vec{a}\times\vec{b}-\vec{a}\times\vec{c}=\vec{c}\times\vec{d}-\vec{b}\times\vec{d}$

$\Rightarrow \vec{a}\times(\vec{b}-\vec{c})=(\vec{c}-\vec{b})\times\vec{d}$

$\Rightarrow \vec{a}\times(\vec{b}-\vec{c})-(\vec{c}-\vec{b})\times\vec{d}=0$

$\Rightarrow \vec{a}\times(\vec{b}-\vec{c})-\vec{d}\times(\vec{b}-\vec{c})=0$

$\Rightarrow (\vec{a}-\vec{d})\times(\vec{b}-\vec{c})=0 \Rightarrow (\vec{a}-\vec{d})\|(\vec{b}-\vec{c})$

$\therefore (\vec{a}-\vec{d})\cdot(\vec{b}-\vec{c})\neq0 \Rightarrow \vec{a}\cdot\vec{b}+\vec{d}\cdot\vec{c}\neq\vec{d}\cdot\vec{b}+\vec{a}\cdot\vec{c}$

Sol 5: Since, $\hat{v}$ is unit vector along the incident ray and $\hat{w}$ is the unit vector along the reflected ray.

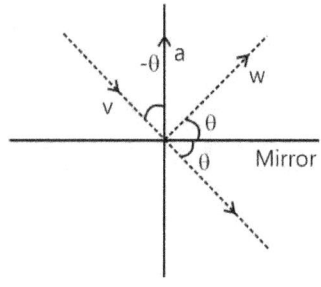

Hence, $\hat{a}$ is a unit vector along the external bisector of $\hat{v}$ and $\hat{w}$.

$\therefore \hat{w}-\hat{v}=\lambda\,\hat{a}$

On squaring both sides, we get

$\Rightarrow 1+1-\hat{w}\cdot\hat{v}=\lambda^2 \Rightarrow 2-2\cos2\theta=\lambda^2 \Rightarrow \lambda=2\sin\theta$

where 2θ is the angle between $\hat{v}$ and $\hat{w}$.

Hence, $\hat{w}-\hat{v}=2\sin\theta\cdot\hat{a}$

$= 2\cos(90°-\theta)\hat{a} = -(2\hat{a}\cdot\hat{v})\hat{a}$

$\hat{w}=\hat{v}-2(\hat{a}\cdot\hat{v})\hat{a}$

Sol 6: (B, D) Let vector $\overrightarrow{AO}$ be parallel to line of intersection of planes P_1 and P_2 through origin

Normal to plane P_1 is

$\overrightarrow{n_1}=[(2\hat{j}+3\hat{k})\times(4\hat{j}-3\hat{k})]=-18\hat{i}$

Normal to plane P_2 is

$\overrightarrow{n_2}=(\hat{j}-\hat{k})\times(3\hat{i}+3\hat{j})=3\hat{i}-3\hat{j}-\hat{k}$

$\therefore \overrightarrow{OA}$ is parallel to $\pm\left(\overrightarrow{n_1}\times\overrightarrow{n_2}\right)=54\hat{j}-54\hat{k}$

$\therefore$ Angle between $54(\hat{j}-\hat{k})$ and $\left(2\hat{i}+\hat{j}-2\hat{k}\right)$ is

$\cos\theta=\pm\left(\dfrac{54+108}{3\cdot54\cdot\sqrt{2}}\right)=\pm\dfrac{1}{\sqrt{2}}$ $\therefore \theta=\dfrac{\pi}{4},\dfrac{3\pi}{4}$

Sol 7: (A, D) Let, $\vec{a}=\hat{i}+\hat{j}+2\hat{k}$, $\vec{b}=\hat{i}+2\hat{j}+\hat{k}$ and $\vec{c}=\hat{i}+\hat{j}+\hat{k}$

$\therefore$ A vector coplanar to $\vec{a}$ and $\vec{b}$, and perpendicular to $\vec{c}$

$\vec{r}=\lambda(\vec{a}\times\vec{b})\times\vec{c}=\lambda\{(\vec{a}\cdot\vec{c})\vec{v}-(\vec{b}\cdot\vec{c})\vec{a}\}$

$= \lambda\{(1+1+4)(\hat{i}+2\hat{j}+\hat{k})-(1+2+1)(\hat{i}+\hat{j}+2\hat{k})\}$

$= \lambda\{6\hat{i}+12\hat{j}+6\hat{k}-6\hat{i}-6\hat{j}-12\hat{k}\}=\lambda\{6\hat{j}-6\hat{k}\}=6\lambda\{\hat{j}-\hat{k}\}$

For $\lambda=\dfrac{1}{6} \Rightarrow$ (a) is correct.

and $\lambda=-\dfrac{1}{6} \Rightarrow$ (d) is correct.

Sol 8: From the given information, it is clear that

$\vec{a}=\dfrac{\hat{i}-2\hat{j}}{\sqrt{5}}$

$\Rightarrow |\vec{a}|=1, |\vec{b}|=1, |\vec{a}\cdot\vec{b}|=0$

Now, $(2\vec{a}+\vec{b})\cdot[(\vec{a}\times\vec{b})\times(\vec{a}-2\vec{b})]$

$= (2\vec{a}+\vec{b})\cdot[a^2\vec{b}-(\vec{a}\cdot\vec{b})\cdot\vec{a}+2b^2\cdot\vec{a}-2(\vec{b}\cdot\vec{a})\cdot a]$

$= [2\vec{a}+\vec{b}]\cdot[\vec{b}+2\vec{a}]=4\vec{a}^2+\vec{b}^2=4\cdot1+1=5$ [as $\vec{a}.\vec{b}=0$]

Sol 9: $\vec{r} \times \vec{b} = \vec{c} \times \vec{b}$

$\Rightarrow (\vec{r} - \vec{c}) \times \vec{b} = 0 \Rightarrow \vec{r} - \vec{c} = \lambda \vec{b}$

or $\vec{r} = \vec{c} + \lambda \vec{b}$ ...(i)

Given, $\vec{r} \cdot \vec{a} = 0$, taking dot product with $\vec{a}$ for Eq. (i)

$\Rightarrow \vec{r} \cdot \vec{a} = \vec{a} \cdot \vec{c} + \lambda \vec{a} \cdot \vec{b}$

$\therefore \lambda = \dfrac{-\vec{a} \cdot \vec{c}}{\vec{a} \cdot \vec{b}} \quad (\because \vec{r} \cdot \vec{a} = 0)$ ...(ii)

From Eqs. (i) and (ii), we get

$\vec{r} = \vec{c} - \dfrac{\vec{a} \cdot \vec{c}}{\vec{a} \cdot \vec{b}} \vec{b}$, taking dot with $\vec{b}$, we get

$\vec{r} \cdot \vec{b} = \vec{c} \cdot \vec{b} - \dfrac{\vec{a} \cdot \vec{c}}{\vec{a} \cdot \vec{b}} (\vec{b} \cdot \vec{b})$

$= (-1 + 2) - \dfrac{(-1 - 3)}{(1)} (1 + 1)$ where,

$\begin{bmatrix} \vec{a} = -\hat{i} - \hat{k} \\ \vec{b} = -\hat{i} + \hat{j} \\ \vec{c} = \hat{i} + 2\hat{j} + 3\hat{k} \end{bmatrix} = 1 + 8 = 9$

Sol 10: (B, C) Given: $w \cdot (\hat{u} \times \vec{v}) = 1$

$\Rightarrow |w| |(\hat{u} \times \vec{v})| \cos\theta = 1 \Rightarrow \cos\theta = 1$

$w \perp \hat{u} \times \hat{v} \Rightarrow w \perp \hat{u}$ and $w \perp \hat{v}$ and $|\hat{u} \times \hat{v}| = 1$

Angle between $\hat{u}$ and $\hat{v}$ can change to have initially many of vectors $\hat{v}$ as $\hat{w} \perp \hat{v}$

If $\hat{u}$ lies in xy plane then $\hat{u} = u_1 \hat{i} + u_2 \hat{j}$

$\Rightarrow \hat{w} \cdot \hat{u} = 0 \Rightarrow u_1 + u_2 = 0 \Rightarrow |u_1| = |u_2|$

Sol 11: (A, C, D) In $\triangle PQR$

$-\vec{a} = \vec{b} + \vec{c}$

$\Rightarrow \vec{a} \cdot \vec{a} = (\vec{b} + \vec{c}) \cdot (\vec{b} + \vec{c})$

$\Rightarrow |\vec{a}|^2 = |\vec{b}|^2 + |\vec{c}|^2 + 2\vec{b} \cdot \vec{c}$

$\Rightarrow |\vec{a}|^2 = (4\sqrt{3})^2 + |\vec{c}|^2 + 2 \times 24$

$\Rightarrow (12)^2 = (4\sqrt{3})^2 + |\vec{c}|^2 + 2 \times 24$

$\Rightarrow |\vec{c}|^2 = 144 - 96$

$\Rightarrow \boxed{|\vec{c}| = 4\sqrt{3}}$

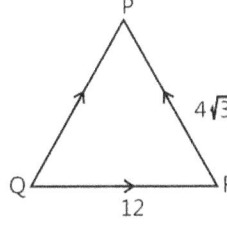

$\Rightarrow \dfrac{|\vec{c}|^2}{2} - |\vec{a}| = \dfrac{48}{2} - 12 = 24 - 12 = 12$

Given $\vec{b} \cdot \vec{c} = 24$

$-|\vec{b}| |\vec{c}| \cos p = 24$

$-4\sqrt{3} \times 4\sqrt{3} \cos p = 24$

$\cos p = \dfrac{-1}{2}$

Since $|\vec{b}| = |\vec{c}|$

$\angle PQR = \angle PRQ \Rightarrow \angle QPR = 120^0$

and $\angle PQR = \angle PRQ = 30^0 \Rightarrow |\vec{a} \times \vec{b} + \vec{c} \times \vec{a}| = 48\sqrt{3}$

And $\vec{a} \times \vec{b} = -72$

Sol 12: $\vec{r} \times \vec{b} = \vec{c} \times \vec{b}$

Taking cross with a

$\Rightarrow \vec{a} \times (\vec{r} \times \vec{b}) = \vec{a} \times (\vec{c} \times \vec{b})$

$\Rightarrow (\vec{a} \cdot \vec{b}) \vec{r} - (\vec{a} \cdot \vec{r}) \vec{b} = \vec{a} \times (\vec{c} \times \vec{b})$

$\Rightarrow \vec{r} = -3\hat{i} + 6\hat{j} + 3\hat{k}$

$\vec{r} \cdot \vec{b} = 3 + 6 = 9$

Sol 13: $(2\vec{a} + \vec{b}) \cdot [(\vec{a} \times \vec{b}) \times (\vec{a} - 2\vec{b})]$

$\vec{a} = \dfrac{\hat{i} - 2\hat{j}}{\sqrt{5}}, \vec{b} = \dfrac{2\hat{i} + \hat{j} + 3\hat{k}}{\sqrt{14}}$

$\Rightarrow (2\vec{a} + \vec{b})[\vec{a} \times (\vec{a} \times \vec{b}) - (\vec{a} \times \vec{b}) \times \vec{b}]$

$= (2\vec{a} \times \vec{b})[(\vec{a} \cdot \vec{a})\vec{b} - (\vec{a} \cdot \vec{b})\vec{a} + 2(\vec{a} \cdot \vec{b})\vec{b} + 2(\vec{b} \vec{b})\vec{a}]$

$= (2\vec{a} + \vec{b}) \cdot [\vec{b} + 2\vec{a}] \quad \{\vec{a} \cdot \vec{b} = 0\}$

$= 2\vec{a} \vec{b} + 4|\vec{a}|^2 + |\vec{b}|^2 - 2 \vec{b} \cdot \vec{a}$

$= 4|\vec{a}|^2 + |\vec{b}|^2$

$= 4 + 1 = 5$

Sol 14: Given: $\vec{a} \times \vec{b} + \vec{b} \times \vec{c} = p\vec{a} + q\vec{b} + r\vec{c}$

$\Rightarrow \vec{a} \cdot (\vec{a} \times \vec{b}) + \vec{a} \cdot (\vec{b} \times \vec{c}) = p + q(\vec{a} \cdot \vec{b}) + r(\vec{a} \cdot \vec{c})$

$\Rightarrow \vec{a} \cdot (\vec{b} \times \vec{c}) = P + \dfrac{q}{2} + \dfrac{r}{2}$ ...(i)

Similarly, $\vec{b} \cdot (\vec{a} \times \vec{b}) + \vec{b} \cdot (\vec{b} \times \vec{c}) = \dfrac{p}{2} + q + \dfrac{r}{2}$

$\Rightarrow \dfrac{p}{2} + q + \dfrac{r}{2} = 0$ ...(ii)

and $\dfrac{p}{2} + \dfrac{q}{2} + r = a(b \times c)$ ...(iii)

From (i), (ii) and (iii)

$P = -q = r \Rightarrow \dfrac{p^2 + 2q^2 + r^2}{q^2} = 4$

Sol 15: (C) $\Rightarrow \vec{x} + \vec{y} = 3\hat{i} + \hat{j} - 2\hat{k}$

and $\vec{x} - \vec{y} = \hat{i} - 3\hat{j} - 4\hat{k}$

On solving we get

$\vec{x} = 2i - j - 3k$

$\vec{y} = i + 2j + k$

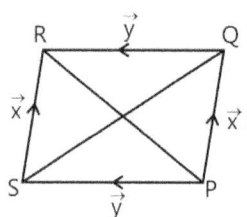

Volume of parallelopiped

$= \begin{vmatrix} 2 & -1 & -3 \\ 1 & 2 & 1 \\ 1 & 2 & 3 \end{vmatrix}$

$= 2(6-2) + 1(3-1) - 3(2-0)$

$= 8 + 2 = 10$

Sol 16: (B, D) Vector perpendicular to ℓ_1 and ℓ_2 is given by

$\begin{vmatrix} \hat{i} & \hat{j} & \hat{k} \\ 2 & 2 & 1 \\ 1 & 2 & 2 \end{vmatrix}$

$= \hat{j}(4-2) - \hat{j}(4-1) + \hat{k}(4-2)$

$= 2\hat{i} - 3\hat{j} + 2\hat{k}$

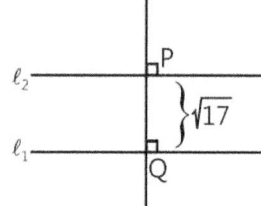

The eq. of line $\perp$ to ℓ_1 and ℓ_2

$\dfrac{x-0}{2} = \dfrac{y-0}{-3} = \dfrac{z-0}{2} = \gamma$

$\Rightarrow Q \equiv (2\gamma, -3\gamma, 2\gamma)$

The point Q lies on ℓ_2, then $= \dfrac{2\gamma - 3}{1} = \dfrac{-3\gamma + 1}{2} = \dfrac{2\gamma - 4}{2}$

$\Rightarrow \gamma = 1$

$\Rightarrow Q \equiv (2, -3, 2)$

Distance of P from Q in $\sqrt{17}$

$PQ^2 = 17 = (2-3-2)^2 + (-3-3-25)^2 + (2-5-2)^2$

$\Rightarrow S = -2, \dfrac{-10}{9}$

$\Rightarrow P \equiv (-1, -1, 0)$ and $\left(\dfrac{7}{9}, \dfrac{7}{9}, \dfrac{8}{9} \right)$

Sol 17: $V = a\hat{i} + b\hat{j} + c\hat{k} : a, b, c \in \{-1, 1\}$

Total number of selection $= 8_{c_3}$

No. of coplanar vectors $= 6 \times 4 = 24$

Total number of non co-planar vet

$= {}^8c_3 - 24 = 32 = 2^p$

$= P = 5$

Sol 18: (C) (i) $\left[2\vec{a} \times b \;\; 3\vec{b} \times \vec{c} \;\; \vec{c} \times a \right] = 6\left[\vec{a} \;\; \vec{b} \;\; \vec{c} \right]^2$

$= 6(2)^2 = 24\left[\vec{a}. \vec{b}. \vec{c} = 2 \right]$

(ii) $\left[3(\vec{a} + \vec{b}) \;\; \vec{b} + \vec{c} \;\; 2\vec{c} + \vec{a} \right] = 6\left[\vec{a} + \vec{b} \;\; \vec{b} + \vec{c} \;\; \vec{c} + \vec{a} \right]$

$= 6\left(\vec{a} + \vec{b}. \left[(\vec{b} + \vec{c}) \times (\vec{c} + \vec{a}) \right] \right)$

$= 6(\vec{a} \times \vec{b})\left[\vec{b} \times \vec{c} + \vec{b} \times \vec{a} + \vec{c} \times \vec{c} + \vec{c} \times \vec{a} \right] = 12a.(b \times c)$

$= 12\left[\vec{a} \;\; \vec{b} \;\; \vec{c} \right] = 12 \times 5 = 60$ $\left(\left[\vec{a} \;\; \vec{b} \;\; \vec{c} \right] = 5 \right)$

(iii) $\dfrac{1}{2} \left| (2\vec{a} + 3\vec{b}) \times (\vec{a} - \vec{b}) \right|$

$= \dfrac{1}{2} \left| (2\vec{a} \times \vec{a} - 2\vec{a} \times \vec{b} + 3\vec{b} \times \vec{a} - 3\vec{b} \times b) \right|$

$= \dfrac{1}{2} \left| 5\vec{a} \times \vec{b} \right| = \dfrac{1}{2} \times 5 \times 40 = 100$ $\left[\dfrac{1}{2} \left| \vec{a} \times \vec{b} \right| = 20 \right]$

(iii) $\left| (\vec{a} \times \vec{b}) \times \vec{a} \right| = \left| \vec{a} \times \vec{a} + \vec{b} \times \vec{a} \right| = \left| \vec{b} \times \vec{a} \right|$

$= 30$ $\left[\left| \vec{a} \times \vec{b} \right| = 30 \right]$

Sol 19: $\left| \vec{a} - \vec{b} \right|^2 + \left| \vec{b} - \vec{c} \right|^2 + \left| \vec{c} - \vec{a} \right|^2 = 9$

$\Rightarrow 3\left(\left| \vec{a} \right|^2 + \left| \vec{b} \right|^2 + \left| \vec{c} \right|^2 \right) - \left| \vec{a} + \vec{b} + \vec{c} \right|^2 = 9$

$\Rightarrow 3(1 + 1 + 1) - \left| \vec{a} + \vec{b} + \vec{c} \right|^2 = 9$

$\Rightarrow \left| \vec{a} + \vec{b} + \vec{c} \right|^2 = 0 \Rightarrow \vec{a} + \vec{b} + \vec{c} = 0$

$\Rightarrow \vec{b} + \vec{c} = -\vec{a}$

Now, $\left| 2\vec{a} + 5\vec{b} + 5\vec{c} \right| = \left| 2\vec{a} + 5(\vec{b} + \vec{c}) \right|$

$= \left| 2\vec{a} - 5\vec{a} \right| = 3\left| \vec{a} \right| = 3$

Sol 20: (C) Let $\vec{c} = 2\hat{i} + 3\hat{j} + 4\hat{k}$

$\vec{a} \times \vec{c} = \vec{c} \times b \Rightarrow (\vec{a} + \vec{b}) \times \vec{c} = 0$

$\Rightarrow (\vec{a} + \vec{b}) \parallel \vec{c}$

Let $(\vec{a} + \vec{b}) = \lambda \vec{c}$

$\Rightarrow |\vec{a} + \vec{b}| = |\lambda| \, |\vec{c}|$

$\Rightarrow \sqrt{29} = |\lambda| . \sqrt{29} \Rightarrow \lambda = \pm 1$

$\therefore \vec{a} + \vec{b} = \pm (2\hat{i} + 3\hat{j} + 4\hat{k})$

Now, $(\vec{a} + \vec{b}).(-7\hat{i} + 2\hat{j} + 3\hat{k}) = \pm(-14 + 6 + 12) = \pm 4$

Sol 21: (C) Any vectors $\vec{v}$ coplanar with $\vec{a}$ and $\vec{b}$ is given by

$\vec{v} = m\vec{a} + n\vec{b}$

$= m(\hat{i} + \hat{j} + \hat{k}) + n(\hat{i} + \hat{j} - \hat{k})$

$= (m + n)\hat{i} + (m - n)\hat{j} + (m + n)\hat{k}$ ... (i)

Projection to $\vec{v}$ on $\vec{c}$ is given by $\dfrac{\vec{v}.\vec{c}}{|\vec{c}|} = \dfrac{1}{\sqrt{3}}$

$\Rightarrow (m + n) - (m - n) - (m + n) = 1$

$\Rightarrow m + n - m + n - m - n = 1$

$\Rightarrow m + 1 = n$

$\Rightarrow m = n - 1$

Substituting in (i)

$(2n - 1)\hat{i} - \hat{j} + 2n - 1\hat{k}$

for $n = 2$

$3\hat{i} - \hat{j} + 3\hat{k}$

Sol 22: (A, D) Let $\vec{r}$ the vector coplanar with $i + j + 2k$ and $i + 2j + k$ then

$\vec{r} = m(i + j + 2k) + n(i + 2j + k)$

$= (m + n)i + (m + 2n)j + (2m + n)k$

$\vec{r} \perp \vec{c}$, then

$m + n + m + 2n + 2m + n = 0$

$\Rightarrow m + n = 0$

$\Rightarrow \vec{r} = (0)i + (0 + n)j + (n + 0)k$

$= nj + mk$

2. 3D GEOMETRY

1. COORDINATE OF A POINT IN SPACE

Let P be a point in the space. If a perpendicular from that point is dropped to the xy-plane, then the algebraic length of this perpendicular is considered as z-coordinate. From the foot of the perpendicular, drop a perpendicular to x and y axes, and algebraic lengths of perpendicular are considered as y and x coordinates, respectively.

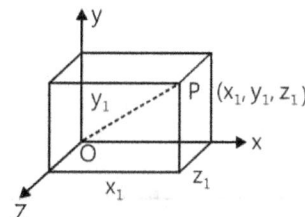

Figure 27.1

2. VECTOR REPRESENTATION OF A POINT IN SPACE

If (x, y, z) are the coordinates of a point P in space, then the position vector of the point P w.r.t. the same origin is $\overrightarrow{OP} = x\hat{i} + y\hat{i} + z\hat{k}$.

3. DISTANCE FORMULA

If (x_1, y_1, z_1) and (x_2, y_2, z_2) are any two points, then the distance between them can be calculated by the following formula: $\sqrt{(x_1 - x_2)^2 + (y_1 - y_2)^2 + (z_1 - z_2)^2}$

3.1 Vector Method

If OA and OB are the position vectors of two points A (x_1, y_1, z_1) and B (x_2, y_2, z_2), then $AB = |\overrightarrow{OB} - \overrightarrow{OA}|$

$\Rightarrow AB = |(x_2 i + y_2 j + z_2 k) - (x_1 i + y_1 j + z_1 k)|$　　$\Rightarrow AB = \sqrt{(x_2 - x_1)^2 + (y_2 - y_1)^2 + (z_2 - z_1)^2}$

3.2 Distance of a Point from Coordinate Axes

Let PA, PB and PC be the distances of the point P(x, y, z) from the coordinates axes OX, OY and OZ, respectively. Then $PA = \sqrt{y^2 + z^2}$, $PB = \sqrt{z^2 + x^2}$, $PC = \sqrt{x^2 + y^2}$

Illustration 1: Show that the points (0, 7, 10), (–1, 6, 6) and (–4, 9, 6) form a right-angled isosceles triangle.

(JEE MAIN)

Sol: By using distance formula we can find out length of sides formed by these points and if it satisfies Pythagoras theorem then these points form a right angled triangle.

Let $A \equiv (0, 7, 10)$, $B \equiv (-1, 6, 6)$, $C \equiv (-4, 9, 6) AB^2 = (0 + 1)^2 + (7 - 6)^2 + (10 - 6)^2 = 18$

$\therefore \quad AB = 3\sqrt{2}$ Similarly $BC = 3\sqrt{2}$ and $AC = 6$; Clearly $\quad AB^2 + BC^2 = AC^2$ and $AB = BC$

Hence, $\triangle ABC$ is isosceles right angled.

Illustration 2: Find the locus of a point which moves such that the sum of its distance from points A(0, 0, −α) and B(0, 0, α) is constant. **(JEE MAIN)**

Sol: Consider the point whose locus is required be P(x, y, z). As sum of its distance from point A and B is constant therefore PA + PB = constant = 2a.

Let P(x, y, z) be the variable point whose locus is required

Given that PA + PB = constant = 2a(say)

$$\therefore \quad \sqrt{(x-0)^2 + (y-0)^2 + (z+\alpha)^2} + \sqrt{(x-0)^2 + (y-0)^2 + (z-\alpha)^2} = 2a$$

$$\Rightarrow \sqrt{x^2 + y^2 + (z+\alpha)^2} = 2a - \sqrt{x^2 + y^2 + (z-\alpha)^2}$$

$$\Rightarrow x^2 + y^2 + z^2 + \alpha^2 + 2z\alpha = 4a^2 + x^2 + y^2 + z^2 + \alpha^2 - 2z\alpha - 4a\sqrt{x^2 + y^2 + (z-\alpha)^2}$$

$$\Rightarrow 4z\alpha - 4a^2 = -4a\sqrt{x^2 + y^2 + (z-\alpha)^2} \Rightarrow \frac{z^2\alpha^2}{a^2} + a^2 - 2z\alpha = x^2 + y^2 + z^2 + \alpha^2 - 2z\alpha \Rightarrow \frac{x^2 + y^2}{a^2 - \alpha^2} + \frac{z^2}{a^2} = 1$$

4. SECTION FORMULA

If a point P divides the distance between the points $A(x_1, y_1, z_1)$ and $B(x_2, y_2, z_2)$ in the ratio of m:n, then the

coordinates of P are $\left(\dfrac{mx_2 + nx_1}{m+n}, \dfrac{my_2 + ny_1}{m+n}, \dfrac{mz_2 + nz_1}{m+n} \right)$

Note: Midpoint $\left(\dfrac{x_1 + x_2}{2}, \dfrac{y_1 + y_2}{2}, \dfrac{z_1 + z_2}{2} \right)$

5. DIRECTION COSINES AND DIRECTION RATIOS

(a) Direction cosines: If α, β, γ are the angles which the line makes with the positive directions of the axes x, y and z coordinates, respectively, then $\cos\alpha$, $\cos\beta$, $\cos\gamma$ are called the direction cosines (d.c.s) of the line. The direction cosines are usually denoted by (l, m, n), where $l = \cos\alpha$, $m = \cos\beta$ and $n = \cos\gamma$.

(b) If l, m, n are the direction cosines of a line, then $l^2 + m^2 + n^2 = 1$

(c) Direction ratios: If the intercepts a, b, c are proportional to the direction cosines l,

(d) m, n, then a, b, c are called the direction ratios (d.r.s).

(e) If l, m, n are the direction cosines and a, b, c are the direction ratios of a vector, then

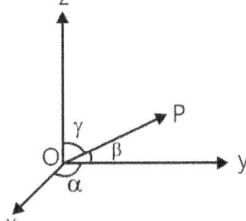

Figure 27.2

$$l = \frac{a}{\sqrt{a^2 + b^2 + c^2}}, m = \frac{b}{\sqrt{a^2 + b^2 + c^2}}, n = \frac{c}{\sqrt{a^2 + b^2 + c^2}} \quad \text{or}$$

$$l = \frac{-a}{\sqrt{a^2 + b^2 + c^2}}, m = \frac{-b}{\sqrt{a^2 + b^2 + c^2}}, n = \frac{-c}{\sqrt{a^2 + b^2 + c^2}}$$

(f) If $OP = r$, where O is the origin and l, m, n are the direction cosines of OP, then the coordinates of P are (lr, mr, nr) If direction cosines of the line AB are l, m, n, $|AB| = r$, and the coordinates of A is (x_1, y_1, z_1), then the coordinates of B are $(x_1 + rl, y_1 + rm, z_1 + rn)$

Illustration 3: Let α, β, γ be the angles made with the coordinate axes. Prove that $\sin^2\alpha + \sin^2\beta + \sin^2\gamma = 2$

(JEE ADVANCED)

Sol: Here line makes angles α, β, γ with the co-ordinates axes, hence by using its direction cosine we can prove given equation.

Since a line makes angles α, β, γ with the coordinates axes, $\cos\alpha$, $\cos\beta$, $\cos\gamma$, are direction cosines.

$\therefore \quad \cos^2\alpha + \cos^2\beta + \cos^2\gamma = 1$

$\Rightarrow (1 - \sin^2\alpha) + (1 - \sin^2\beta) + (1 - \sin^2\gamma) = 1 \Rightarrow \sin^2\alpha + \sin^2\beta + \sin^2\gamma = 2$

Illustration 4: Find the direction cosines l, m, n of a line using the following relations: $l + m + n = 0$ and $2mn + 2ml - nl = 0$.

(JEE ADVANCED)

Sol: By solving these two equations simultaneously, we will be get $l : m : n$.

Given, $l + m + n = 0$ ... (i) $2mn + 2ml - nl = 0$ ...(ii)

From equation (i), $n = -(l + m)$ Substituting $n = -(l + m)$ in equation (ii), we get,

$-2m(l + m) + 2ml + (l + m)l = 0$ $\Rightarrow -2ml - 2m^2 + 2ml + l^2 + ml = 0$

$\Rightarrow l^2 + ml - 2m^2 = 0$ $\Rightarrow \left(\dfrac{l}{m}\right)^2 + \left(\dfrac{l}{m}\right) - 2 = 0 \Rightarrow \dfrac{l}{m} = \dfrac{-1 \pm \sqrt{1+8}}{2} = \dfrac{-1 \pm 3}{2} = 1, -2$

Case I: When $\dfrac{l}{m} = 1$:   In this case $m = l$ From equation (1), $2l + n = 0$

$\Rightarrow n = -2l$

$\therefore l : m : n = 1 : 1 : -2$ $\therefore$ Direction ratios of the line are 1, 1, -2

$\therefore$ Direction cosines are $\pm\dfrac{1}{\sqrt{1^2 + 1^2 + (-2)^2}}, \pm\dfrac{1}{\sqrt{1^2 + 1^2 + (-2)^2}}, \pm\dfrac{-2}{\sqrt{1^2 + 1^2 + (-2)^2}} = \dfrac{1}{\sqrt{6}}, \dfrac{1}{\sqrt{6}}, \dfrac{-2}{\sqrt{6}}$ or $-\dfrac{1}{\sqrt{6}}, -\dfrac{1}{\sqrt{6}}, \dfrac{2}{\sqrt{6}}$

Case II: When $\dfrac{l}{m} = -2$: In this case $l = -2m$

From equation (i), $-2m + m + n = 0$ Þ $n = m$ $\therefore$ $l : m : n = -2m : m : m$

$\therefore$ Direction ratios of the line are -2, 1, 1

$\therefore$ Direction cosines are given by

$\dfrac{-2}{\sqrt{(-2)^2 + 1^2 + 1^2}}, \dfrac{1}{\sqrt{(-2)^2 + 1^2 + 1^2}}, \dfrac{1}{\sqrt{(-2)^2 + 1^2 + 1^2}} = \dfrac{-2}{\sqrt{6}}, \dfrac{1}{\sqrt{6}}, \dfrac{1}{\sqrt{6}}$ or $\dfrac{2}{\sqrt{6}}, \dfrac{-1}{\sqrt{6}}, \dfrac{-1}{\sqrt{6}}$

6. ANGLE BETWEEN TWO LINE SEGMENTS

If a_1, b_1, c_1 and a_2, b_2, c_2 are the direction ratios of any two lines, respectively, then $a_1 i + b_1 j + c_1 k$ and $a_2 i + b_2 j + c_2 k$ are the two vectors parallel to the lines, and the angle between them is given by the following formula:

$$\cos\theta = \dfrac{a_1 a_2 + b_1 b_2 + c_1 c_2}{\sqrt{a_1^2 + b_1^2 + c_1^2}\sqrt{a_2^2 + b_2^2 + c_2^2}}$$

(a) The lines are perpendicular if $a_1 a_2 + b_1 b_2 + c_1 c_2 = 0$

(b) The lines are parallel if $\dfrac{a_1}{a_2} = \dfrac{b_1}{b_2} = \dfrac{c_1}{c_2}$

(c) Two parallel lines have same direction cosines, i.e. $l_1 = l_2$, $m_1 = m_2$, $n_1 = n_2$

Illustration 5: Prove that the lines, whose direction cosines given by the relations $a^2 l + b^2 m + c^2 n = 0$ and $mn + nl + lm = 0$, are perpendicular if $\dfrac{1}{a^2} + \dfrac{1}{b^2} + \dfrac{1}{c^2} = 0$ and parallel, if $a \mp b \pm c = 0$ **(JEE ADVANCED)**

Sol: Here if two lines are perpendicular then, $l_1 l_2 + m_1 m_2 + n_1 n_2 = 0$ and if they are parallel then,

$l_1 = l_2, \ m_1 = m_2, \ n_1 = n_2$

Given that, $a^2 l + b^2 m + c^2 n = 0$ ... (i)

and $mn + nl + lm = 0$ ... (ii)

Eliminating m from equations (i) and (ii), we have $-\dfrac{1}{b^2}(a^2 l + c^2 n)n + nl - \dfrac{1}{b^2}(a^2 l + c^2 n)l = 0$

$\Rightarrow a^2 nl + c^2 n^2 - b^2 nl + a^2 l^2 + c^2 ln = 0 \qquad \Rightarrow a^2 \dfrac{l}{n} + c^2 - b^2 \dfrac{l}{n} + a^2 \dfrac{l^2}{n^2} + c^2 \dfrac{l}{n} = 0$

$\Rightarrow a^2 \left(\dfrac{l}{n}\right)^2 + \left(a^2 - b^2 + c^2\right)\left(\dfrac{l}{n}\right) + c^2 = 0$ ... (iii)

Let $\dfrac{l_1}{n_1}, \dfrac{l_2}{n_2}$ be the roots of the equation (iii).

$\therefore$ Product of roots $\dfrac{l_1}{n_1} \cdot \dfrac{l_2}{n_2} = \dfrac{c^2}{a^2} \qquad \Rightarrow \dfrac{l_1 l_2}{1/a^2} = \dfrac{n_1 n_2}{1/c^2} \Rightarrow \dfrac{l_1 l_2}{1/a^2} = \dfrac{m_1 m_2}{1/b^2} = \dfrac{n_1 n_2}{1/c^2}$ [By symmetry]

$\Rightarrow \dfrac{l_1 l_2}{1/a^2} = \dfrac{m_1 m_2}{1/b^2} = \dfrac{n_1 n_2}{1/c^2} = \dfrac{l_1 l_2 + m_1 m_2 + n_1 n_2}{1/a^2 + 1/b^2 + 1/c^2}$

For perpendicular lines $\quad l_1 l_2 + m_1 m_2 + n_1 n_2 = 0$

$\Rightarrow \dfrac{1}{a^2} + \dfrac{1}{b^2} + \dfrac{1}{c^2} = 0 \qquad$ Two lines are parallel if $l_1 = l_2, \ m_1 = m_2, n_1 = n_2$

$\Rightarrow \dfrac{l_1}{n_1} = \dfrac{l_2}{n_2} \qquad \Rightarrow$ roots of equation (iii) are equal $\Rightarrow (a^2 - b^2 + c^2)^2 - 4a^2 c^2 = 0 \Rightarrow a^2 - b^2 + c^2 = \pm 2ac$

$\Rightarrow a^2 + c^2 \pm 2ac = b^2 \Rightarrow (a \pm c)^2 = b^2 \qquad \Rightarrow (a \pm c) = \pm b \qquad \Rightarrow a \mp b \pm c = 0$

Note: In the above result, the two signs are independent of each other. So, the total cases would be $(a + b + c = 0, a + b - c = 0, a - b + c = 0, a - b - c = 0)$.

7. PROJECTION OF A LINE SEGMENT ON A LINE

If (x_1, y_1, z_1) and (x_2, y_2, z_2) are the coordinates of P and Q, respectively, then the projection of the line segments PQ on a line having direction cosines l, m, n is $|l(x_2 - x_1) + m(y_2 - y_1) + n(z_2 - z_1)|$.

Vector form: Projection of a vector $\vec{a}$ on another vector $\vec{b}$ is $= \dfrac{\vec{a}\vec{b}}{|\vec{b}|}$. In the above case, we replace $2\sqrt{6}$ with

$\vec{PQ}$ as $(x_2 - x_1)\hat{i} + (y_2 - y_1)\hat{j} + (z_2 - z_1)\hat{k}$ and $\Rightarrow (x - 1)^2 = 11$ with $l\hat{i} + m\hat{j} + n\hat{k}$.

where $l|r|, m|r| \& n|r|$ are the projections of r in the coordinate axes OX, OY and OZ, respectively.

$r = |r|(l\hat{i} + m\hat{j} + n\hat{k})$

Illustration 6: Find the projection of the line joining the coordinates (1, 2, 3) and (−1, 4, 2) on line having direction ratios 2, 3, −6. **(JEE MAIN)**

Sol: Here projection of line joining (1, 2, 3) and (–1, 4, 2) on the line having direction ratios 2, 3, –6 is given by $\ell(x_2 - x_1) + m(y_2 - y_1) + n(z_2 - z_1)$.

Let $A \equiv (1,2,3), B \equiv (-1,4,2)$. Direction ratios of the given line PQ are 2, 3, –6

$\therefore$ Direction cosines of PQ are $\dfrac{2}{7}, \dfrac{3}{7}, -\dfrac{6}{7}$

Projection of AB on PQ = $\ell(x_2 - x_1) + m(y_2 - y_1) + n(z_2 - z_1)$

$= \dfrac{2}{7}(-1-1) + \dfrac{2}{7}(4-2) - \dfrac{6}{7}(2-3) = \dfrac{-4+6+6}{7} = \dfrac{8}{7}$

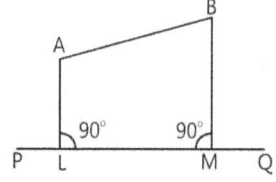

Figure 27.3

8. PLANE

If a line joining any two points on a surface entirely lies on it or if a line joining any two points on a surface is perpendicular to some fixed straight line, then the surface is called a plane. This fixed line is called the normal to the plane.

8.1 Equation of a Plane

(a) **Normal form:** The equation of a plane is given by lx + my + nz = p, where ℓ, m, n are the direction cosines of the normal to the plane and p is the distance of the plane from the origin.

(b) **General form:** The equation of a plane is given by ax + by + cz + d = 0, where a, b, c are the direction ratios of the normal to the plane.

(c) The equation of a plane passing through the point (x_1, y_1, z_1) is given by $a(x - x_1) + b(y - y_1) + c(z - z_1) = 0$, where a, b, c are the direction ratios of the normal to the plane.

(d) **Plane through three points:** The equation of a plane through three noncollinear points is given by

(x_1, y_1, z_1), (x_2, y_2, z_2), (x_3, y_3, z_3) is $\begin{vmatrix} x & y & z & 1 \\ x_1 & y_1 & z_1 & 1 \\ x_2 & y_2 & z_2 & 1 \\ x_3 & y_3 & z_3 & 1 \end{vmatrix} = 0 \equiv \begin{vmatrix} x-x_3 & y-y_3 & z-z_3 \\ x_1-x_3 & y_1-y_3 & z_1-z_3 \\ x_2-x_3 & y_2-y_3 & z_2-z_3 \end{vmatrix} = 0$

(e) **Intercept form:** The equation of a plane cutting the intercepts a, b, c on the axes is given by $\dfrac{x}{a} + \dfrac{y}{b} + \dfrac{z}{c} = 1$.

(f) **Vector form:** The equation of a plane passing through a point having a position vector $\vec{a}$ and unit vector normal to plane is $(\vec{r} - \vec{a}) \cdot \hat{n} = 0 \Rightarrow \vec{r}.\hat{n} = \vec{a}.\hat{n}$

(g) The equation of any plane parallel to the given plane ax + by + cz + d = 0 is given by ax + by + cz + λ = 0 (same direction ratios), where λ is any scalar.

(h) The equation of a plane passing through a given point $\vec{a}$ and parallel to two vectors $\vec{b}$ and $\vec{c}$ is given by $\vec{r} \cdot (\vec{b} \times \vec{c}) = \vec{a} \cdot (\vec{b} \times \vec{c})$ where $\vec{r}$ is a position vector of any point on the plane.

8.2 Plane Parallel to a Given Plane

The general equation of the plane parallel to the plane ax + by + cz + d = 0 is ax + by + cz + k = 0, where k is any scalar.

Distance between two parallel planes ax + by + cz + d_1 = 0 and ax + by + cz + d_2 = 0 is given by $\dfrac{|d_1 - d_2|}{\sqrt{a^2 + b^2 + c^2}}$

Illustration 7: Find the distance between the planes 2x – y + 2z = 4 and 6x – 3y + 6z = 2.　　　　**(JEE MAIN)**

Sol: Here if two planes are parallel then the distance between them is equal to $\dfrac{|d_1 - d_2|}{\sqrt{a^2 + b^2 + c^2}}$.

Given planes are $2x - y + 2z - 4 = 0$ ... (i)

and $6x - 3y + 6z - 2 = 0$ ... (ii)

We find that $\dfrac{a_1}{a_2} = \dfrac{b_1}{b_2} = \dfrac{c_1}{c_2}$. Hence, planes (i) and (ii) are parallel

Plane (ii) may be written as $2x - y + 2z - 2/3 = 0$ ... (iii)

$\therefore$ Required distance between the planes $= \dfrac{\left|4 - (2/3)\right|}{\sqrt{2^2 + (-1)^2 + 2^3}} = \dfrac{10}{3.3} = \dfrac{10}{9}$

8.3 Plane Passing Through the Line of Intersection of Planes

Let π_1 and π_2 be the two planes represented by equations $\vec{r} \cdot \hat{n}_1 = d_1$ and $\vec{r} \cdot \hat{n}_2 = d_2$, respectively. The position vector of any point on the line of intersection must satisfy both equations.

If $\vec{t}$ is the position vector of a point on the line, then $\vec{t} \cdot \hat{n}_1 = d_1$ and $\vec{t} \cdot \hat{n}_2 = d_2$

Therefore, for all real values of λ. we have $\vec{t} \cdot (\hat{n}_1 + \lambda \hat{n}_2) = d_1 + \lambda d_2$

Because $\vec{t}$ is arbitrary, it satisfies for any point on the line. Hence, the equation $\vec{r} \cdot (\hat{n}_1 + \lambda \hat{n}_2) = d_1 + \lambda d_2$ represents a plane π_3 which is such that if any vector X satisfies the equations of both the planes π_1 and π_2, it also satisfies the equation of plane π_3.

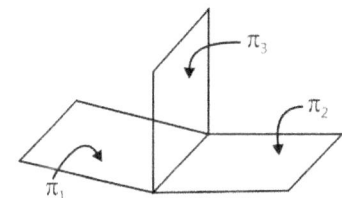

Figure 27.4

8.4 Cartesian Form

In a Cartesian system, let $n_1 = A_1\hat{i} + B_1\hat{j} + C_1\hat{k}$, $n_2 = A_2\hat{i} + B_2\hat{j} + C_2\hat{k}$ and $r = x\hat{i} + y\hat{j} + z\hat{k}$.

On substituting above values in vector equation we get,

$x(A_1 + \lambda A_2) + y(B_1 + \lambda B_2) + z(C_1 + \lambda C_2) = d_1 + \lambda d_2$ or $(A_1 x + B_1 y + C_1 z - d_1) + \ell(A_2 x + B_2 y + C_2 z - d_2) = 0$

Illustration 8: Show that the points $(0, -1, 0)$, $(2, 1, -1)$, $(1, 1, 1)$, $(3, 3, 0)$ are coplanar. **(JEE MAIN)**

Sol: Equation of any plane passing through (x_1, y_1, z_1) is given by $a(x - x_1) + b(y - y_1) + c(z - z_1) = 0$, by using this formula we can obtain respective equation of plane.

Let $A \equiv (0, -1, 0)$, $B \equiv (2, 1, -1)$, $C \equiv (1, 1, 1)$ and $D \equiv (3, 3, 0)$

Equation of a plane through $A(0, -1, 0)$ is $a(x - 0) + b(y + 1) + c(z - 0) = 0$

$\Rightarrow ax + by + cz + b = 0$ ...(i)

If plane (i) passes through $B(2, 1, -1)$ and $C(1, 1, 1)$

Then $2a + 2b - c = 0$ ...(2) and $a + 2b + c = 0$ ...(iii)

From equations (ii) and (iii), we have $\dfrac{a}{2 + 2} = \dfrac{b}{-1 - 2} = \dfrac{c}{4 - 2}$ or $\dfrac{a}{4} = \dfrac{b}{-3} = \dfrac{c}{2} = k\,(\text{say})$

Substituting values of a, b, c in equation (i), equation of the required plane is $4kx - 3k(y + 1) + 2kz = 0$

$\Rightarrow 4x - 3y + 2z - 3 = 0$ ...(iv)

Thus, point $D(3, 3, 0)$ lies on plane (iv).

Because the points on the plane passes through A, B, C, the points A, B, C and D are coplanar.

Illustration 9: Find the equation of the plane upon which the length of normal from the origin is 10 and direction ratios of this normal are 3, 2, 6. **(JEE ADVANCED)**

Sol: Let p be the length of perpendicular from the origin to the plane and ℓ, m, n be the direction cosines of this normal. The equation is given by

$$\ell x + my + nz = p \qquad \qquad \text{... (i)}$$

From the data provided, p = 10 and the direction ratios of the normal to the plane are 3, 2, 6.

$\therefore$ Direction cosines of normal to the required plane are $\ell = \dfrac{3}{7}, m = \dfrac{2}{7}, n = \dfrac{6}{7}$

Substituting values of ℓ, m, n, p in equation (i), equation of the required plane is $\dfrac{3}{7}x + \dfrac{2}{7}y + \dfrac{6}{7}z = 10$

$\Rightarrow 3x + 2y + 6z = 70$

Illustration 10: A point P moves on a plane $\dfrac{x}{a} + \dfrac{y}{b} + \dfrac{z}{c} = 1$. A plane through P and perpendicular to OP meets the coordinate axes in A, B and C. If the planes through A, B and C parallel to the planes x = 0, y = 0, z = 0 intersect in Q, find the locus of Q. **(JEE ADVANCED)**

Sol: Similar to above problem.

Given plane is $\dfrac{x}{a} + \dfrac{y}{b} + \dfrac{z}{c} = 1$ $\qquad \qquad \text{... (i)}$

Let $P \equiv (h, k, \ell)$, Then, $\dfrac{h}{a} + \dfrac{k}{b} + \dfrac{\ell}{c} = 1$ $\qquad \qquad \text{... (ii)}$

$(OP) = \sqrt{h^2 + k^2 + \ell^2}$

Direction cosines of OP are $\dfrac{h}{\sqrt{h^2 + k^2 + \ell^2}}, \dfrac{k}{\sqrt{h^2 + k^2 + \ell^2}}, \dfrac{\ell}{\sqrt{h^2 + k^2 + \ell^2}}$

$\therefore$ Equation of the plane through P and normal to OP is

$$\dfrac{h}{\sqrt{h^2 + k^2 + \ell^2}}x + \dfrac{k}{\sqrt{h^2 + k^2 + \ell^2}}y + \dfrac{\ell}{\sqrt{h^2 + k^2 + \ell^2}} = \sqrt{h^2 + k^2 + \ell^2}$$

$\Rightarrow hx + ky + \ell z = (h^2 + k^2 + \ell^2)$, $A \equiv \left(\dfrac{h^2 + k^2 + \ell^2}{h}, 0, 0\right)$, $B \equiv \left(0, \dfrac{h^2 + k^2 + \ell^2}{k}, 0\right)$, $C \equiv \left(0, 0, \dfrac{h^2 + k^2 + \ell^2}{\ell}\right)$

$\Rightarrow A = \left(\dfrac{h^2 + k^2 + \ell^2}{h}, 0, 0\right)$, $B = \left(0, \dfrac{h^2 + k^2 + \ell^2}{k}, 0\right)$, $C = \left(0, 0, \dfrac{h^2 + k^2 + \ell^2}{\ell}\right)$

Let $Q \equiv (\alpha, \beta, \gamma)$, then $\alpha = \dfrac{h^2 + k^2 + \ell^2}{h}, \beta = \dfrac{h^2 + k^2 + \ell^2}{k}, \gamma = \dfrac{h^2 + k^2 + \ell^2}{\ell}$ $\qquad \text{... (iii)}$

Now $\dfrac{1}{\alpha^2} + \dfrac{1}{\beta^2} + \dfrac{1}{\gamma^2} = \dfrac{h^2 + k^2 + \ell^2}{(h^2 + k^2 + \ell^2)^2} = \dfrac{1}{(h^2 + k^2 + \ell^2)}$ $\qquad \text{... (iv)}$

From equation (iii), $h = \dfrac{h^2 + k^2 + \ell^2}{\alpha}$

$\therefore \dfrac{h}{a} = \dfrac{h^2 + k^2 + \ell^2}{a\alpha}$ Similarly $\dfrac{k}{b} = \dfrac{h^2 + k^2 + \ell^2}{b\beta}$ and $\dfrac{\ell}{c} = \dfrac{h^2 + k^2 + \ell^2}{c\gamma}$

$\therefore \dfrac{h^2 + k^2 + \ell^2}{a\alpha} + \dfrac{h^2 + k^2 + \ell^2}{b\beta} + \dfrac{h^2 + k^2 + \ell^2}{c\gamma} = \dfrac{h}{a} + \dfrac{k}{b} + \dfrac{\ell}{c} = 1$ [from equation (ii)] or, $\dfrac{1}{a\alpha} + \dfrac{1}{b\beta} + \dfrac{1}{c\gamma} = \dfrac{1}{h^2 + k^2 + \ell^2} = \dfrac{1}{\alpha^2} + \dfrac{1}{\beta^2} + \dfrac{1}{\gamma^2}$ [from equation (iv)]

$\therefore$ Required locus of $Q(\alpha, \beta, \gamma)$ is $\dfrac{1}{ax} + \dfrac{1}{by} = \dfrac{1}{cz} = \dfrac{1}{x^2} + \dfrac{1}{y^2} + \dfrac{1}{z^2}$.

8.5 Plane and a Point

(a) A plane divides the three-dimensional space into two equal segments. Two points $A(x_1, y_1, z_1)$ and $B(x_2, y_2, z_3)$ lie on the same sides of the plane $ax + by + cz + d = 0$ if the two expressions $ax_1 + by_1 + cz_1 + d$ and $ax_2 + by_2 + cz_3 + d$ are of same sign, and lie on the opposite sides of plane if both of these expressions are of opposite sign.

(b) Perpendicular distance of the point (x', y', z') from the plane $ax + by + cz + d = 0$ is given by

$\dfrac{ax' + by' + cz' + d}{\sqrt{a^2 + b^2 + c^2}}$.

(c) The length of the perpendicular from the point having a position vector $\vec{a}$ to the plane $\vec{r}.\vec{n} = d$ is given by

$P = \dfrac{|\vec{a}.\vec{n} - d|}{|\vec{n}|}$

(d) The coordinates of the foot of the perpendicular from the point (x_1, y_1, z_1) to the plane $ax + by + cz + d = 0$ are

given by $\dfrac{x' - x_1}{a} = \dfrac{y' - y_1}{b} = \dfrac{z' - z_1}{c} = -\dfrac{(ax_1 + by_1 + cz_1 + d)}{a^2 + b^2 + c^2}$

(e) If $P'(x', y', z')$ is the image of a point $P(x_1, y_1, z_1)$ w.r.t. the plane $ax + by + cz + d = 0$, then

$\dfrac{x' - x_1}{a} = \dfrac{y' - y_1}{b} = \dfrac{z' - z_1}{c} = -2\dfrac{(ax_1 + by_1 + cz_1 + d)}{a^2 + b^2 + c^2}$

NOMORECLASS CONCEPTS

The distance between two parallel planes $ax + by + cx + d = 0$ and $ax + by + cx + d' = 0$ is given by

$\dfrac{|d - d'|}{\sqrt{a^2 + b^2 + c^2}}$

If a variable point P moves so that $PA^2 - PB^2 = K$, where K is a constant and A and B are the two points, then the locus of P is a plane.

Illustration 11: Show that the points (1, 2, 3) and (2, −1, 4) lie on the opposite sides of the plane $x + 4y + z − 3 = 0$.

(JEE MAIN)

Sol: Substitute given points in to the given equation of plane, if their values are in opposite sign then the points are on opposite sides of the plane.

Since $1 + 4 \times 2 + 3 − 3 = 9$ and $2 − 4 + 4 − 3 = −1$ are of opposite sign, the points are on opposite sides of the plane.

8.6 Angle Between Two Planes

Let us consider two planes $ax + by + cz + d = 0$ and $a'x + b'y + c'z + d = 0$. Angle between these planes is the angle between their normals. Let (a, b, c) and (a', b', c') be the direction ratios of their normals of the two planes, respectively, and the angle θ between them is given by

$$\cos\theta = \frac{aa' + bb' + cc'}{\sqrt{a^2 + b^2 + c^2}\sqrt{a'^2 + b'^2 + c'^2}}.$$

The planes are perpendicular if $aa' + bb' + cc' = 0$ and the planes are parallel if $\dfrac{a}{a'} = \dfrac{b}{b'} = \dfrac{c}{c'}$.

In vector form, if θ is the angle between the planes $\vec{r}.\vec{n} = d_1$ and $\vec{r}.\vec{n}_2 = d_2$, then $\cos\theta = \dfrac{\vec{n}_1.\vec{n}_2}{|\vec{n}_1||\vec{n}_2|}$. The planes are perpendicular if $\vec{n}_1.\vec{n}_2 = 0$ and the planes are parallel if $\vec{n}_1 = \lambda\vec{n}_2$.

8.6.1 Angle Bisectors

(a) Equations of the planes bisecting the angle between the two given planes $a_1x + b_1y + c_1z + d_1 = 0$ and $a_2x + b_2y + c_2z + d_2 = 0$ are

$$\frac{a_1x + b_1y + c_1z + d_1}{\sqrt{a_1^2 + b_1^2 + c_1^2}} = \pm\frac{a_2x + b_2y + c_2z + d_2}{\sqrt{a_2^2 + b_2^2 + c_2^2}}$$

(b) Equation of bisector of the angle containing the origin is given by

$$\frac{a_1x + b_1y + c_1z + d_1}{\sqrt{a_1^2 + b_1^2 + c_1^2}} = \frac{a_2x + b_2y + c_2z + d_2}{\sqrt{a_2^2 + b_2^2 + c_2^2}} \quad \text{[where } d_1 \text{ and } d_2 \text{ are positive]}$$

(c) In order to find the bisector of acute/obtuse angle, both the constant terms should be positive. If

$a_1 a_2 + b_1 b_2 + c_1 c_2 > 0$ Þ   then the origin lies in the obtuse angle

$a_1 a_2 + b_1 b_2 + c_1 c_2 < 0$ Þ   then the origin lies in the acute angle

now apply step (ii) according to the question.

8.7 Family of Planes

(a) The equation of any plane passing through the line of intersection of nonparallel planes or through the given line is

$a_1x + b_1y + c_1z + d_1 = 0$ and $a_2x + b_2y + c_2z + d_2 = 0$, i.e. $P_1 = 0$ and $P_2 = 0$

$a_1x + b_1y + c_1z + d_1 + \lambda(a_2x + b_2y + c_2z + d_2) = 0$, i.e. $P_1 + \lambda P_2 = 0$

(b) The equation of plane passing through the intersection of the planes $\vec{r}.\vec{n}_1 = d_1$ & $\vec{r}.\vec{n}_2 = d_2$ is $r.(n_1 + \lambda n_2) = d_1 + \lambda d_2$, where λ is an arbitrary scalar.

Illustration 12: The plane $x - y - z = 4$ is rotated through 90° about its line of intersection with the plane $x + y + 2z = 4$. Find its equation in the new position. **(JEE MAIN)**

Sol: As the required plane passes through the line of intersection of given planes, therefore its equation may be taken as $x + y + 2z - 4 + k(x - y - z - 4) = 0$

$\Rightarrow (1 + k)x + (1 - k)y + (2 - k)z - 4 - 4k = 0$ ... (iii)

Thus, planes (i) and (iii) are mutually perpendicular.

$\therefore (1 + k) - (1 - k) - (2 - k) = 0 \Rightarrow 1 + k - 1 + k - 2 + k = 0 \quad \Rightarrow k = 2/3$

Substituting $k = 2/3$ in equation (iii), we get, $5x + y + 4z = 20$. This is the required equation of the plane in its new position.

Illustration 13: Find the equation of the plane through the point (1, 1, 1) and passing through the line of intersection of the planes x + y + z = 6 and 2x + 3y + 4z + 5 = 0   **(JEE MAIN)**

Sol: Similar to above illustration.

Given planes are x + y + z − 6 = 0   ... (i)

and 2x + 3y + 4z + 5 = 0   ... (ii)

Given point is P(1, 1, 1).

Equation of any plane passing through the line of intersection of the planes (i) and (ii) is

x + y + z − 6 + k (2x + 3y + 4z + 5) = 0   ... (iii)

If plane (iii) passes through a point P, then 1 + 1 + 1 − 6 + k (2 + 3 + 4 + 5) = 0

$\Rightarrow k = \dfrac{3}{14}$

From equation (i), the required plane is 20x + 23y + 26z − 69 = 0

Illustration 14: If the planes x − cy − bz = 0, cx − y + az = 0 and bx + ay − z = 0 pass through a straight line, then find the value of $a^2 + b^2 + c^2 + 2abc$.   **(JEE ADVANCED)**

Sol: Here the plane passing through the line of intersection of planes x − cy − bz = 0 and cx − y + az = 0 is same as the plane bx + ay − z = 0. Hence by using family of planes we can obtain required result.

Given planes are x − cy − bz = 0   ... (i)

cx − y + az = 0   ... (ii)

bx + ay − z = 0   ... (iii)

Equation of any plane passing through the line of intersection of the planes (i) and (ii) may be written as

x − cy − bz + λ(cx − y + az) = 0 $\Rightarrow$ x(1 + lc) − y(c + λ) + z(−b + aλ) = 0   ... (iv)

If planes (3) and (4) are the same, then equations (iii) and (iv) will be identical.

$$\therefore \quad \underset{\text{(i)}}{\dfrac{1 + c\lambda}{b}} = \underset{\text{(ii)}}{\dfrac{-(c + \lambda)}{a}} = \underset{\text{(iii)}}{\dfrac{-b + a\lambda}{-1}} ;$$

From equations (i) and (ii), a + acλ = −bc − bλ

$$\Rightarrow \lambda = -\dfrac{(a + bc)}{(ac + b)}$$ ... (v)

From equations (ii) and (iii), c + λ = −ab + a²λ

$$\Rightarrow \lambda = -\dfrac{(ab + c)}{1 - a^2}$$ ... (vi)

From equations (v) and (vi), we have, $\dfrac{-(a + bc)}{ac + b} = \dfrac{-(ab + c)}{(1 - a^2)}$

$\Rightarrow a − a^3 + bc − a^2bc = a^2bc + ac^2 + ab^2 + bc$ $\Rightarrow 2a^2bc + ac^2 + ab^2 + a^3 − a = 0$

$\Rightarrow a^2 + b^2 + c^2 + 2abc = 1$

Illustration 15: Through a point P(h, k, ℓ), a plane is drawn at right angles to OP to meet the coordinate axes in A, B and C. If OP = p, show that the area of ΔABC is $\dfrac{p^5}{2hk\ell}$.   **(JEE ADVANCED)**

Sol: Here line OP is normal to the plane, therefore ℓx + my nz = p. where ℓ, m and n are direction cosines of given plane.

$$OP = \sqrt{h^2 + k^2 + \ell^2} = p$$

Direction cosines of OP are $\dfrac{h}{\sqrt{h^2 + k^2 + \ell^2}}, \dfrac{k}{\sqrt{h^2 + k^2 + \ell^2}}, \dfrac{\ell}{\sqrt{h^2 + k^2 + \ell^2}}$

Since OP is the normal to the plane, therefore, equation of the plane will be

$$\dfrac{h}{\sqrt{h^2 + k^2 + \ell^2}}x + \dfrac{k}{\sqrt{h^2 + k^2 + \ell^2}}y + \dfrac{\ell}{\sqrt{h^2 + k^2 + \ell^2}}z = \sqrt{h^2 + k^2 + \ell^2}$$

$$\Rightarrow hx + ky + \ell z = h^2 + k^2 + \ell^2 = p^2$$

$$\therefore \quad A = \left(\dfrac{p^2}{h}, 0, 0\right), B = \left(0, \dfrac{p^2}{k}, 0\right), C = \left(0, 0, \dfrac{p^2}{\ell}\right)$$

Thus, area of $\triangle ABC = \dfrac{\left|\overrightarrow{AB} \times \overrightarrow{AC}\right|}{2}$

$$= \dfrac{\left|\left(\dfrac{p^2}{h}\hat{i} - \dfrac{p^2}{k}\hat{j}\right) \times \left(\dfrac{p^2}{h}\hat{i} - \dfrac{p^2}{\ell}\hat{k}\right)\right|}{2} = \dfrac{\left|\left(\dfrac{p^4}{h\ell}\hat{j} + \dfrac{p^4}{kh}\hat{k} + \dfrac{p^4}{k\ell}\hat{i}\right)\right|}{2}$$

$$= \dfrac{1}{2}\sqrt{p^8\left(\dfrac{1}{h^2\ell^2} + \dfrac{1}{h^2k^2} + \dfrac{1}{k^2\ell^2}\right)} = \dfrac{1}{2}\sqrt{\dfrac{p^8}{h^2\ell^2 k^2}\left(\ell^2 + h^2 + k^2\right)} = \dfrac{p^5}{2hk\ell}$$

9. TETRAHEDRON

Volume of a tetrahedron given the coordinates of its vertices $A(x_1, y_1, z_1)$, $B(x_2, y_2, z_2)$, $C(x_3, y_3, z_3)$ and $D(x_4, y_4, z_4)$ can be calculated by

$$V = \dfrac{1}{6}\begin{Vmatrix} x_1 & y_1 & z_1 & 1 \\ x_2 & y_2 & z_2 & 1 \\ x_3 & y_3 & z_3 & 1 \\ x_4 & y_4 & x_4 & 1 \end{Vmatrix} = \dfrac{1}{6}\begin{Vmatrix} x_2 - x_1 & y_2 - y_1 & z_2 - z_1 \\ x_3 - x_1 & y_3 - y_1 & z_3 - z_1 \\ x_4 - x_1 & y_4 - y_1 & z_4 - z_1 \end{Vmatrix}$$

NOMORECLASS CONCEPTS

Four points (x_r, y_r, z_r); $r = 1, 2, 3, 4$; will be coplanar if the volume of the tetrahedron with the points as vertices is zero. Therefore, the condition of coplanarity of the points

(x_1, y_1, z_1), (x_2, y_2, z_2), (x_3, y_3, z_3) and (x_4, y_4, z_4) is $\begin{vmatrix} x_2 - x_1 & y_2 - y_1 & z_2 - z_1 \\ x_3 - x_1 & y_3 - y_1 & z_3 - z_1 \\ x_4 - x_1 & y_4 - y_1 & z_4 - z_1 \end{vmatrix} = 0.$

Centroid of a Tetrahedron

Let $A(x_1, y_1, z_1)$, $B(x_2, y_2, z_2)$, $C(x_3, y_3, z_3)$ and $D(x_4, y_4, z_4)$ be the vertices of a tetrahedron.

The coordinate of its centroid (G) is given as $\left(\dfrac{\Sigma x_i}{4}, \dfrac{\Sigma y_i}{4}, \dfrac{\Sigma z_i}{4} \right)$

Illustration 16: If two pairs of opposite edges of a tetrahedron are mutually perpendicular, show that the third pair will also be mutually perpendicular. **(JEE MAIN)**

Sol: If two lines are perpendicular then summation of product of their respective direction ratios is equals to zero.

Let OABC be the tetrahedron where O is the origin and coordinate of A, B, C be (x_1, y_1, z_1), (x_2, y_2, z_2), (x_3, y_3, z_3), respectively.

Let OA $\perp$ BC and OB $\perp$ CA. We have to prove that OC $\perp$ BA

Direction ratios of OA are $x_1 - 0$, $y_1 - 0$, $z_1 - 0$ or x_1, y_1, z_1

Direction ratios of BC are $(x_3 - x_2)$, $(y_3 - y_2)$, $(z_3 - z_2)$

OA $\perp$ BC

$\Rightarrow x_1(x_3 - x_2) + y_1(y_3 - y_2) + z_1(z_3 + z_2) = 0$ ... (i)

Similarly, OB $\perp$ CA

$\Rightarrow x_2(x_1 - x_3) + y_2(y_1 - y_3) + z_2(z_1 - z_3) = 0$ ... (ii)

Figure 27.5

On adding equations (1) and (2), we obtain the following equation:

$x_3(x_1 - x_2) + y_3(y_1 - y_2) + z_3(z_1 - z_2) = 0$

$\therefore$ OC $\perp$ BA [$\because$ direction ratios of OC are x_3, y_3, z_3 and that of BA are $(x_1 - x_2)$, $(y_1 - y_2)$, $(z_1 - z_2)$]

10. LINE

10.1 Equation of a line

A straight line in space will be determined if it is the intersection of two given nonparallel planes and therefore, the equation of a straight line is present as a solution of the system constituted by the equations of the two planes, $a_1x + b_1y + c_1z + d_1 = 0$ and $a_2x + b_2y + c_2z + d_2 = 0$. This form is also known as non-symmetrical form.

(a) The equation of a line passing through the point (x_1, y_1, z_1) with a, b, c as direction ratios is $\dfrac{x - x_1}{a} = \dfrac{y - y_1}{b} = \dfrac{z - z_1}{c} = r$. This form is called symmetrical form. A general point on the line is given by $(x_1 + ar, y_1 + br, z_1 + cr)$.

(b) Vector equation of a straight line passing through a fixed point with position vector $\vec{a}$ and parallel to a given vector $\vec{b}$ is $\vec{r} = \vec{a} + \lambda\vec{b}$, where λ is a scalar.

(c) The equation of the line passing through the points (x_1, y_1, z_1) and (x_2, y_2, z_2) is

(d) $\dfrac{x - x_1}{x_2 - x_1} = \dfrac{y - y_1}{y_2 - y_1} = \dfrac{z - z_1}{z_2 - z_1}$.

(e) Vector equation of a straight line passing through two points with position vectors $\vec{a}$ and $\vec{b}$ is $\vec{r} = \vec{a} + \lambda(\vec{b} - \vec{a})$.

(f) Reduction of Cartesian form of equation of a line to vector form and vice versa is as

(g) $\dfrac{x - x_1}{a} = \dfrac{y - y_1}{b} = \dfrac{z - z_1}{c}$ $\Leftrightarrow$ $r = (x_1\hat{i} + y_1\hat{j} + z_1\hat{k}) + \lambda(a\hat{i} + b\hat{j} + c\hat{k})$

Straight lines parallel to coordinate axes

Straight lines	Equations
(i) Through origin	$y = mx$, $z = nx$
(ii) x-axis	$y = 0$, $z = 0$
(iii) y-axis	$x = 0$, $z = 0$
(iv) z-axis	$x = 0$, $y = 0$
(v) Parallel to x-axis	$y = p$, $z = q$
(vi) Parallel to y-axis	$x = h$, $z = q$
(vii) Parallel to z-axis	$x = h$, $y = p$

The number of lines which are equally inclined to the coordinate axes are 4.

Illustration 17: Find the equation of the line passing through the points (3, 4, –7) and (1, –1, 6) in vector form as well as in Cartesian form. **(JEE MAIN)**

Sol: Here line in vector form is given by $r = (x_1\hat{i} + y_1\hat{j} + z_1\hat{k}) + \lambda(a\hat{i} + b\hat{j} + c\hat{k})$ and in Cartesian form is given by

$\dfrac{x - x_1}{a} = \dfrac{y - y_1}{b} = \dfrac{z - z_1}{c}$.

Let $A \equiv (3, 4, -7)$, $B \equiv (1, -1, 6)$; Now, $\vec{a} = \overrightarrow{OA} = 3\hat{i} + 4\hat{j} - 7\hat{k}, \vec{b} = \overrightarrow{OB} = \hat{i} - \hat{j} + 6\hat{k}$

Equation (in vector form) of the line passing through $A(a)$ and $B(\vec{b})$ is $r = a + t(\vec{b} - \vec{a})$

$\Rightarrow \vec{r} = 3\vec{i} + 4\vec{j} - 7\vec{k} + t(-2\vec{i} - 5\vec{j} + 13\vec{k})$ ... (i)

Equation in Cartesian form is $\dfrac{x-3}{3-1} = \dfrac{y-4}{4+1} = \dfrac{z+7}{-7-6} \Rightarrow \dfrac{x-3}{2} = \dfrac{y-4}{5} = \dfrac{z+7}{-13}$

Illustration 18: Show that the two lines $\dfrac{x-1}{2} = \dfrac{y-2}{3} = \dfrac{z-3}{4}$ and $\dfrac{x-4}{5} = \dfrac{y-1}{2} = z$ intersect. Find also the point of intersection of these lines. **(JEE MAIN)**

Sol: The given lines will intersect if any point on respective lines coincide for some value of λ and r.

Given lines are $\dfrac{x-1}{2} = \dfrac{y-2}{3} = \dfrac{z-3}{4}$ .... (i)

and $\dfrac{z-4}{5} = \dfrac{y-1}{2} = \dfrac{z-0}{1}$ .... (ii)

Any point on line (1) is $P(2r + 1, 3r + 2, 4r + 3)$ and any point on line (2) is $Q(5\lambda + 4, 2\lambda + 1, \lambda)$

Lines (i) and (ii) will intersect if P and Q coincide for some value of λ and r.

$\Rightarrow 2r + 1 = 5\lambda + 4 \qquad \Rightarrow 2r - 5\lambda = 3$ ... (iii)

$\Rightarrow 3r + 2 = 2\lambda + 1 \qquad \Rightarrow 3r - 2\lambda = -1$ ... (iv)

$\Rightarrow 4r + 3 = \lambda \qquad \Rightarrow 4r - \lambda = -3$ ... (v)

Solving equations (iii) and (iv), we get $r = -1$, $\lambda = -1$; these obtained values of r and λ clearly satisfy equation (v)

$\Rightarrow P \equiv (-1, -1, -1)$. Hence, lines (i) and (ii) intersect at $(-1, -1, -1)$

Illustration 19: Find the angle between the lines $x - 3y - 4 = 0$, $4y - z + 5 = 0$ and $x + 3y - 11 = 0$, $2y - z + 6 = 0$.

(JEE MAIN)

Sol: $\dfrac{x-4}{3} = \dfrac{y-0}{1} = \dfrac{z-5}{4}$ ... (i)

$\dfrac{x-11}{-3} = \dfrac{y-0}{1} = \dfrac{z-6}{2}$ ... (ii)

$a = 3, b = 1, c = 4$ $\qquad\qquad \therefore \qquad a^1 = -3, b^1 = 1, c^1 = 2$

$aa^1 + bb^1 + cc^1 = -9 + 1 + 8 = 0 \Rightarrow \cos\theta = 0 \qquad\qquad \theta = 90$

Illustration 20: Find the equation of the line drawn through point $(1, 0, 2)$ to meet at right angle with the line $\dfrac{x+1}{3} = \dfrac{y-2}{-2} = \dfrac{z+1}{-1}$.

(JEE ADVANCED)

Sol: If two lines are perpendicular then summation of product of their direction ratios are equal to zero. Hence by obtaining direction ratio of these line, we will be get the result.

Given line is $\dfrac{x+1}{3} = \dfrac{y-2}{-2} = \dfrac{z+1}{-1}$ ... (i)

Let $P \equiv (1, 0, 2)$; coordinates of any point on line (i) may be written as $Q \equiv (3r - 1, -2r + 2, -r - 1)$.

Direction ratios of PQ are $3r - 2, -2r + 2, -r - 3$

Direction ratios of the line $\dfrac{x+1}{3} = \dfrac{y-2}{-2} = \dfrac{z+1}{-1}$ are $3, -2, -1$

Since PQ $\perp$ to the line $\Rightarrow 3(3r - 2) - 2(-2r + 2) - 1(-r - 3) = 0$

$\Rightarrow 9r - 6 + 4r - 4 + r + 3 = 0 \Rightarrow 14r = 7 \Rightarrow r = \dfrac{1}{2} \qquad \therefore$ Direction ratios of PQ are $-\dfrac{1}{2}, 1, -\dfrac{7}{2}$

Hence, equation of the line PQ is $\dfrac{x-1}{-1} = \dfrac{y-0}{2} = \dfrac{z-2}{-7}$

Illustration 21: Find the equation of the line of intersection of the planes $4x + 4y - 5z = 12$, $8x + 12y - 13z = 32$ in the symmetric form

(JEE ADVANCED)

Sol: Consider the line of intersection meet the xy-plane at $P(\alpha, \beta, 0)$, therefore obtain the value of α and β and direction ratios of line of intersection to solve the problem.

Given planes are $4x + 4y - 5z - 12 = 0$ ... (i)

and $8x + 12y - 13z - 32 = 0$ ... (ii)

Let ℓ, m, n be the direction ratios of the line of intersection. Then,

$4\ell + 4m - 5n = 0$ and $8\ell + 12m - 13n = 0 \Rightarrow \dfrac{\ell}{-52+60} = \dfrac{m}{-40+52} = \dfrac{n}{48-32} \Rightarrow \dfrac{\ell}{2} = \dfrac{m}{3} = \dfrac{n}{4}$

Direction ratios of the line of intersection are $2, 3, 4$.

Let the line of intersection meet the xy-plane at $P(\alpha, \beta, 0)$

Then P lies on planes (i) and (i) $\Rightarrow 4\alpha + 4\beta - 12 = 0 \Rightarrow \alpha + \beta - 3 = 0$

and $8\alpha + 12\beta - 32 = 0$ ... (v)

or $2\alpha + 3\beta - 8 = 0$ ... (vi)

Solving equations (v) and (vi), we get $\dfrac{\alpha}{-8+9} = \dfrac{\beta}{-6+8} = \dfrac{1}{3-2} \Rightarrow \alpha = 1, \beta = 2$

Hence, equation of the line of intersection in symmetrical form is $\dfrac{x-1}{2} = \dfrac{y-2}{3} = \dfrac{z-0}{4}$

10.1 Coplanar Lines

Coplanar lines are lines that entirely lie on the same plane.

(i) If $\dfrac{x-\alpha}{\ell} = \dfrac{y-\beta}{m} = \dfrac{z-\gamma}{n}$ and $\dfrac{x-\alpha'}{\ell'} = \dfrac{x-\beta'}{m'} = \dfrac{z-\gamma'}{n'}$, are the lines, then the condition for intersection/coplanarity

is $\begin{vmatrix} \alpha-\alpha' & \beta-\beta' & \gamma-\gamma' \\ \ell & m & n \\ \ell' & m' & n' \end{vmatrix} = 0$ and the plane containing the aforementioned lines is $\begin{vmatrix} x-\alpha & y-\beta & z-\gamma \\ \ell & m & n \\ \ell' & m' & n' \end{vmatrix} = 0.$

(ii) Condition of coplanarity if both lines are in general form.

Let the lines be $ax + by + cz + d = 0 = a'x + b'y + c'z + d'$ and $ax + by + gz + \delta = 0 = \alpha'x + \beta'y + \gamma'z + \delta' = 0$

If $\Delta = \begin{vmatrix} a & b & c & d \\ a' & b' & c' & d' \\ \alpha & \beta & \gamma & \delta \\ \alpha' & \beta' & \gamma' & \delta' \end{vmatrix} = 0$, then they are coplanar.

10.2 Skew Lines

Skew lines are two lines that do not intersect and are not parallel.

If $\Delta = \begin{vmatrix} \alpha'-\alpha & \beta'-\beta & \gamma'-\gamma \\ \ell & m & n \\ \ell' & m' & n' \end{vmatrix} \neq 0$, then the lines are skew.

Shortest distance

Let the equation of the lines be $\dfrac{x-\alpha}{\ell} = \dfrac{y-\beta}{m} = \dfrac{z-\gamma}{n}$ and $\dfrac{x-\alpha'}{\ell'} = \dfrac{x-\beta'}{m'} = \dfrac{z-y'}{n'}$

S.D. $= \dfrac{(\alpha-\alpha')(mn'-m'n) + (\beta-\beta')(n\ell'-n'\ell) + (\gamma-\ell')(\ell m'-\ell'm)}{\sqrt{\sum(mn'-m'n)^2}} = \begin{vmatrix} \alpha'-\alpha & \beta'-\beta & \gamma'-\gamma \\ \ell & m & n \\ \ell' & m' & n' \end{vmatrix} \Big/ \sqrt{\sum(nm'-m'n)^2}$

Vector form

For lines $\vec{a}_1 + \lambda\vec{b}_1$ and $\vec{a}_2 + \lambda\vec{b}_2$ to be skew, the following condition should be satisfied: $(\vec{b}_1 \times \vec{b}_2) \cdot (\vec{a}_1 - \vec{a}_2) \neq 0$

NOMORECLASS CONCEPTS

Shortest distance between two skew lines is perpendicular to both the lines.

10.3 Intersecting Lines

Two or more lines that intersect at a point are called intersecting lines, and their shortest distance between the two lines is

zero, i.e. $\left|\dfrac{(\vec{b}_1 \times \vec{b}_2) \cdot (\vec{a}_2 - \vec{a}_1)}{|\vec{b}_1 \times \vec{b}_2|}\right| = 0$ $\Rightarrow (\vec{b}_1 \times \vec{b}_2) \cdot (\vec{a}_1 - \vec{a}_2) = 0$ $\Rightarrow [\vec{b}_1 \vec{b}_2 (\vec{a}_2 - \vec{a}_1)] = 0$ $\Rightarrow \begin{vmatrix} x_2 - x_1 & y_2 - y_1 & z_2 - z_1 \\ a_1 & b_1 & c_1 \\ a_2 & a_2 & c_2 \end{vmatrix} = 0$

10.4 Parallel Lines

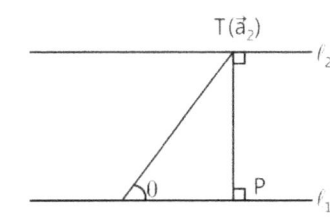

Parallel lines are lines that never intersect, and are coplanar.

Let $\left|-\dfrac{8}{3}, \dfrac{1}{3}, \dfrac{16}{3}\right|$ and $\left|\dfrac{8}{3}, \dfrac{1}{3}, \dfrac{16}{3}\right|$ be the two parallel lines.

Let the lines be given by $\vec{r} = \vec{a}_1 + \lambda \vec{b}$ ... (i)

$\vec{r} = \vec{a}_2 + \mu \vec{b}$ ... (ii)

Figure 27.6

where $\vec{a}_1$ is the position vector of a point S on $\dfrac{x}{2} = \dfrac{y-1}{3} = \dfrac{z}{5}$ and $\bar{a}_2$ is the position vector of a point T on ℓ_2. As ℓ_1 and ℓ_2 are coplanar, if the foot of the perpendicular from T on the line ℓ_1 is P, then the distance between the lines ℓ_1 and ℓ_2 = |TP|. Let θ be the angle between the vectors $\vec{ST}$ and $\vec{b}$.

Then $\vec{b} \times \vec{ST} = (|\vec{b}||\vec{ST}|\sin\theta)\hat{n}$ ... (iii)

where $\hat{n}$ is the unit vector perpendicular to the plane of the lines $a\ell + bm + cn = 0$ and $ax_1 + by_1 + cz_1 + d \neq 0$. But $\vec{ST} = \vec{a}_2 - \vec{a}_1$

Therefore, from equation (iii), we get $\vec{b} \times (\vec{a}_2 - \vec{a}_1) = |\vec{b}|PT\hat{n}$ (as PT = ST sin θ)

$\Rightarrow |\vec{b} \times (\vec{a}_2 - \vec{a}_1)| = |\vec{b}|PT \cdot 1 (as |\hat{n}| = 1)$ Hence, the distance between the given parallel lines is $d = |\vec{PT}| = \left|\dfrac{\vec{b} \times (\vec{a}_2 - \vec{a}_1)}{|\vec{b}|}\right|$.

10.5 Angular Bisector

If $a(x - \alpha) + b(y - \beta)$, m_1, n_1 and $+c(z - \gamma) = 0$, , m_2, n_2 are the direction cosines of the two lines inclined to each other at an angle θ, then the direction cosines of the

(a) Internal bisectors of the angle between the lines are $\dfrac{\ell_1 + \ell_2}{2\cos(\theta/2)}, \dfrac{m_1 + m_2}{2\cos(\theta/2)}$ and $\dfrac{n_1 + n_2}{2\cos(\theta/2)}$.

(b) External bisectors of the angle between the lines are $\dfrac{\ell_1 - \ell_2}{2\sin(\theta/2)}, \dfrac{m_1 - m_2}{2\sin(\theta/2)}$ and $\dfrac{n_1 - n_2}{2\sin(\theta/2)}$.

10.6 Reduction to Symmetric Form

Let the line in nonsymmetrical form be represented as $a_1x + b_1y + c_1z + d_1 = 0$, $a_2x + b_2y + c_2z + d_2 = 0$.

To find the equation of the line in symmetrical form, (i) its direction ratios and (ii) coordinate of any point on it must be known.

(a) **Direction ratios:** Let ℓ, m, n be the direction ratios of the line. Since the line lies on both planes, it must be perpendicular to the normal of both planes. So $a_1\ell + b_1m + c_1n = 0$, $a_2\ell + b_2m + c_2n = 0$. From these equations, proportional values of ℓ, m, n can be found by using the method of cross-multiplication, i.e.

$$\dfrac{\ell}{b_1c_2 - b_2c_1} = \dfrac{m}{c_1a_2 - c_2a_1} = \dfrac{n}{a_1b_2 - a_2b_1}$$

Alternate method

The vector $\begin{vmatrix} i & j & k \\ a_1 & b_1 & c_1 \\ a_2 & b_2 & c_2 \end{vmatrix} = i(b_1c_2 - b_2c_1) + j(c_1a_2 - c_2a_1) + k(a_1b_2 - a_2b_1)$ will be parallel to the line of intersection of the

two given planes. Hence, $\ell : m : n = (b_1c_2 - b_2c_1) : (c_1a_2 - c_2a_1) : (a_1b_2 - a_2b_1)$.

(b) **Coordinate of any point on the line:** Note that as ℓ, m, n cannot be zero simultaneously, so at least one must be nonzero. Let $a_1b_2 - a_2b_1 \neq 0$, so that the line cannot be parallel to xy-plane, and will intersect it. Let it intersect xy-plane at the point $(x_1, y_1, 0)$. These $a_1x_1 + b_1y_1 + d_1 = 0$ and $a_2x_1 + b_2y_1 + d_2 = 0$. Solving these, we get a point on the line. Thus, we get the following equation:

$$\frac{x - x_1}{b_1c_2 - b_2c_1} = \frac{y - y_1}{c_1a_2 - c_2a_1} = \frac{z - 0}{a_1b_2 - a_2b_1} \quad \text{or} \quad \frac{x - (b_1d_2 - b_2d_1 / a_1b_2 - a_2b_1)}{b_1c_2 - b_2c_1} = \frac{y - (d_1a_2 - d_2a_1 / a_1b_2 - a_2b_1)}{c_1a_2 - c_2a_1} = \frac{z - 0}{a_1b_2 - a_2b_1}$$

Note: If $\ell \neq 0$, take a point on yz-plane as $(0, y_1, z_1)$ and if $m \neq 0$, take a point on xz-plane as $(x_1, 0, z_1)$.

Alternate method:

If $\dfrac{a_1}{a_2} \neq \dfrac{b_1}{b_2}$, then put z = 0 in both equations and solve the equation $a_1x + b_1y + d_1 = 0$, $a_2x + b_2y + d_2 = 0$ or put y = 0

and solve the equation $a_1x + c_1z + d_1 = 0$ and $a_2x + c_2z + d_2 = 0$.

10.7 Point and Line

Foot Length and Equation of Perpendicular from a Point to a Line

Cartesian form: Let equation of the line be $\dfrac{x - a}{\ell} = \dfrac{y - b}{m} = \dfrac{z - c}{n} = r \, (\text{say})$ ...(i)

and $A(\alpha, \beta, \gamma)$ be the point. Any point on line (i) is $P(\ell r + a, mr + b, nr + c)$. If it is the foot of the perpendicular from point A on the line, then AP is perpendicular to the line.

$\Rightarrow \ell(\ell r + a - \alpha) + m(mr + b - \beta) + n(nr + c - \gamma) = 0$, i.e. $r = [(\alpha - a)\ell + (\beta - b)m + (\gamma - c)n]/\, \ell^2 + m^2 + n^2$

Using this value of r, we get the foot of the perpendicular from point A on the given line. Because the foot of the perpendicular P is known, the length of the perpendicular $AP = \sqrt{(\ell r + a - \alpha^2) + (mr + b - \beta)^2 + (nr + c - \gamma)^2}$ is given by the equation of perpendicular as $\dfrac{x - \alpha}{\ell r + a - \alpha} = \dfrac{y - \beta}{mr + b - \beta} = \dfrac{z - \gamma}{nr + c - \gamma}$

Illustration 22: Find the coordinates of the foot of the perpendicular drawn from the point A(1, 2, 1) to the line joining B(1, 4, 6) and C(5, 4, 4). **(JEE ADVANCED)**

Sol: Using section formula we will get co-ordinates of the foot D, and as AD is perpendicular to BC therefore $\overrightarrow{AD}.\overrightarrow{BC} = 0.$

Let D be the foot of the perpendicular drawn from A on BC,

and let D divide BC in the ratio k:1. Then, the coordinates

of D are $\left(\dfrac{5k + 1}{k + 1}, \dfrac{4k + 4}{k + 1}, \dfrac{4k + 6}{k + 1} \right)$ ...(i)

Now, $\overrightarrow{AD}$ = Position vector of D – Position vector of A

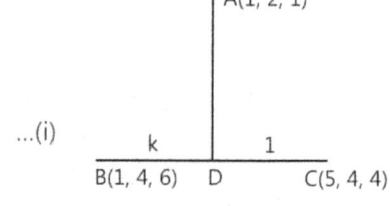

Figure 27.7

$$= \left(\frac{5k+1}{k+1} - 1\right)\hat{i} + \left(\frac{4k+4}{k+1} - 2\right)\hat{j} + \left(\frac{4K+6}{K+1} - 1\right)\hat{K} = \left(\frac{4k}{k+1}\right)\hat{i} + \left(\frac{2k+2}{k+1}\right)\hat{j} + \left(\frac{3k+5}{k+1}\right)\hat{k}$$

and $\vec{BC}$ = Position vector of C − Position vector of B = $(5\hat{i} + 4\hat{j} + 4\hat{k}) - (\hat{i} + 4\hat{j} + 6\hat{k}) = 4\hat{i} + 0\hat{j} - 2\hat{k}$

since $\vec{AD} \perp \vec{BC}$ $\Rightarrow$ $\vec{AD}.\vec{BC} = 0$ $\Rightarrow$ $\left[\left(\frac{4k}{k+1}\right)\hat{i} + \left(\frac{2k+2}{k+1}\right)\hat{j} + \left(\frac{3k+5}{k+1}\right)\hat{k}\right].(4\hat{i} + 0\hat{j} - 2\hat{k}) = 0$

$\Rightarrow 4\left(\frac{4k}{k+1}\right) + 0\left(\frac{2k+2}{k+1}\right) - 2\left(\frac{3k+5}{k+1}\right) = 0$ $\Rightarrow$ $\frac{16k}{k+1} + 0 - 2\frac{(3k+5)}{k+1} = 0$

$\Rightarrow$ $16k - 6k - 10 = 0$ $\Rightarrow$ $k = 1$

Substituting k = 1 in equation (i), we obtain the coordinates of D as (3, 4, 5).

10.8 Vector Form

Equation of a line passing through a point having position vector $\bar{a}$ and perpendicular to the lines $\vec{r_1} = \vec{a_1} + \lambda\vec{b_1}$
and $\vec{r_1} = \vec{a_2} + \lambda\vec{b_2}$ is parallel to $\vec{b_1} \times \vec{b_2}$. So the vector equation of such line is $\vec{r} = \vec{a} + \lambda(\vec{b_1} \times \vec{b_2})$. The equation of the

perpendicular passing through $\vec{a}$ is $\vec{r} = \vec{a} + \mu\left((\vec{a} - \vec{a}) - \left(\frac{(\vec{a} - \vec{a})\vec{b}}{|\vec{b}|^2}\right)\vec{b}.\right)$

10.9 Image w.r.t. the Line

Let $L \equiv \frac{x - x_2}{a} = \frac{y - y_2}{b} = \frac{z - z_2}{c}$ be the given line.
Let (x', y', z') be the image of the point $P(x_1, y_1, z_1)$ w.r.t the line L. Then

(i) $a(x_1 - x') + b(y_1 - y') + c(z_1 - z') = 0$

(ii) $\frac{\frac{x_1 - x'}{2} - x_2}{a} = \frac{\frac{y_1 - y'}{2} - y_2}{b} = \frac{\frac{z_1 - z'}{2} - z_2}{c} = \lambda$

From (ii), the value of x', y', z' in terms of λ can be obtained as x' = 2aλ + 2x_2 − x_1, y' = $2b\lambda - 2y_2 - y_1$, z' = 2cλ + 2z_2 − z_1
On substituting values of x', y', z' in (i), we get λ and on re-substituting value of λ. we get (x' y' z').

Illustration 23: Find the length of the perpendicular from P(2, −3, 1) to the line $\frac{x+1}{2} = \frac{y-3}{3} = \frac{z+2}{-1}$ **(JEE MAIN)**

Sol: Here Co-ordinates of any point on given line may be taken as Q ≡ (2r − 1, 3r + 3, −r − 2), therefore by using distance formula we can obtain required length.

Given line is $\frac{x+1}{2} = \frac{y-3}{3} = \frac{z+2}{-1}$ ...(i) P = (2, −3, 1)

Coordinates of any point on line (i) may be written as Q ≡ (2r − 1, 3r + 3, −r − 2)

Direction ratios of PQ are 2r − 3, 3r + 6, −r − 3.

Direction ratios of AB are 2, 3, −1.

Since PQ ⊥ AB $\Rightarrow$ 2(2r − 3) + 3(3r + 6) − 1(−r − 3) = 0 $\Rightarrow$ $r = \frac{-15}{14}$

$\Rightarrow$ $Q = \left(\frac{-22}{7}, \frac{-3}{14}, \frac{-13}{14}\right)$ $\Rightarrow$ $PQ = \frac{\sqrt{531}}{14}$ units

10.10 Plane Passing Through a Given Point and Line

Let the plane pass through the given point A($\vec{a}$) and line $\vec{r} = \vec{b} + \lambda\vec{c}$. For any position of point R (r) on the plane, vectors $\overrightarrow{AB}, \overrightarrow{RA}$ and $\vec{c}$ are coplanar.

Then $\begin{bmatrix} \vec{r}-\vec{a} & \vec{b}-\vec{a} & \vec{c} \end{bmatrix} = 0$, which is the required equation of the plane.

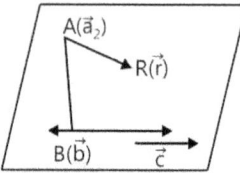

Figure 27.8

Angle between a plane and a line:

Angle between a line and a plane is complementary to the angle made by the line with the normal of plane. Hence, if θ is the angle between the line $\dfrac{x-x_1}{\ell} = \dfrac{y-y_1}{m} = \dfrac{z-z_1}{n}$ and the plane

$ax + by + cz + d = 0$, then $\sin\theta = \left(\dfrac{a\ell + bm + cn}{\sqrt{(a^2+b^2+c^2)}\sqrt{(\ell^2+m^2+n^2)}} \right)$

Vector form:

If θ is the angle between the line $\vec{r} = (\vec{a} + \lambda\vec{b})$ and $\vec{r}\cdot\vec{n} = d$ then $\sin\theta = \left[\dfrac{\vec{b}\cdot\vec{n}}{|\vec{b}||\vec{n}|} \right]$

Line and plane are perpendicular if $\dfrac{\ell}{a} = \dfrac{m}{b} = \dfrac{n}{c}$, i.e. $\vec{b}\times\vec{n} = 0$.

Line and plane are parallel if $a\ell + bm + cn = 0$, i.e. $\vec{b}\cdot\vec{n} = 0$.

NOMORECLASS CONCEPTS

Condition for a Line to Lie on a Plane

(i) **Cartesian form:** Line $\dfrac{x-x_1}{\ell} = \dfrac{y-y_1}{m} = \dfrac{z-z_1}{n}$ would lie on a plane $ax + by + cz + d = 0$ if,

$ax_1 + by_1 + cz_1 + d = 0$ and $a\ell + bm + cn = 0$

(ii) **Vector form:** Line $\vec{r} = \vec{a} + \lambda\vec{b} +$ would lie on the plane $\vec{r}\cdot\hat{n} = d$ if $\vec{b}\cdot\hat{n} = 0$ & $\vec{a}\cdot\hat{n} = d$.

The number of lines which are equally inclined to the coordinate axes is 4.

If ℓ, m, n are the d.c.s of a line, then the maximum value of lmn $= \dfrac{1}{3\sqrt{3}}$.

Illustration 24: Find the shortest distance and the vector equation of the lines of shortest distance between the lines given by $\vec{r} = 3\vec{i} + 8\vec{j} + 3\vec{k} + \lambda(3\vec{i} - \vec{j} + \vec{k})$ and $\vec{r} = -3\vec{i} - 7\vec{j} + 6\vec{k} + \mu(-3\vec{i} + 2\vec{j} + 4\vec{k})$.　　　　**(JEE ADVANCED)**

Sol: Consider LM is the shortest distance between given lines therefore LM is perpendicular to these lines, hence by obtaining their direction ratios and using perpendicular formula we will get the result.

Given lines are $\vec{r} = 3\vec{i} + 8\vec{j} + 3\vec{k} + \lambda(3\vec{i} - \vec{j} + \vec{k})$　　　　　　　　... (i)

and $\vec{r} = -3\vec{i} - 7\vec{j} + 6\vec{k} + \mu(-3\vec{i} + 2\vec{j} + 4\vec{k})$　　　　　　　　... (ii)

Equations of lines (i) and (ii) in Cartesian form are

$AB : \dfrac{x-3}{3} = \dfrac{y-8}{-1} = \dfrac{z-3}{1} = \lambda$ and $CD : \dfrac{x+3}{-3} = \dfrac{y+7}{2} = \dfrac{z-6}{4} = \mu$

Let $L \equiv (3\lambda + 3, -\lambda + 8, \lambda + 3)$ and $M \equiv (-3\mu - 3, 2\mu - 7, 4\mu + 6)$

Direction ratios of LM are $3\lambda + 3\mu + 6, -\lambda - 2\mu + 15, \lambda - 4\mu - 3$ since LM $\perp$ AB

$\Rightarrow \ 3(3\lambda + 3\mu + 6) - 1(-\lambda - 2\mu + 15) + 1(\lambda - 4\mu - 3) = 0$

or, $11\lambda + 7\mu = 0$ ... (v)

Again LM $\perp$ CD

$\therefore \ -3(3\lambda + 3\mu + 6) + 2(-\lambda - 2\mu + 15) + 4(\lambda - 4\mu - 3) = 0$

or, $-7\lambda - 29\mu = 0$ ... (vi)

Solving equations (v) and (vi), we get $\lambda = 0, \mu = 0 \ \Rightarrow \ L \equiv (3, 8, 3), M \equiv (-3, -7, 6)$

Hence, the shortest distance LM $= \sqrt{(3+3)^2 + (8+7)^2 + (3-6)^2} = \sqrt{270} = 3\sqrt{30}$ units

Vector equation of LM is $\vec{r} = 3\vec{i} + 8\vec{j} + 3\vec{k} + t\,(-6\vec{i} + 15\vec{j} - 3\vec{k})$.

11. SPHERE

11.1 General Equation

The general equation of a sphere is given by $x^2 + y^2 + z^2 + 2ux + 2vy + 2wz + d = 0$, where $(-u, -v, -w)$ is the center and $\sqrt{u^2 + u^2 + w^2 - d}$ is the radius of the sphere.

11.2 Diametric Form

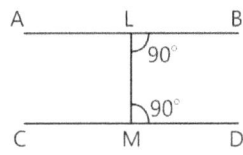

Figure 27.9

If (x_1, y_1, z_1) and (x_2, y_2, z_2) are the coordinates of the extremities of a diameter of a sphere, then its equation is given by $(x - x_1)(x - x_2) + (y - y_1)(y - y_2) + (z - z_1)(z - z_2) = 0$.

11.3 Plane and Sphere

If the perpendicular distance of the plane from the center of the sphere is equal to the radius of the sphere, then the plane touches the sphere. The plane $lx + my + nz = p$ touches the sphere $x^2 + y^2 + z^2 + 2ux + 2vy + 2wz + d = 0$, if $(u\ell + vm + wn + p)^2 = (l^2 + m^2 + n^2)(u^2 + v^2 + w^2 - d)$.

11.4 Intersection of Straight Line and a Sphere

Let the equations of the sphere and the straight line be

$x^2 + y^2 + z^2 + 2ux + 2vy + 2wz + d = 0$ ... (i)

and $\dfrac{x - \alpha}{\ell} = \dfrac{y - \beta}{m} = \dfrac{z - \gamma}{n} = r$, (say) ... (ii)

Any point on the line (ii) is $(\alpha + \ell r, \beta + mr, \gamma + nr)$. If this point lies on the sphere (i), then we have

$(\alpha + \ell r)^2 + (\beta + mr)^2 + (\gamma + nr)^2 + 2u(\alpha + \ell r) + 2v(\beta + mr) + 2w(\gamma + nr) + d = 0$

$\Rightarrow r^2[\ell^2 + m^2 + n^2] + 2r[\ell(u + \alpha) + m(v + \beta) + n(w + \gamma)] + (\alpha^2 + \beta^2 + \gamma^2 + 2u\alpha + 2v\beta + 2w\gamma + d) = 0$ ... (iii)

This is a quadratic equation in r and thus two values of r are obtained. Therefore, the line (ii) intersects the sphere (i) at two points which may be real, coincident and imaginary, according to roots of (iii).

If ℓ, m, n are the actual d.c.s of the line, then $l^2 + m^2 + n^2 = 1$ and then the equation (iii) can be simplified.

11.5 Orthogonality of Two Spheres

Let the equation of the two spheres be $x^2 + y^2 + z^2 + 2ux + 2vy + 2wz + d = 0$ ... (i)

and $x^2 + y^2 + z^2 + 2u'x + 2v'y + 2w'z + d' = 0$ ... (ii)

If the sphere (i) and (ii) cut orthogonally, then $2uu' + 2vv' + 2ww' = d + d'$, which is the required condition. If the spheres $x^2 + y^2 + z^2 = a^2$ and $x^2 + y^2 + z^2 + 2ux + 2vy + 2wz + d = 0$ cut orthogonally, then $d = a^2$.

Two spheres of radii r_1 and r_2 cut orthogonally, then the radius of the common circle is $\dfrac{r_1 r_2}{\sqrt{r_1^2 + r_2^2}}$.

Illustration 25: A plane passes through a fixed point (a, b, c). Show that the locus of the foot of the perpendicular to it from the origin is the sphere $x^2 + y^2 + z^2 - ax - by - cz = 0$ **(JEE ADVANCED)**

Sol: Consider $P(\alpha, \beta, \gamma)$ be the foot of perpendicular from origin to, therefore by getting the direction ratios of OP we will get the required result.

Let the equation of the variable plane be

$\ell x + my + nz + d = 0$ ... (i)

Plane passes through the fixed point (a, b, c)

$\therefore \ell a + mb + nc + d = 0$ ... (ii)

Let $P(\alpha, \beta, \gamma)$ be the foot of the perpendicular from the origin to plane (i)

Direction ratios of OP are $\alpha - 0, \beta - 0, \gamma - 0$, i.e. α, β, γ

From equation (i), it is clear that the direction ratios of the normal to the plane, i.e. OP are ℓ, m, n, and α, β, γ are the direction ratio of the same line OP $\therefore \dfrac{\alpha}{\ell} = \dfrac{\beta}{m} = \dfrac{\gamma}{n} = \dfrac{1}{k}$ (say); $\quad \ell = k\alpha, m = k\beta, n = k\gamma$ ... (iii)

Substituting values of ℓ, m, n from equation (iii) in equation (ii), we get, $ka\alpha + kb\beta + kc\gamma + d = 0$ ... (iv)

Since α, β, γ lies on plane (i) $\therefore \ell\alpha + m\beta + n\gamma + d = 0$ ... (v)

Substituting values of ℓ, m, n from equation (iii) in equation (v), we get $k\alpha^2 + k\beta^2 + k\gamma^2 + d = 0$ ... (vi)

[substituting value of d from equation (iv) in equation (vi)] or $\alpha^2 + \beta^2 + \gamma^2 - a\alpha - b\beta - c\gamma = 0$

Therefore, locus of foot of the perpendicular from the point $P(\alpha, \beta, \gamma)$ is $x^2 + y^2 + z^2 - ax - by - cz = 0$

Illustration 26: Find the equation of the sphere if it touches the plane $\vec{r} \cdot (2\hat{i} - 2\hat{j} - \hat{k}) = 0$ and the position vector of its center is $3\hat{i} + 6\hat{j} - 4\hat{k}$. **(JEE ADVANCED)**

Sol: Here equation of the required sphere is $|\vec{r} - \vec{c}| = a$ where a is the radius of the sphere.

Given plane is $\vec{r} \cdot (2\vec{i} - 2\vec{j} - \vec{k}) = 0$ ... (i)

Let H be the center of the sphere, then $\overrightarrow{OH} = 3\vec{i} + 6\vec{j} - 4\vec{k} = c$ (say)

Radius of the sphere = length of perpendicular from H to plane (i)

$= \dfrac{|c \cdot (2i - 2j - k)|}{|2\vec{i} - 2\vec{j} - \vec{k}|} = \dfrac{|(3i + 6j - 4k) \cdot (2i - 2j - k)|}{(2\vec{i} - 2\vec{j} - \vec{k})} = \dfrac{|6 - 12 + 4|}{3} = \dfrac{2}{3} = a$ (say)

Equation of the required sphere is $|\vec{r} - \vec{c}| = a$

$\Rightarrow |x\vec{i} + y\vec{j} + z\vec{k} - (3\vec{i} + 6\vec{j} - 4\vec{k})| = \dfrac{2}{3}$ or $|(x - 3)\vec{i} + (y - 6)\vec{j} + (z + 4)\vec{k}|^2 = \dfrac{4}{9} \Rightarrow (x - 3)^2 + (y - 6)^2 + (z + 4)^2$

$= 4/9$ or $9(x^2 + y^2 + z^2 - 6x - 12y + 8z + 61) = 4 \Rightarrow 9x^2 + 9y^2 + 9z^2 - 54x - 108y + 72z + 545 = 0$

Illustration 27: Find the equation of the sphere with the points (1, 2, 2) and (2, 3, 4) as the extremities of a diameter. Find the coordinates of its center. **(JEE MAIN)**

Sol: Equation of the sphere having (x_1, y_1, z_1) and (x_2, y_2, z_2) as the extremities of a diameter is $(x - x_1)(x - x_2) + (y - y_1)(y - y_2) + (z - z_1)(z - z_2) = 0$.

Let $A \equiv (1, 2, 2)$, $B \equiv (2, 3, 4)$

Equation of the sphere having (x_1, y_1, z_1) and (x_2, y_2, z_2) as the extremities of a diameter is

$(x - x_1)(x - x_2) + (y - y_1)(y - y_2) + (z - z_1)(z - z_2) = 0$

Here $x_1 = 1$, $x_2 = 2$, $y_1 = 2$, $y_2 = 3$, $z_1 = 2$, $z_2 = 4$

$\therefore$ Required equation of the sphere is $(x - 1)(x - 2) + (y - 2)(y - 3) + (z - 2)(z - 4) = 0$ or

$x^2 + y^2 + z^2 - 3x - 5y - 6z + 16 = 0$

Center of the sphere is the midpoint of AB

$\therefore$ Center is $\left(\dfrac{3}{2}, \dfrac{5}{2}, 3\right)$.

Illustration 28: Find the equation of the sphere passing through the points (3, 0, 0), (0, −1, 0), (0, 0, −2) and whose center lies on the plane $3x + 2y + 4z = 1$ **(JEE ADVANCED)**

Sol: Consider the equation of the sphere be $x^2 + y^2 + z^2 + 2ux + 2vy + 2wz + d = 0$.

As the sphere passes through these given points hence these points will satisfy equation of sphere. ... (i)

Let $A \equiv (3, 0, 0)$, $B \equiv (0, -1, 0)$, $C \equiv (0, 0, -2)$ since sphere (i) passes through A, B and C.

$\therefore 9 + 6u + d = 0$ ... (ii)

$1 - 2v + d = 0$ ... (iii)

$4 - 4w + d = 0$ ... (iv)

Since the center $(-u, -v, -w)$ of the sphere lies on plane $3x + 2y + 4z = 1 \therefore -3u - 2v - 4w = 1$ ....(v)

(ii) − (iii) $\Rightarrow 6u + 2v = -8$ ... (vi)

(iii) − (iv) $\Rightarrow -2v + 4w = 3$ ... (vii)

From (vi), $u = \dfrac{-2v - 8}{6}$ ... (viii)

From (vii), $4w = 3 + wv$

Substituting the values of u, v and w in equation (v), we get $\dfrac{2v + 8}{2} + 2v - 3 - 2v = 1$

$\Rightarrow 2v + 8 - 4v - 6 - 4v = 2 \Rightarrow v = 0$

From equation (viii), $u = \dfrac{0 - 8}{6} = -\dfrac{4}{3}$;   From equation (ix), $4w = 3$ $\therefore w = 3/4$

From equation (iii), $d = 2v - 1 = 0 - 1 = -1$ From equation (i), the equation of the required sphere is

$x^2 + y^2 + z^2 - \dfrac{0 - 8}{6} - \dfrac{8}{3}x + \dfrac{3}{2}z - 1 = 0$ or $6x^2 + 6y^2 + 6z^2 - 16x + 9z - 6 = 0$

FORMULAE SHEET

(a) Distance between the points (x_1, y_1, z_1) and (x_2, y_2, z_2) is $\sqrt{(x_2 - x_1)^2 + (y_2 - y_1)^2 + (z_2 - z_1)^2}$

(b) Coordinates of the point dividing the distance between the points (x_1, y_1, z_1) and (x_2, y_2, z_2) in the ratio m:n are

$$\left(\frac{mx_2 + nx_1}{m + n}, \frac{my_2 + ny_1}{m + n}, \frac{mz_2 + nz_1}{m + n} \right)$$

(c) If $A(x_1, y_1, z_1)$, $B(x_2, y_2, z_2)$ and $C(x_3, y_3, z_3)$ are vertices of a triangle, then its centroid is

$$\left(\frac{x_1 + x_2 + x_3}{3}, \frac{y_1 + y_2 + y_3}{3}, \frac{z_1 + z_2 + z_3}{3} \right)$$

(d) If $A(x_1, y_1, z_1)$ and $B(x_2, y_2, z_2)$ are the two points, the point which divides the line segment AB in ratio $\lambda:1$ is

$$\left(\frac{\lambda x_2 + x_1}{\lambda + 1}, \frac{\lambda y_2 + y_1}{\lambda + 1}, \frac{\lambda z_2 + z_1}{\lambda + 1} \right)$$

(e) If (x_1, y_1, z_1) and (x_2, y_2, z_2) are the two points on the line with $x_2 - x_1, y_2 - y_1, z_2 - z_1$ as direction ratios, then their d.c.s are

$$\pm \frac{x_2 - x_1}{\sqrt{\Sigma(x_2 - x_1)^2}}, \pm \frac{y_2 - y_1}{\sqrt{\Sigma(x_2 - x_1)^2}}, \pm \frac{z_2 - z_1}{\sqrt{\Sigma(x_2 - x_1)^2}}$$

(f) If ℓ, m, n are d.c.s of a line, then $l^2 + m^2 + n^2 = 1$. Thus, if a line makes angles α, β, γ with axes, then $\cos^2 \alpha + \cos^2 \beta + \cos^2 \gamma = 1$ and $\sin^2 \alpha + \sin^2 \beta + \sin^2 \gamma = 2$

(g) If a, b, c are the d.r.s of a line, then the d.c.s of the line are $\pm \dfrac{a}{\sqrt{\Sigma a^2}}, \pm \dfrac{b}{\sqrt{\Sigma a^2}}, \pm \dfrac{c}{\sqrt{\Sigma a^2}}$

(h) If $p(x, y, z)$ is a point in space such that $\overrightarrow{OP} = \vec{r}$ has d.c.s ℓ, m, n, then

(a) $\ell|\vec{r}|, m|\vec{r}|, n|\vec{r}|$ are the projections on x-axis, y-axis and z-axis, respectively.

(b) $x = \ell|\vec{r}|, y = m|\vec{r}|, z = n|\vec{r}|$

(c) $\vec{r} = |\vec{r}|(l\hat{i} + m\hat{j} + n\hat{k})$ and $\hat{r} = \ell\hat{i} + m\hat{j} + n\hat{k}$

Moreover, if a, b, c are d.r.s of a vector $\vec{r}$, then $\vec{r} = \dfrac{|\vec{r}|}{\sqrt{a^2 + b^2 + c^2}}(a\hat{i} + b\hat{j} + c\hat{k})$.

(i) Length of projection of the line segment joining (x_1, y_1, z_1) and (x_2, y_2, z_2) on a line with d.c.s ℓ, m, n is $|\ell(x_2 - x_1) + m(y_2 - y_1) + n(z_2 - z_1)|$

(j) If θ is the angle between two lines having direction ratios a_1, b_1, c_1 and a_2, b_2, c_2 then

$$\cos \theta = \pm \frac{a_1 a_2 + b_1 b_2 + c_1 c_2}{\sqrt{\Sigma a_1^2}\sqrt{\Sigma a_2^2}}$$

(k) Two lines are parallel if $\dfrac{a_1}{a_2} = \dfrac{b_1}{b_2} = \dfrac{c_1}{c_2}$ and two lines are perpendicular if $a_1 a_2 + b_1 b_2 + c_1 c_2 = 0$

(l) Cartesian equations of a line passing through (x_1, y_1, z_1) and having direction ratios a, b, c are

$$\frac{x - x_1}{a} = \frac{y - y_1}{b} = \frac{z - z_1}{c} = t$$

(m) Vector equation of a line passing through the point $A(\vec{a})$ and parallel to vector $\vec{b}$ is $\vec{r} = \vec{a} + \lambda \vec{b}$ for scalar λ.

(n) Cartesian equation of a line passing through two points having coordinates (x_1, y_1, z_1) and (x_2, y_2, z_2) is

$$\frac{x - x_1}{x_2 - x_1} = \frac{y - y_1}{y_2 - y_1} = \frac{z - z_1}{z_2 - z_1} .$$

(o) Vector equation of a line passing through two points having position vectors $\vec{a}$ and $\vec{b}$ is $\vec{r} = \vec{a} + \lambda(\vec{b} - \vec{a})$

(p) Distance between the parallel lines $\vec{r} = \vec{a}_1 + \lambda \vec{b}$ and $\vec{r} = \vec{a}_2 + \mu \vec{b}$ is $\dfrac{|\vec{b} \times (\vec{a}_2 - \vec{a}_1)|}{|\vec{b}|}$

(q) Shortest distance (S.D.) between two lines with equations; $\vec{r} = \vec{a}_1 + \lambda \vec{b}_1$ and $\vec{r} = \vec{a}_2 + \mu \vec{b}_2$ is

$\dfrac{|(\vec{b}_1 \times \vec{b}_2) \cdot (\vec{a}_2 - \vec{a}_1)|}{|\vec{b}_1 \times \vec{b}_2|}$. If θ is the angle between the lines, then $\cos\theta = \dfrac{\vec{b}_1 \cdot \vec{b}_2}{|\vec{b}_1||\vec{b}_2|}$

(r) The length of perpendicular from the point (α, β, γ) to the line $\dfrac{x - x_1}{l} = \dfrac{y - y_1}{m} = \dfrac{z - z_1}{n}$ (l, m, n being d.cs)

is given by $\sqrt{(\alpha - x_1)^2 + (\beta - y_1)^2 + (\gamma - z_1)^2 - [l(\alpha - x_1) + m(\beta - y_1) + n(\gamma - z_1)]^2}$

(s) If $\vec{a}$ and $\vec{b}$ are the unit vectors along the sides of an angle, then $\vec{a} + \vec{b}$ and $\vec{a} - \vec{b}$ are the vectors, respectively, along the internal and external bisector of the angle. In fact, the bisectors of the angles between the lines,

$\vec{r} = x\vec{a}$ and $\vec{r} = y\vec{b}$ are given by $\vec{r} = \lambda\left(\dfrac{\vec{a}}{|\vec{a}|} + \dfrac{\vec{b}}{|\vec{b}|}\right); \lambda \in R$

(t) Equation of plane passing through the point (x_1, y_1, z_1) is $a(x - x_1) + b(y - y_1) + c(z - z_1) = 0$.

(u) Equation of plane passing through three points (x_1, y_1, z_1), (x_2, y_2, z_2) and (x_3, y_3, z_3) is

$$\begin{vmatrix} x - x_1 & y - y_1 & z - z_1 \\ x_2 - x_1 & y_2 - y_1 & z_2 - z_1 \\ x_3 - x_1 & y_3 - y_1 & z_3 - z_1 \end{vmatrix} = 0$$

(v) Equation of a plane making intercepts a, b, c on axes is $\dfrac{x}{a} + \dfrac{y}{b} + \dfrac{z}{c} = 1$

(w) Vector equation of a plane through the point $\vec{a}$ and perpendicular to the unit vector $\hat{n}$ is $(\vec{r} - \vec{a}) \cdot \hat{n} = 0$

(x) If θ is the angle between the two planes $\vec{r} \cdot \hat{n}_1 = d_1$ and $\vec{r} \cdot \hat{n}_2 = d_2$, then $\cos\theta = \dfrac{\hat{n}_1 \cdot \hat{n}_2}{|\hat{n}_1||\hat{n}_2|}$

(y) Equation of a plane containing the line $\dfrac{x - x_1}{a} = \dfrac{y - y_1}{b} = \dfrac{z - z_1}{c}$ and passing through the point (x_2, y_2, z_2) not

on the line is $\begin{vmatrix} x - x_1 & y - y_1 & z - z_1 \\ a & b & c \\ x_2 - x_1 & y_2 - y_1 & z_2 - z_1 \end{vmatrix} = 0$

(z) Equation of a plane through the line $\dfrac{x - x_1}{a_1} = \dfrac{y - y_1}{b_1} = \dfrac{z - z_1}{c_1}$ and parallel to the line $\dfrac{x - x_2}{a_2} = \dfrac{y - y_2}{b_2} = \dfrac{z - z_2}{c_2}$

is $\begin{vmatrix} x - x_1 & y - y_1 & z - z_1 \\ a_1 & b_1 & c_1 \\ a_2 & b_2 & c_2 \end{vmatrix} = 0$

(aa) If θ is the angle between the line $\dfrac{x - x_1}{a} = \dfrac{y - y_1}{b} = \dfrac{z - z_1}{c}$ and the plane $Ax + By + Cz + D = 0$, then

$$\sin\theta = \frac{|\,aA + bB + cC\,|}{\sqrt{a^2 + b^2 + c^2}\,\sqrt{A^2 + B^2 + C^2}}$$

(ab) Length of perpendicular from (x_1, y_1, z_1) to the plane $ax + by + cz + d = 0$ is $\dfrac{|\,ax_1 + by_1 + cz_1 + d\,|}{\sqrt{a^2 + b^2 + c^2}}$

(ac) The equation of a sphere with center at the origin and radius 'a' is $|\vec{r}| = a$ or $x^2 + y^2 + z^2 = a^2$

(ad) Equation of a sphere with center (α, β, γ) and radius 'a' is $(x - \alpha)^2 + (y - \beta)^2 + (z - \gamma)^2 = a^2$

(ae) Vector equation of the sphere with center $\vec{c}$ and radius 'a' is $|\vec{r} - \vec{c}| = a$ or $(\vec{r} - \vec{c})\cdot(\vec{r} - \vec{c}) = a^2$

(af) General equation of sphere is $x^2 + y^2 + z^2 + 2ux + 2vy + 2wz + d = 0$ whose center is $(-u, -v, -w)$ and radius is $\sqrt{u^2 + v^2 + w^2 - d}$

(ag) Equation of a sphere concentric with $x^2 + y^2 + z^2 + 2ux + 2vy + 2wz + d = 0$ is $x^2 + y^2 + z^2 + 2ux + 2vy + 2wz + \lambda = 0$, where λ is a real number.

Solved Examples

JEE Main/Boards

Example 1: Find the coordinates of the point which divides the join of P(2, –1, 4) and Q(4, 3, 2) in the ratio 2 : 3 (i) internally (ii) externally

Sol: By using section formula we can obtain the result.

Let R(x, y, z) be the required point

(i) $x = \dfrac{2 \times 4 + 3 \times 2}{2 + 3} = \dfrac{14}{5}$; $y = \dfrac{2 \times 3 + 3 \times (-1)}{2 + 3} = \dfrac{3}{5}$

$z = \dfrac{2 \times 2 + 3 \times 4}{2 + 3} = \dfrac{16}{5}$

So, the required point is $R\left(\dfrac{14}{5}, \dfrac{3}{5}, \dfrac{16}{5}\right)$

(ii) $x = \dfrac{2 \times 4 - 3 \times 2}{2 - 3} = -2$; $y = \dfrac{2 \times 3 - 3 \times (-1)}{2 - 3} = -9$

$z = \dfrac{2 \times 2 - 3 \times 4}{2 - 3} = 8$

Therefore, the required point is R(–2. –9, 8)

Example 2: Find the points on X-axis which are at a distance of $2\sqrt{6}$ units from the point (1, –2, 3)

Sol: Consider required point is P(x, 0, 0), therefore by using distance formula we can obtain the result.

Let P(x, 0, 0) be a point on X-axis such that distance of P from the point (1, –2, 3) is $2\sqrt{6}$

$\Rightarrow \sqrt{(1 - x)^2 + (-2 - 0)^2 + (3 - 0)^2} = 2\sqrt{6}$

$\Rightarrow (x - 1)^2 + 4 + 9 = 24 \qquad \Rightarrow (x - 1)^2 = 11$

$\Rightarrow x - 1 = \pm\sqrt{11} \qquad\qquad \Rightarrow x = 1 \pm \sqrt{11}$

Example 3: If a line makes angles α, β, γ with OX, OY, OZ, respectively, prove that $\sin^2\alpha + \sin^2\beta + \sin^2\gamma = 2$

Sol: Same as illustration 2.

Let ℓ, m, n be the d.c.'s of the given line, then

$\ell = \cos\alpha, m = \cos\beta, n = \cos\gamma$

$\cos^2\alpha + \cos^2\beta + \cos^2\gamma = 1$

$\Rightarrow (1 - \sin^2\alpha) + (1 - \sin^2\beta) + (1 - \sin^2\gamma) = 1$

$\Rightarrow \sin^2\alpha + \sin^2\beta + \sin^2\gamma = 2$

Example 4: Projections of a line segment on the axes are 12, 4 and 3 respectively. Find the length and direction cosines of the line segment.

Sol: Let ℓ, m, n be the direction cosines and r be the length of the given segment, then $\ell r, mr, nr$ are the projections of the segment on the axes.

Let l, m, n be the direction cosines and r be the length of the given segment, then lr, mr, nr are the projections of the segment on the axes; therefore $lr = 12, mr = 4, nr = 3$

Squaring and adding, we get

$$r^2(l^2 + m^2 + n^2) = 12^2 + 4^2 + 3^2 \Rightarrow r^2 = 169$$

$\Rightarrow r = 13 \Rightarrow$ length of segment $= 13$

And direction cosines of segment are

$$l = \frac{12}{r} = \frac{12}{13}, \ m = \frac{4}{r} = \frac{4}{13} \text{ and } n = \frac{3}{r} = \frac{3}{13}$$

Example 5: Find the length of the perpendicular from the point $(1, 2, 3)$ to the line through $(6, 7, 7)$ and having direction ratios $(3, 2, -2)$.

Sol: By using distance formula i.e. $| l (x_2 - x_1) + m(y_2 - y_1) + n(z_2 - z_1)|$, we can obtain required length.

Direction cosines of the line are

$$\frac{3}{\sqrt{3^2 + 2^2 + (-2)^2}}, \frac{2}{\sqrt{3^2 + 2^2 + (-2)^2}}, \frac{-2}{\sqrt{3^2 + 2^2 + (-2)^2}}$$

i.e. $\dfrac{3}{\sqrt{17}}, \dfrac{2}{\sqrt{17}}, \dfrac{-2}{\sqrt{17}}$

$\therefore$ AN = Projection of AP on AB

$$= (6 - 1)\frac{3}{\sqrt{17}} + (7 - 2)\frac{2}{\sqrt{17}} + (7 - 3)\frac{(-2)}{\sqrt{17}}$$

$$= \frac{15 + 10 - 8}{\sqrt{17}} = \frac{17}{\sqrt{17}} = \sqrt{17}$$

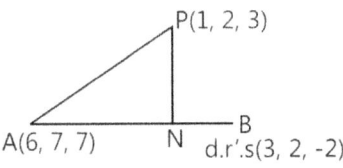

Also, $AP = \sqrt{(6 - 1)^2 + (7 - 2)^2 + (7 - 3)^2}$

$$= \sqrt{25 + 25 + 16} = \sqrt{66}$$

$\therefore PN = \sqrt{AP^2 - AN^2} = \sqrt{66 - 17} = \sqrt{49} = 7$ unit

Example 6: Find the equation of the plane through the points A(2, 2, −1), B(3, 4, 2) and C(7, 0, 6)

Sol: As we know, equation of a plane passing through the point (x_1, y_1, z_1) is given by

$a(x - x_1) + b(y - y_1) + c \quad (z - z_1) = 0$.

The general equation of a plane through (2, 2, −1) is
$a(x - 2) + b(y - 2) + c(z + 1) = 0$ .... (i)

It will pass through B(3, 4, 2) and C(7, 0, 6) if
$a(3 - 2) + b(4 - 2) + c(2 + 1) = 0$ or

$a + 2b + 3c = 0$ .... (ii)

& $a(7 - 2) + b(0 - 2) + c(6 + 1) = 0$ or

$5a - 2b + 7c = 0$ .... (iii)

Solving (ii) and (iii) by cross multiplication, we get

$$\frac{a}{14 + 6} = \frac{b}{15 - 7} = \frac{c}{-2 - 10} \text{ or } \frac{a}{5} = \frac{b}{2} = \frac{c}{-3} = \lambda \text{ (say)}$$

$\Rightarrow a = 5\lambda, b = 2\lambda, c = -3\lambda$

Substituting the values of a, b and c in (i), we get
$5\lambda(x - 2) + 2\lambda(y - 2) - 3\lambda(z + 1) = 0$

or, $5(x - 2) + 2(y - 2) - 3(z + 1) = 0$

$\Rightarrow 5x + 2y - 3z = 17$,

Which is the required equation of the plane.

Example 7: Find the angle between the planes $x + y + 2z = 9$ and $2x - y + z = 15$

Sol: By using formula $\cos\theta = \dfrac{a_1 a_2 + b_1 b_2 + c_1 c_2}{\sqrt{a_1^2 + b_1^2 + c_1^2}\sqrt{a_2^2 + b_2^2 + c_2^2}}$

we can obtain the result.

The angle between $x + y + 2z = 9$ and $2x - y + z = 15$ is given by

$$\cos\theta = \frac{a_1 a_2 + b_1 b_2 + c_1 c_2}{\sqrt{a_1^2 + b_1^2 + c_1^2}\sqrt{a_2^2 + b_2^2 + c_2^2}}$$

$$\cos\theta = \frac{(1)(2) + 1(-1) + (2)(1)}{\sqrt{1^2 + 1^2 + 2^2}\sqrt{2^2 + (-1)^2 + 1^2}} = \frac{1}{2} \Rightarrow \theta = \frac{\pi}{3}$$

Example 8: Find the distance between the parallel planes $2x - y + 2z + 3 = 0$ and $4x - 2y + 4z + 5 = 0$

Sol: By making the coefficient of x, y and z as unity we will be get required result.

Let $P(x_1, y_1, z_1)$ be any point on $2x - y + 2z + 3 = 0$, then,

$2x_1 - y_1 + 2z_1 + 3 = 0$ ... (i)

The length of the perpendicular from

$P(x_1, y_1, z_1)$ to $4x - 2y + 4z + 5 = 0$ is

$$= \left|\frac{4x_1 - 2y_1 + 4z_1 + 5}{\sqrt{4^2 + (-2)^2 + 4^2}}\right| = \left|\frac{2(2x_1 - y_1 + 2z_1) + 5}{\sqrt{36}}\right|$$

$$= \left|\frac{2(-3) + 5}{6}\right| = \frac{1}{6}$$

Example 9: The equation of a line are $6x - 2 = 3y + 1 = 2z - 2$. Find its direction ratios and its equation in symmetric form.

Sol: The given line is $6x - 2 = 3y + 1 = 2z - 2$

$$\Rightarrow 6\left(x - \frac{1}{3}\right) = 3\left(y + \frac{1}{3}\right) = 2(z - 1)$$

$$\Rightarrow \frac{x - \frac{1}{3}}{1} = \frac{y + \frac{1}{3}}{2} = \frac{z - 1}{3}$$

[We make the coefficients of x, y and z as unity]

This equation is in symmetric form. Thus the direction ratios of the line are 1, 2 and 3 and this line passes through the point $\left(\frac{1}{3}, -\frac{1}{3}, 1\right)$.

Example 10: Find the image of the point $(3, -2, 1)$ in the plane $3x - y + 4z = 2$.

Sol: Consider Q be the image of the point P(3, -2, 1) in the plane $3x - y + 4z = 2$. Then PQ is normal to the plane hence direction ratios of PQ are 3, -1, 4.

Let Q be the image of the point P(3, -2, 1) in the plane $3x - y + 4z = 2$. Then PQ is normal to the plane. Therefore direction ratios of PQ are 3, -1, 4. Since PQ passes through P(3, -2, 1) and has direction ratios 3, -1, 4. Therefore equation of PQ is

$$\frac{x - 3}{3} = \frac{y + 2}{-1} = \frac{z - 1}{4} = r \qquad \text{(say)}$$

Let the coordination of Q be $(3r + 3, -r, -2. 4r + 1)$. Let R be the mid-point of PQ. Then R lies on the plane $3x - y + 4z = 2$. The coordinates of R are

$$\left(\frac{3r + 3 + 3}{2}, \frac{-r - 2 - 2}{2}, \frac{4r + 1 + 1}{2}\right)$$

or $\left(\frac{3r + 6}{2}, \frac{-r - 4}{2}, 2r + 1\right)$

$$3\left(\frac{3r + 6}{2}\right) - \left(\frac{-r - 4}{2}\right) + 4(2r + 1) = 2$$

$$\Rightarrow \quad 13r = -13 \quad \Rightarrow \quad r = -1$$

So, the coordinates of Q are $(0, -1, -3)$

JEE Advanced/Boards

Example 1: Find the equations of the bisector planes of the angles between the planes $2x - y + 2z + 3 = 0$ and $3x - 2y + 6z + 8 = 0$ and specify the plane bisecting the acute angle and the plane bisecting obtuse angle.

Sol: As we know, Equation of the planes bisecting the angle between two given planes $a_1x + b_1y + c_1z + d_1 = 0$ and $a_2x + b_2y + c_2z + d_2 = 0$ are

$$\frac{a_1x + b_1y + c_1z + d_1}{\sqrt{a_1^2 + b_1^2 + c_1^2}} = \pm \frac{a_2x + b_2y + c_2z + d_2}{\sqrt{a_2^2 + b_2^2 + c_2^2}}$$

The two given planes are

$2x - y + 2z + 3 = 0$ ... (i)

and $3x - 2y + 6z + 8 = 0$ ... (ii)

The equations of the planes bisecting the angles between (i) and (ii) are

$$\frac{2x - y + 2z + 3}{\sqrt{2^2 + (-1)^2 + 2^2}} = \pm \frac{3x - 2y + 6z + 8}{\sqrt{3^2 + (-2)^2 + 6^2}}$$

$$\Rightarrow 14x - 7y + 14z + 21 = \pm(9x - 6y + 18z + 24)$$

Hence the two bisector planes are

$5x - y - 4z - 3 = 0$ ... (iii)

and $23x - 13y + 32z + 45 = 0$ ... (iv)

Now we find angle θ between (i) & (iii)

We have,

$$\cos\theta = \frac{5(2) + (-1)(-1) + 2(-4)}{\sqrt{2^2 + (-1)^2 + 2^2} \sqrt{5^2 + (-1)^2 + (-4)^2}} = \frac{1}{\sqrt{42}}$$

Thus the angle between (i) & (iii) is more than $\frac{\pi}{4}$. Therefore, (iii) is the bisector of obtuse angle between (i) and (ii) and hence (iv) bisects acute angle between them.

Example 2: Find the distance of the point $(1, -2, 3)$ from the plane $x - y + z = 5$ measured parallel to the line whose direction cosines are proportional to 2, 3, -6.

Sol: By using distance formula we can obtain required length,

Equation of line through $(1, -2, 3)$ parallel to the line with d.r.'s 2, 3, -6 is

$$\frac{x - 1}{2} = \frac{y + 2}{3} = \frac{z - 3}{-6} = r \qquad \text{...(i)}$$

Any point on it is $(1 + 2r, -2 + 3r, 3 - 6r)$

Line (i) meets the plane $x - y + z = 5$.

If $1 + 2r - (-2 + 3r) + (3 - 6r) = 5$; i.e. if $r = \frac{1}{7}$

∴ Point of intersection is $\left(\dfrac{9}{7}, -\dfrac{11}{7}, \dfrac{15}{7}\right)$

whose distance from $(1, -2, 3)$ is

$$\sqrt{\left(\dfrac{9}{7} - 1\right)^2 + \left(-\dfrac{11}{7} + 2\right)^2 + \left(\dfrac{15}{7} - 3\right)^2}$$

$$= \sqrt{\dfrac{4}{49} + \dfrac{9}{49} + \dfrac{36}{49}} = \sqrt{\dfrac{49}{49}} = 1$$

Example 3: Show that the lines

$$\dfrac{x-1}{3} = \dfrac{y+1}{2} = \dfrac{z-1}{5} \qquad \dots \text{(i)}$$

and $\dfrac{x+2}{4} = \dfrac{y-1}{3} = \dfrac{z+1}{-2} \qquad \dots \text{(ii)}$

do not intersect. Also find the shortest distance between them.

Sol: If $\begin{vmatrix} x_2 - x_1 & y_2 - y_1 & z_2 - z_1 \\ \ell_1 & m_1 & n_1 \\ \ell_2 & m_2 & n_2 \end{vmatrix} \neq 0$

then the lines do not intersect each other. And using distance formula we will be get required shortest distance.

Points on (i) and (ii) are $(1, -1, 1)$ and $(-2, 1, -1)$

respectively and their d.c.'s are $\dfrac{3}{\sqrt{38}}, \dfrac{2}{\sqrt{38}}, \dfrac{5}{\sqrt{38}}$

and $\dfrac{4}{\sqrt{29}}, \dfrac{3}{\sqrt{29}}, \dfrac{-2}{\sqrt{29}}$ respectively.

$$\therefore \begin{vmatrix} x_2 - x_1 & y_2 - y_1 & z_2 - z_1 \\ \ell_1 & m_1 & n_1 \\ \ell_2 & m_2 & n_2 \end{vmatrix} = \begin{vmatrix} -2-1 & 1+1 & -1-1 \\ \dfrac{3}{\sqrt{38}} & \dfrac{2}{\sqrt{38}} & \dfrac{5}{\sqrt{38}} \\ \dfrac{4}{\sqrt{29}} & \dfrac{3}{\sqrt{29}} & \dfrac{-2}{\sqrt{29}} \end{vmatrix}$$

$$= \dfrac{1}{\sqrt{38}} \times \dfrac{1}{\sqrt{29}} \begin{vmatrix} -3 & 2 & -2 \\ 3 & 2 & 5 \\ 4 & 3 & -2 \end{vmatrix}$$

$$= \dfrac{1}{\sqrt{38}} \times \dfrac{1}{\sqrt{29}}, [-3(-4-15) + 2(20+6) - 2(9-8)] \neq 0$$

Hence the given lines do not intersect.

Any point P on (i) is $(1 + 3r_1, 2r_1, -1, 5r_1 + 1)$ and a point on (ii) is $Q(4r_2 - 2, 3r_2 + 1, -2r_2 - 1)$

∴ Direction ratios of PQ are

$(4r_2 - 3r_1 - 3, 3r_2 - 2r_1 + 2, -2r_2 - 5r_1 - 2)$

If PQ is perpendicular to (i) and (ii), we have

$3(4r_2 - 3r_1 - 3) + 2(3r_2 - 2r_1 + 2) + 5(-2r_2 - 5r_1 - 2) = 0$

& $4(4r_2 - 3r_1 - 3) + 3(3r_2 - 2r_1 + 2) - 2(-2r_2 - 5r_1 - 2) = 0$

i.e. $8r_2 - 38r_1 - 15 = 0$ & $29r_2 - 8r_1 - 2 = 0$

Solving them, $\dfrac{r_2}{76 - 120} = \dfrac{r_1}{-435 + 16} = \dfrac{1}{1038}$

$\Rightarrow r_2 = -\dfrac{44}{1038}, r_1 = -\dfrac{419}{1038}$

∴ Points P and Q are $\left(-\dfrac{1257}{1038} + 1, -\dfrac{838}{1038} - 1, -\dfrac{2095}{1038} + 1\right)$

and $\left(-\dfrac{176}{1038} - 2, -\dfrac{132}{1038} + 1, \dfrac{88}{1038} - 1\right)$

We can find the distance PQ by distance formula which is the shortest distance.

Example 4: Find the angle between the lines whose direction ratios satisfy the equations :

$3\ell + m + 5n = 0, 6mn - 2n\ell + 5\ell m = 0$

Sol: Here, $\cos\theta = \dfrac{aa' + bb' + cc'}{\sqrt{a^2 + b^2 + c^2}\sqrt{a'^2 + b'^2 + c'^2}}$.

The given equations are $3\ell + m + 5n = 0$ $\qquad \dots \text{(i)}$

and $6mn - 2n\ell + 5\ell m = 0$ $\qquad \dots \text{(ii)}$

From (i), we have $m = -3\ell - 5n$ $\qquad \dots \text{(iii)}$

Putting in (ii), we get

$6(-3\ell - 5n)n - 2n\ell + 5\ell(-3\ell - 5n) = 0$

$\Rightarrow 30n^2 + 45\ell n + 15\ell^2 = 0$

$\Rightarrow 2n^2 + 3\ell n + \ell^2 = 0 \qquad \Rightarrow (n + \ell)(2n + \ell) = 0$

$\Rightarrow$ Either $\ell = -n$ or $\ell = -2n$

If $\ell = -n$, then from (iii), $m = -2n$

If $\ell = -2n$, then from (iii), $m = n$

Thus the direction ratios of two lines are

$-n, -2n, n$ and $-2n, n, n$

i.e. $1, 2, -1$ and $-2, 1, 1$

∴ If θ is the angle between the lines, then

$\cos\theta = \dfrac{1.(-2) + 2.1 + (-1).1}{\sqrt{1+4+1}.\sqrt{4+1+1}} = \dfrac{-2 + 2 - 1}{\sqrt{6}.\sqrt{6}} = \dfrac{-1}{6}$.

Example 5: Find the equation of the plane through the intersection planes $2x + 3y + 4z = 5$, $3x - y + 2z = 3$ and parallel to the straight line having direction cosines $(-1, 1, -1)$.

Sol: By using formula of family of plane, we will get the result.

Equation of plane through the given planes is
$2x + 3y + 4z - 5 + \lambda(3x - y + 2z - 3) = 0$

i.e. $(2 + 3\lambda)x + (3 - \lambda)y + (4 + 2\lambda)z + (-5 - 3\lambda) = 0$

This plane is parallel to the given straight line.

$\Rightarrow -1(2 + 3\lambda) + 1(3 - \lambda) + (-1)(4 + 2\lambda) = 0$

$\Rightarrow -2 - 3\lambda + 3 - \lambda - 4 - 2\lambda = 0$

$\Rightarrow 6\lambda = -3 \qquad \Rightarrow \lambda = -\dfrac{1}{2}$

$\therefore$ Equation of required plane is

$\dfrac{1}{2}x + \dfrac{7}{2}y + 3z - \dfrac{7}{2} = 0 \qquad \Rightarrow x + 7y + 6z = 7$

Example 6: Prove that the lines

$\dfrac{x+1}{3} = \dfrac{y+3}{5} = \dfrac{z+5}{7}$ and $\dfrac{x-2}{1} = \dfrac{y-4}{4} = \dfrac{z-6}{7}$

are coplanar. Also, find the plane containing these two lines.

Sol: As similar to example 3.

We know the lines $\dfrac{x - x_1}{\ell_1} = \dfrac{y - y_1}{m_1} = \dfrac{z - z_1}{n_1}$

and $\dfrac{x - x_2}{\ell_2} = \dfrac{y - y_2}{m_2} = \dfrac{z - z_2}{n_2}$ are coplanar if

$$\begin{vmatrix} x_2 - x_1 & y_2 - y_1 & z_2 - z_1 \\ \ell_1 & m_1 & n_1 \\ \ell_2 & m_2 & n_2 \end{vmatrix} = 0$$

and the equation of the plane containing these two lines is

$$\begin{vmatrix} x - x_1 & y - y_1 & z - z_1 \\ \ell_1 & m_1 & n_1 \\ \ell_2 & m_2 & n_2 \end{vmatrix} = 0$$

Here $x = -1, y_1 = -3, z_1 = -5, x_2 = 2, y_2 = 4, z_2 = 6,$

$\ell_1 = 3, m_1 = 5, n_1 = 7, \ell_2 = 1, m_2 = 4, n_2 = 7$

$$\begin{vmatrix} x_2 - x_1 & y_2 - y_1 & z_2 - z_1 \\ \ell_1 & m_1 & n_1 \\ \ell_2 & m_2 & n_2 \end{vmatrix} = \begin{vmatrix} 3 & 7 & 11 \\ 3 & 5 & 7 \\ 1 & 4 & 7 \end{vmatrix}$$

$= 21 - 98 + 77 = 0$

So, the given lines are coplanar, The equation of the plane containing the lines is

$$= \begin{vmatrix} x+1 & y+3 & z+5 \\ 3 & 5 & 7 \\ 1 & 4 & 7 \end{vmatrix} = 0$$

or $(x + 1)(35 - 28) - (y + 3)(21 - 7) + (z + 5)(12 - 5) = 0$ or

$x - 2y + z = 0$

Example 7: Find the equation of the plane passing through the lines of intersection of the planes $2x - y = 0$ and $3z - y = 0$ and perpendicular to the plane $4x + 5y - 3z = 8$.

Sol: Here by using the family of plane and formula of two perpendicular plane we will get the result.

The plane $2x - y + k(3z - y) = 0$

$\Leftrightarrow 2x - (1 + k)y + 3kz = 0$ is perpendicular to the plane $4x + 5y - 3z = 8$

$\Rightarrow 2.4 - (1 + k).5 + 3k(-3) = 0 \qquad \Rightarrow 14k = 3$

$\Rightarrow k = \dfrac{3}{14}$

Thus the required equation is

$2x - y + \left(\dfrac{3}{14}\right)(3z - y) = 0 \Leftrightarrow 28x - 17y + 9z = 0$

Example 8: Show that the lines

$\dfrac{x+5}{3} = \dfrac{y+4}{1} = \dfrac{z-7}{-2}, 3x + 2y + z - 2 = 0 = x - 3y + 2z - 13$

are coplanar and find the equation to the plane in which they lie.

Sol: By using the condition of coplanarity of line, we will get given lines are coplanar or not. And after that by using general equation of the plane we can obtain required equation of plane.

The general equation of the plane through the second line is

$3x + 2y + z - 2 + k(x - 3y + 2z - 13) = 0$

$\Leftrightarrow x(3 + k) + y(2 - 3k) + z(1 + 2k) - 2 - 13k = 0;$

K being the parameter

This contains the first line only if

$3(3 + k) + (2 - 3k) - 2(1 + 2k) = 0 \Rightarrow k = \dfrac{9}{4}$

Hence the equation of the plane which contains the two lines is

$21x - 19y + 22x - 125 = 0$

This plane clearly passes through the point $(-5, -4, 7)$

Exercise 1

Q.1 Direction cosines of a line are $\frac{3}{7}, \frac{-2}{7}, \frac{6}{7}$, find its direction ratios.

Q.2 Find the direction ratios of a line passing through the points (2, 1, 0) and (1, –2, 3).

Q.3 Find the angle between the lines
$\frac{x}{1} = \frac{y}{2} = \frac{z}{0}$ and $\frac{x-1}{3} = \frac{y+5}{2} = \frac{z-3}{1}$.

Q.4 Find the equation of a line parallel to the vector $3\hat{i} - \hat{j} - 3\hat{k}$ and passing through the point (–1,1,1).

Q.5 Write the vector equation of a line whose Cartesian equation is $\frac{x+3}{2} = \frac{y-1}{4} = \frac{z+1}{5}$.

Q.6 Write the Cartesian equation of a line whose vector equations is $\vec{r} = (3\hat{i} + 2\hat{j} - 5\hat{k}) + \lambda(-2\hat{i} + \hat{j} + 3\hat{k})$.

Q.7 Find the value of p, such that the line

$\frac{x}{1} = \frac{y}{3} = \frac{z}{2p}$ and $\frac{x}{-3} = \frac{y}{5} = \frac{z}{2}$

are perpendicular to each other.

Q.8 Write the Cartesian equation of the plane $\vec{r}(3\hat{i} + 2\hat{j} + 5\hat{k}) = 7$.

Q.9 Write the vector equation of plane $3x - y - 4z + 7 = 0$.

Q.10 Find the vector, normal to the plane $\vec{r}.(3\hat{i} - 7\hat{k}) + 5 = 0$.

Q.11 Find the direction ratios of a line, normal to the plane $7x + y - 2z = 1$.

Q.12 Find the angle between the line $\frac{x+2}{4} = \frac{y-1}{-5} = \frac{z}{7}$ and the plane $3x - 2z + 4 = 0$.

Q.13 Find the distance of the plane $x + y + 3z + 7 = 0$ from origin.

Q.14 Find the distance of the plane $3x - 3y + 3z = 0$ from the point (1, 1, 1).

Q.15 Find the intercepts cut by the plane $3x - 2y + 4z - 12 = 0$ on axes.

Q.16 Direction ratios of a line are 1, 3, –2. Find its direction cosines.

Q.17 Find the direction cosines of y-axis.

Q.18 Find the direction ratio of the line

$\frac{x+2}{1} = \frac{2y-1}{3} = \frac{3-z}{5}$.

Q.19 Find the angle between the planes

$r.(\hat{i} - 2\hat{j} - 2\hat{k}) = 1$ and $r.(3\hat{i} - 6\hat{j} + 2\hat{k}) = 0$.

Q.20 Find the angle between the line $\vec{r} = (2\hat{i} - \hat{j} + 3\hat{k}) + \lambda(3\hat{i} - 6\hat{j} + 2\hat{k})$ and the plane $\vec{r}.(\hat{i} + \hat{j} + \hat{k}) + 3$

Q.21 Find the direction cosines of the two lines which are connected by the relations $\ell - 5m + 3n = 0$ and $7\ell^2 + 5m^2 - 3n^2 = 0$.

Q.22 Prove that, the line passing through the point (1, 2, 3) and (–1, –2, –3) is perpendicular to the line passing through the points (–2, 1, 5) and (3, 3, 2).

Q.23 Find the coordinates of the foot of the perpendicular drawn from the point (1, 2, 1) to the line joining the points (1, 4, 6) and (5, 4, 4).

Q.24 If a variable line in two adjacent positions has direction cosines ℓ, m, n and $\ell + \delta\ell, m + \delta m, n + \delta n$, prove that the small angle $\delta\theta$ between two position is given by $(\delta\theta)^2 = (\delta\ell)^2 + (\delta m)^2 + (\delta n)^2$.

Q.25 Verify that $\frac{\ell_1 + \ell_2 + \ell_3}{\sqrt{3}}, \frac{m_1 + m_2 + m_3}{\sqrt{3}}, \frac{n_1 + n_2 + n_3}{\sqrt{3}}$

can be taken as direction cosines of a line equally inclined to three mutually perpendicular lines with direction cosines $\ell_1, m_1, n_1; \ell_2, m_2, n_2$ and ℓ_3, m_3, n_3

Q.26 Find the equations of line through the point (3, 0, 1) and parallel to the planes $x + 2y = 0$ and $3y - z = 0$.

Q.27 Find the equations of the planes through the intersection of the planes $x + 3y + 6 = 0$ and $3x - y - 4z = 0$ whose perpendicular distance from the origin is equal to 1.

Q.28 Find the equation of the plane through the points $(-1, 1, 1)$ and $(1, -1, 1)$ and perpendicular to the plane $x + 2y + 2z = 5$.

Q.29 Find the distance of the point $(-1, -5, -10)$ from the plane $x - y + z = 5$ measured parallel to the line

$$\frac{x-2}{3} = \frac{y+1}{4} = \frac{z-2}{12}.$$

Q.30 Find the vector and Cartesian forms of the equation of the plane passing through $(1, 2, -4)$ and parallel to the line $\vec{r} = \hat{i} + 2\hat{j} - 4\hat{k} + \lambda(2\hat{i} + 3\hat{j} + 6\hat{k})$ and $\vec{r} = \hat{i} - 3\hat{j} + 5\hat{k} + \mu(\hat{i} + \hat{j} - \hat{k})$.

Q.31 If straight line having direction cosines given by $a\ell + bm + cn = 0$ and $fmn + gn\ell + h\ell m = 0$ are perpendicular, then prove that $\frac{f}{a} + \frac{g}{b} + \frac{h}{c} = 0$.

Q.32 Prove that, the lines $x = ay + b$, $z = cy + d$ and $x = a'y + b'$, $z = c'y + d'$ are perpendicular to each other, if $aa' + cc' + 1 = 0$.

Q.33 Find the equation of the plane passing through the intersection of the planes $4x - y + z = 10$ and $x + y - z = 4$ and parallel to the line with direction ratios 2, 1, 1. Find also the perpendicular distance of $(1, 1, 1)$ from this plane.

Q.34 The foot of the perpendicular drawn from the origin to the plane is $(2, 5, 7)$. Find the equation of plane.

Q.35 Find the equation of a plane through $(-1, -1, 2)$ and perpendicular to the planes $3x + 2y - 3z = 1$ and $5x - 4y + z = 5$.

Q.36 Find the angle between the lines whose direction cosines are given by equations $\ell + m + n = 0$; $\ell^2 + m^2 - n^2 = 0$

Q.37 Find the equation of the line which passes through $(5, -7, -3)$ and is parallel to the line of intersection of the planes $x - 3y - 5 = 0$ and $9y - z + 16 = 0$.

Q.38 Prove that, the plane through the points $(1, 1, 1)$, $(1, -1, 1)$ and $(-7, 3, -5)$ is perpendicular to xz-plane.

Q.39 Find the length and coordinates of the foot of perpendicular from points $(1, 1, 2)$ to the plane $2x - 2y + 4z + 5 = 0$.

Q.40 Find the vector equation in the scalar product form, of the plane passing through the points $(1, 0, -1)$, $(3, 2, 2)$ and parallel to line

$$r = \hat{i} + \hat{j} + \lambda(\hat{i} - 2\hat{j} + 3\hat{k}).$$

Q.41 Find the distance between the parallel planes $2x - y + 3z - 4 = 0$ and $6x - 3y + 9z + 13 = 0$.

Q.42 Prove that, the equation of a plane. Which meets the axes in A, B, and C and the given centroid of triangle ABC is the point (α, β, γ), is $\frac{x}{\alpha} + \frac{y}{\beta} + \frac{z}{\gamma} = 3$.

Q.43 Find the equation of the plane passing through the origin and the line of intersection of the planes $x - 2y + 3z + 4 = 0$ and $x - y + z + 3 = 0$.

Q.44 Prove that, the line $2x + 2y - z - 6 = 0$, $2x + 3y - z - 8 = 0$ is parallel to the plane $y = 0$. Find the coordinates of the point where this line meets the plane $x = 0$.

Q.45 Find the equation of the plane through the line $ax + by + cz + d = 0$, $a'x + b'y + c'z + d' = 0$ and parallel to the line $\frac{x}{\ell} = \frac{y}{m} = \frac{z}{n}$.

Q.46 Find the equation of a plane parallel to x-axis and has intercepts 5 and 7 on y and z-axis, respectively.

Q.47 A variable plane at a constant distance p from origin meets the coordinate axes in points A, B and C, respectively. Through these points, planes are drawn parallel to the coordinate planes, prove that locus of point of intersection is $\frac{1}{x^2} + \frac{1}{y^2} + \frac{1}{z^2} = \frac{1}{p^2}$.

Q.48 Find the value of λ, for which the points with position vectors $\hat{i} - \hat{j} + 3\hat{k}$ and $3\hat{i} + \lambda\hat{j} + 3\hat{k}$ are equidistant from the plane $\vec{r}.(5\hat{i} + 2\hat{j} - 7\hat{k}) + 9 = 0$.

Q.49 Find the equation of a plane which is at a distance of 7 units from the origin and which is normal to the vector $3\hat{i} + 5\hat{j} - 6\hat{k}$

Q.50 Find the vector equation of the plane, $r = \hat{i} - \hat{j} + \lambda(\hat{i} + \hat{j} + \hat{k}) + \mu(4\hat{i} - 2\hat{j} + 3\hat{k})$ in the scalar product from.

Exercise 2

Single Correct Choice Type

Q.1 The sum of the squares of direction cosines of a straight line is

(A) Zero

(B) Two

(C) 1

(D) None of these

Q.2 Which one of the following is best condition for the plane $ax + by + cz + d = 0$ to intersect the x and y axes at equal angle

(A) $|a| = |b|$

(B) $a = -b$

(C) $a = b$

(D) $a^2 + b^2 = 1$

Q.3 The equation of a straight line parallel to the x-axis is given by

(A) $\dfrac{x-a}{1} = \dfrac{y-b}{1} = \dfrac{z-c}{1}$

(B) $\dfrac{x-a}{0} = \dfrac{y-b}{1} = \dfrac{z-c}{1}$

(C) $\dfrac{x-a}{0} = \dfrac{y-b}{0} = \dfrac{z-c}{1}$

(D) $\dfrac{x-a}{1} = \dfrac{y-b}{0} = \dfrac{z-c}{0}$

Q.4 A straight line is inclined to the axes of x and z at angles 45° and 60° respectively, then the inclination of the line to the y-axis is

(A) 30°

(B) 45°

(C) 60°

(D) 90°

Q.5 The coordinates of the point of intersection of the line $\dfrac{x+1}{1} = \dfrac{y+3}{3} = \dfrac{z+2}{-2}$ with the plane $3x + 4y + 5z = 5$

(A) $(5, 15, -14)$

(B) $(3, 4, 5)$

(C) $(1, 3, -2)$

(D) $(3, 12, -10)$

Q.6 Perpendicular is drawn from the point $(0, 3, 4)$ to the plane $2x - 2y + z = 10$. The coordinates of the foot of the perpendicular are

(A) $\left| -\dfrac{8}{3}, \dfrac{1}{3}, \dfrac{16}{3} \right|$

(B) $\left| \dfrac{8}{3}, \dfrac{1}{3}, \dfrac{16}{3} \right|$

(C) $\left| \dfrac{8}{3}, -\dfrac{1}{3}, \dfrac{16}{3} \right|$

(D) $\left| \dfrac{8}{3}, \dfrac{1}{3}, -\dfrac{16}{3} \right|$

Q.7 The equation of the plane through the line of intersection of the planes $2x + y - z - 4 = 0$ and $3x + 5z - 4 = 0$ which cuts off equal intercepts from the x-axis and y-axis is

(A) $3x + 3y - 8z + 8 = 0$

(B) $3x + 3y - 8z - 8 = 0$

(C) $3x - 3y - 8z - 8 = 0$

(D) $x + y - 8z - 8 = 0$

Q.8 The symmetric form of the equation of the line $x + y - z = 1$, $2x - 3y + z = 2$ is

(A) $\dfrac{x+1}{3} = \dfrac{y-2}{-2} = \dfrac{z+1}{-1}$

(B) $\dfrac{x}{2} = \dfrac{y}{3} = \dfrac{z-1}{5}$

(C) $\dfrac{x}{2} = \dfrac{y-1}{3} = \dfrac{z}{5}$

(D) $\dfrac{x-1}{2} = \dfrac{y}{3} = \dfrac{z}{5}$

Q.9 The line $\dfrac{x-1}{1} = \dfrac{y-3}{2} = \dfrac{z-4}{3}$ is parallel to the plane

(A) $2x + y + 2z + 3 = 0$

(B) $2x - y - 2z = 3$

(C) $21x - 12y + z = 0$

(D) $2x + y - 2z = 0$

Q.10 The vertices of the triangle PQR are $(2, 1, 1)$, $(3, 1, 2)$ and $(-4, 0, 1)$. The area of the triangle is

(A) $\dfrac{\sqrt{38}}{2}$

(B) $\sqrt{38}$

(C) 4

(D) 2

Q.11 Equation of straight line which passes through the point $P(1, 0, -3)$ and $Q(-2, 1, -4)$ is

(A) $\dfrac{x-2}{-3} = \dfrac{y+1}{1} = \dfrac{z-4}{-1}$

(B) $\dfrac{x-1}{3} = \dfrac{y}{1} = \dfrac{z+3}{1}$

(C) $\dfrac{x-1/2}{-3} = \dfrac{y-1}{1} = \dfrac{z+4}{-1}$

(D) $\dfrac{x-1}{-3} = \dfrac{y}{1} = \dfrac{z+3}{-1}$

Q.12 A point moves so that the sum of the squares of its distances from the six faces of a cube given by $x = \pm 1$, $y = \pm 1$, $z = \pm 1$ is 10 units. The locus of the point is

(A) $x^2 + y + z^2 = 1$

(B) $x^2 + y^2 + z^2 = 2$

(C) $x + y + z = 1$

(D) $x + y + z = 2$

Q.13 The points $(0, -1, -1)$, $(-4, 4, 4)$, $(4, 5, 1)$ and $(3, 9, 4)$ are

(A) Collinear

(B) Coplanar

(C) Forming a square

(D) None of these

Q.14 The equation of the plane containing the line $\dfrac{x-\alpha}{\ell} = \dfrac{y-\beta}{m} = \dfrac{z-\gamma}{n}$ is $a(x-\alpha) + b(y-\beta) + c(z-\gamma) = 0$, where $a\ell + bm + cn$ is equal to

(A) 1

(B) –1

(C) 2

(D) 0

Q.15 The reflection of the plane $2x + 3y + 4z - 3 = 0$ in the plane $x - y + z - 3 = 0$ is the plane

(a) $4x - 3y + 2z - 15 = 0$

(b) $x - 3y + 2z - 15 = 0$

(c) $4x + 3y - 2z + 15 = 0$

(d) None of these

Previous Years' Questions

Q.1 The value of k such that $\dfrac{x-4}{1} = \dfrac{y-2}{1} = \dfrac{z-k}{2}$ lies in the plane
$2x - 4y + z = 7$, is **(2003)**

(A) 7   (B) –7   (C) No real value   (D) 4

Q.2 If the lines $\vec{r} = \vec{a}_2 + \mu\vec{b}$ and $\dfrac{x-3}{1} = \dfrac{y-k}{2} = \dfrac{z}{1}$
intersect, then the value of k is **(2004)**

(A) $\vec{a}_2$ (B) $\dfrac{9}{2}$ (C) $-\dfrac{2}{9}$ (D) $-\dfrac{3}{2}$

Q.3 A variable plane $\dfrac{x}{a} + \dfrac{y}{b} + \dfrac{z}{c} = 1$ at a unit distance from origin cuts the coordinate axes at A, B and C. Centroid (x, y, z) satisfies the equation $\dfrac{1}{x^2} + \dfrac{1}{y^2} + \dfrac{1}{z^2} = K$.
The value of K is **(2005)**

(A) 9   (B) 3   (C) 1/9   (D) 1/3

Fill in the Blanks for Q.4 and Q.5

Q.4 The area of the triangle whose vertices are A(1, –1, 2), B(2, 1, –1), C(3, –1, 2) is ... **(1983)**

Q.5 The unit vector perpendicular to the plane determined by P(1, –1, 2), Q(2, 0, –1) and R(0, 2, 1) is.............. **(1983)**

Q.6 A plane is parallel to two lines whose direction ratios are (1, 0, –1) and (–1, 1, 0) and it contains the point (1, 1, 1). If it cuts coordinate axes at A, B, C. Then find the volume of the tetrahedron OABC. **(2004)**

Q.7 Find the equation of the plane containing the line $2x - y + z - 3 = 0$, $3x + y + z = 5$ and at a distance of $\dfrac{1}{\sqrt{6}}$ from the point (2, 1, –1). **(2005)**

Q.8 If the line $\dfrac{x-3}{2} = \dfrac{y+2}{-1} = \dfrac{z+4}{3}$ lies in the plane,
$\ell x + my - z = 9$, then $\ell^2 + m^2$ is equal to: **(2016)**

(A) 18   (B) 5   (C) 2   (D) 26

Q.9 The distance of the point $(1, -5, 9)$ from the plane $x - y + z = 5$ measured along the line $x = y = z$ is **(2016)**

(A) $10\sqrt{3}$ (B) $\dfrac{10}{\sqrt{3}}$ (C) $\dfrac{20}{3}$ (D) $3\sqrt{10}$

Q.10 The distance of the point (1, 0, 2) from the point of intersection of the line $\dfrac{x-2}{3} = \dfrac{y+1}{4} = \dfrac{z-2}{12}$ and the plane $x - y + z = 16$, is: **(2015)**

(A) 8   (B) $3\sqrt{21}$ (C) 13   (D) $2\sqrt{14}$

Q.11 The equation of the plane containing the line $2x - 5y + z + 3; x + y + 4z = 5$ and parallel to the plane $x + 3y + 6z = 1$ is **(2015)**

(A) $x + 3y + 6z = -7$ (B) $x + 3y + 6z = 7$

(C) $2x + 6y + 12z = -13$ (D) $2x + 6y + 12z = 13$

Q.12 The number of common tangents to the circles **(2015)**

(A) Meats the curve again in the second in the second quadrant

(B) Meats the curve again in the third quadrant

(C) Meets the curve again in the fourth quadrant

(D) Does not meet the curve again

Q.13 The image of the line $\dfrac{x-1}{3} = \dfrac{y-3}{1} = \dfrac{z-4}{-5}$ in the plane $2x - y + z + 3$ is the **(2014)**

(A) $\dfrac{x+3}{3} = \dfrac{y-5}{1} = \dfrac{z-2}{-5}$ (B) $\dfrac{x+3}{-3} = \dfrac{y-5}{-1} = \dfrac{z-2}{5}$

(C) $\dfrac{x+3}{3} = \dfrac{y-5}{1} = \dfrac{z-2}{-5}$ (D) $\dfrac{x-3}{-3} = \dfrac{y-5}{-1} = \dfrac{z-2}{-5}$

Q.14 The angle between the lines whose direction cosines satisfy the equations $l + m + n = 0$ and $l^2 = m^2 + n^2$ is **(2014)**

(A) $\dfrac{\pi}{3}$ (B) $\dfrac{\pi}{4}$ (C) $\dfrac{\pi}{6}$ (D) $\dfrac{\pi}{2}$

Q.15 If the lines
$\dfrac{x-2}{1} = \dfrac{y-3}{1} = \dfrac{z-4}{-k}$ and $\dfrac{x-1}{k} = \dfrac{y-4}{2} = \dfrac{z-5}{1}$
are coplanar, then k have **(2013)**

(A) Exactly one value   (B) Exactly two value

(C) Exactly three values   (D) Any value

Q.16 An equation of a plane parallel to the plane $x - 2y + 2z - 5 = 0$ and at a unit distance from the origin is **(2012)**

(A) $x - 2y + 2z - 3 = 0$ (B) $x - 2y + 2z + 1 = 0$

(C) $x - 2y + 2z - 1 = 0$ (D) $x - 2y + 2z + 5 = 0$

Q.17 If the angle between the line $x = \dfrac{y-1}{2} = \dfrac{z-3}{\lambda}$ and the plane $x + 2y + 3z = 4$ is $\cos^{-1}\left(\sqrt{\dfrac{5}{14}}\right)$, then λ equals

(2011)

(A) $\dfrac{3}{2}$ (B) $\dfrac{2}{5}$ (C) $\dfrac{5}{3}$ (D) $\dfrac{2}{3}$

Q.18 Statement–I: The point $A(1,0,7)$ is the mirror image of the point $B(1,6,3)$ in the line $\dfrac{x}{1} = \dfrac{y-1}{2} = \dfrac{z-2}{3}$

Statement-II: The line: $\dfrac{x}{1} = \dfrac{y-1}{2} = \dfrac{z-2}{3}$ bisects the line segment joining $A(1,0,7)$ and $B(1,6,3)$ *(2011)*

(A) Statement-I is true, statement-II is true; statement-II is not a correct explanation for statement-I

(B) Statement-I is true, statement-II is false.

(C) Statement-I is false, statement-II is true

(D) Statement-I is true, statement-II is true, statement-II is a correct explanation for statement-I

JEE Advanced/Boards

Exercise 1

Q.1 Points X and Y are taken on the sides QR and RS respectively, of parallelogram PQRS, so that $QX = 4\overrightarrow{XR}$ and $\overrightarrow{RY} = 4\overrightarrow{YS}$. The line XY cuts the line PR at Z. prove that $\overrightarrow{PZ} = \left(\dfrac{21}{25}\right)\overrightarrow{PR}$.

Q.2 Given three points on the xy plane on O(0, 0), A(1, 0) and B(–1, 0). Point P is moving on the plane satisfying the condition $(\overrightarrow{PA}\cdot\overrightarrow{PB}) + 3(\overrightarrow{OA}\cdot\overrightarrow{OB}) = 0$. If the maximum and minimum values of $|\overrightarrow{PA}||\overrightarrow{PB}|$ are M and m, respectively then find the value of $M^2 + m^2$.

Instruction for questions 3 to 6.

Suppose the three vectors, $\vec{a}, \vec{b}, \vec{c}$ on a plane satisfy the condition that

$|\vec{a}| = |\vec{b}| = |\vec{c}| = |\vec{a}+\vec{b}| = 1; c$ is perpendicular to a and $\vec{b}\cdot\vec{c} > 0$, then

Q.3 Find the angle formed by $2\vec{a}+\vec{b}$ and $\vec{b}$.

Q.4 If the vector c is expressed as a linear combination $\lambda\vec{a} + \mu\vec{b}$ then find the ordered pair

$\dfrac{\ell_1 - \ell_2}{2\sin(\theta/2)}, \dfrac{m_1 - m_2}{2\sin(\theta/2)}$ and $\dfrac{n_1 - n_2}{2\sin(\theta/2)}$.

Q.5 For real number x, y the vector $\vec{p} = x\vec{a} + y\vec{c}$ satisfies the condition $0 \le \vec{p}\cdot\vec{a} \le 1$ and $0 \le \vec{p}\cdot\vec{b} \le 1$. Find the maximum value of $\vec{p}\cdot\vec{c}$

Q.6 For the maximum value of x and y, find the linear combination of $\vec{p}$ in terms of $\vec{a}$ and $\vec{b}$.

Q.7 If O be the origin and the coordinates of P be (1, 2, -3), then find the equation of the plane passing through P and perpendicular to OP.

Q.8 Given non zero number x_1, x_2, x_3; y_1, y_2, y_3 and z_1, z_2 and z_3 (i) Can the given numbers satisfy

$$\begin{vmatrix} x_1 & x_2 & x_3 \\ y_1 & y_2 & y_3 \\ z_1 & z_2 & z_3 \end{vmatrix} = 0 \text{ and } \begin{cases} x_1x_2 + y_1y_2 + z_1z_2 = 0 \\ x_2x_3 + y_2y_3 + z_2z_3 = 0 \\ x_3x_1 + y_3y_1 + z_3z_1 = 0 \end{cases}$$

(ii) If $x_1 > 0$ and $y_1 < 0$ for all I = 1, 2, 3 and $P = (x_1, x_2, x_3)$; $Q(y_1, y_2, y_3)$ and O(0, 0, 0) can the triangle POQ be a right angled triangle?

Q.9 ABCD is a tetrahedron with pv's of its angular points as $A(-5, 22, 5); B(1, 2, 3); C(4, 3, 2)$ and $D(-1, 2, -3)$. If the area of the triangle AEF where the quadrilaterals ABDE and ABCF are parallelogram is S, then find the value of S.

Q.10 If x, y are two non-zero and non-collinear vectors satisfying $[(a-2)\alpha^2 + (b-3)\alpha + c]x + [(a-2)\beta^2 + (b-3)\beta + c]y$

$+[(a-2)\gamma^2 + (b-3)\gamma + c](x \times y) = 0$

where α, β, γ are three distinct real numbers, then find the value of $(a^2 + b^2 + c^2)$.

Q.11 Find the distance of the point (-1, -5, -10) from the point of intersection of the line $\vec{r} = 2\hat{i} - \hat{j} + 2\hat{k} + \lambda(3\hat{i} + 4\hat{j} + 2\hat{k})$ and the plane $\vec{r}.(\hat{i} - \hat{j} + \hat{k}) = 5$

Q.12 Find the equations of the straight line passing through the point (1, 2, 3) to intersect the straight line $x + 1 = 2(y-2) = x + 4$ and parallel to the plane $x + 5y + 4z = 0$.

Q.13 Find the equations of the two lines through the origin which intersect the line $\dfrac{x-3}{2} = \dfrac{y-3}{1} = \dfrac{z}{1}$ at an angle of $\dfrac{\pi}{3}$.

Exercise 2

Single Correct Choice Type

Q.1 If P(2, 3, −6) and Q(3, −4, 5) are two points, the direction cosines of line PQ are

(A) $-\dfrac{1}{\sqrt{171}}, -\dfrac{7}{\sqrt{171}}, \dfrac{11}{\sqrt{171}}$ (B) $\dfrac{1}{\sqrt{171}}, -\dfrac{7}{\sqrt{171}}, \dfrac{11}{\sqrt{171}}$

(C) $\dfrac{1}{\sqrt{171}}, \dfrac{7}{\sqrt{171}}, -\dfrac{11}{\sqrt{171}}$ (D) $-\dfrac{1}{\sqrt{171}}, -\dfrac{7}{\sqrt{171}}, \dfrac{11}{\sqrt{171}}$

Q.2 The ratio in which yz-plane divide the line joining the points A(3, 1, −5) and B(1, 4, −6) is

(A) −3 : 1   (B) 3 : 1   (C) −1 : 3   (D) 1 : 3

Q.3 The value of λ for which the lines $3x + 2y + z + 5 = 0 = x + y - 2z - 3$ and $2x - y - \lambda z = 0 = 7x + 10y - 8z$ are perpendicular to each other is

(A) −1   (B) −2   (C) 2   (D) 1

Q.4 The ratio in which yz-plane divides the line joining (2, 4, 5) and (3, 5, 7)

(A) −2 : 3   (B) 2 : 3   (C) 3 : 2   (D) −3 : 2

Q.5 A line makes angle $\alpha, \beta, \gamma, \delta$ with the four diagonals of a cube then $\cos^2\alpha + \cos^2\beta + \cos^2\gamma + \cos^2\delta$ is equal to

(A) 1   (B) 4/3   (C) ¾   (D) 4/5

Q.6 A variable plane passes through a fixed point (a, b, c) and meets the coordinate axes in A, B, C. The locus of the point common to plane through A, B, C parallel to coordinate planes is

(A) ayz + bzx + cxy = xyz   (B) axy + byz + czx = xyz

(C) axy + byz + czx = abc   (D) bcx + acy + abz = abc

Q.7 The equation of the plane bisecting the acute angle between the planes

$2x - y + 2z + 3 = 0$ and $3x - 2y + 6z + 8 = 0$

(A) 23x − 13y + 32z + 45 = 0

(B) 5x − y − 4z = 3

(C) 5x − y − 4z + 45 = 0

(D) 23x − 13y + 32z + 3 = 0

Q.8 The shortest distance between the two straight lines $\dfrac{x-4/3}{2} = \dfrac{y+6/5}{3} = \dfrac{z-3/2}{4}$ and

$\dfrac{5y+6}{8} = \dfrac{2z-3}{9} = \dfrac{3x-4}{5}$ is

(A) $\sqrt{29}$ (B) 3   (C) 0   (D) $6\sqrt{10}$

Q.9 The equation of the straight line through the origin parallel to the line $(b + c)x + (c + a)y + (a + b)z = k$
$= (b - c)x + (c - a)y + (a - b)z$ is

(A) $\dfrac{x}{b^2 - c^2} = \dfrac{y}{c^2 - a^2} = \dfrac{z}{a^2 - b^2}$

(B) $\dfrac{x}{b} = \dfrac{y}{c} = \dfrac{z}{a}$

(C) $\dfrac{x}{a^2 - bc} = \dfrac{y}{b^2 - ca} = \dfrac{z}{c^2 - ab}$

(D) None of these

Assertion Reasoning Type

Q.10 Consider the following statements

Assertion: The plane y + z + 1 = 0 is parallel to x-axis.

Reason: Normal to the plane is parallel to x-axis.

(A) Both A and R are true and R is the correct

(B) Both A and R are true and R is not a correct explanation of A

(C) A is true but R is false

(D) A is false but R is true

Previous Years' Questions

Q.1 A plane passes through $(1, -2, 1)$ and is perpendicular to two planes $2x - 2y + z = 0$ and $x - y + 2z = 4$, then the distance of the plane from the point $(1, 2, 2)$ is *(2006)*

(A) 0   (B) 1   (C) $\sqrt{2}$ (D) $2\sqrt{2}$

Q.2 Let $P(3, 2, 6)$ be a point in space and Q be a point on the line $\vec{r} = (\hat{i} - \hat{j} + 2\hat{k}) + \mu(-3\hat{i} + \hat{j} + 5\hat{k})$. Then the value of μ for which the vector $\overrightarrow{PQ}$ is parallel to the plane $x - 4y + 3z = 1$ is *(2009)*

(A) $\dfrac{1}{4}$ (B) $-\dfrac{1}{4}$ (C) $\dfrac{1}{8}$ (D) $-\dfrac{1}{8}$

Q.3 A line with positive direction cosines passes through the point $P(2, -1, 2)$ and makes equal angles with the coordinate axes. The line meets the plane $2x + y + z = 9$ at point Q. The length of the line segment PQ equals *(2009)*

(A) 1   (B) $\sqrt{2}$ (C) $\sqrt{3}$ (D) 2

For the following question, choose the correct answer from the codes (A), (B), (C) and (D) defined as follows.

(A) Statement-I is true, statement-II is also true; statement-II is the correct explanation of statement-I

(B) Statement-I is true, statement-II is also true; statement-II is not the correct explanation of statement-I.

(C) Statement-I is true; statement-II is false.

(D) Statement-I is false; statement-II is true

Q.4 Consider the planes $3x - 6y - 2z = 15$ and $2x + y -2z = 5$.

Statement-I: The parametric equations of the line of intersection of the given planes are $x = 3 + 14t$, $y = 1 + 2t$, $z = 15t$.

Statement-II: The vectors $14\hat{i} + 2\hat{j} + 15\hat{k}$ is parallel to the line of intersection of the given planes. *(2007)*

Q.5 Consider three planes

$AB : \dfrac{x-3}{3} = \dfrac{y-8}{-1} = \dfrac{z-3}{1} = \lambda$

$CD : \dfrac{x+3}{-3} = \dfrac{y+7}{2} = \dfrac{z-6}{4} = \mu$ and

$L \equiv (3\lambda + 3, -\lambda + 8, \lambda + 3)$

Let L_1, L_2, L_3 be the lines of intersection of the planes P_2 and P_3, P_3 and P_1, P_1 and P_2, respectively.

Statement-I : At least two of the lines L_1, L_2 and L_3 are non-parallel.

Statement-II : The three planes do not have a common point *(2008)*

Paragraph for Q.6 to Q.8

Read the following passage and answer the questions. Consider the lines

$L_1 : \dfrac{x+1}{3} = \dfrac{y+2}{1} = \dfrac{z+1}{2}$, $L_2 : \dfrac{x-2}{1} = \dfrac{y+2}{2} = \dfrac{z-3}{3}$

(2008)

Q.6 The unit vector perpendicular to both L_1 and L_2 is

(A) $\dfrac{-\hat{i} + 7\hat{j} + 7\hat{k}}{\sqrt{99}}$ (B) $\dfrac{-\hat{i} - 7\hat{j} + 5\hat{k}}{5\sqrt{3}}$

(C) $\dfrac{-\hat{i} + 7\hat{j} + 5\hat{k}}{5\sqrt{3}}$ (D) $\dfrac{7\hat{i} - 7\hat{j} - \hat{k}}{\sqrt{99}}$

Q.7 The shortest distance between L_1 and L_2 is

(A) 0   (B) $\dfrac{17}{\sqrt{3}}$ (C) $\dfrac{41}{5\sqrt{3}}$ (D) $\dfrac{17}{5\sqrt{3}}$

Q.8 The distance of the point $(1, 1, 1)$ from the plane passing through the point $(-1, -2, -1)$ and whose normal is perpendicular to both the lines L_1 and L_2 is

(A) $\dfrac{2}{\sqrt{75}}$ (B) $\dfrac{7}{\sqrt{75}}$ (C) $\dfrac{13}{\sqrt{75}}$ (D) $\dfrac{23}{\sqrt{75}}$

Match the Columns

Match the condition/expression in column I with statement in column II.

Q.9 Consider the following linear equations $ax + by + cz = 0$, $bx + cy + az = 0$, $cx + ay + bz = 0$ *(2007)*

Column I	Column II
(A) $a + b + c \neq 0$ and $a^2 + b^2 + c^2 = ab + bc + ca$	(p) The equations represent planes meeting only at a single point

(B) $a+b+c=0$ and $a^2+b^2+c^2 \neq ab+bc+ca$	(q) The equation represent the line $x=y=z$
(C) $a+b+c \neq 0$ and $a^2+b^2+c^2 \neq ab+bc+ca$	(r) The equations represent identical planes
(D) $a+b+c=0$ and $a^2+b^2+c^2 \, ab+bc+ca$	(s) The equations represent the whole of the three dimensional space

Q.10 (i) Find the equation of the plane passing through the points (2, 1, 0), (5, 0, 1) and (4, 1, 1).

(ii) If P is the point (2, 1, 6), then the point Q such that PQ is perpendicular to the plane in (a) and the mid point of PQ lies on it. **(2003)**

Q.11 T is a parallelepiped in which A, B, C and D are vertices of one face and the face just above it has corresponding vertices A', B', C', D', T is now compressed to S with face ABCD remaining same and A', B', C', D' shifted to A", B", C", D" in S. the volume of parallelepiped S is reduced to 90% of T. Prove that locus of A" is a plane. **(2003)**

Q.12 Consider a pyramid OPQRS located in the first octant $(x \geq 0, y \geq 0, z \geq 0)$ with O as origin, and OP and OR along the x-axis and the y-axis, respectively. The base OPQR of the pyramid is a square with $OP = 3$. The point S is directly above the mid-point T of diagonal OQ such that $TS = 3$. Then **(2016)**

(A) The acute angle between OQ and OS is $\dfrac{\pi}{3}$

(B) The equation of the plane containing the triangle OQS is $x - y = 0$

(C) The length of the perpendicular from p to the plane containing the triangle OQS is $\dfrac{3}{\sqrt{2}}$

(D) The perpendicular distance from O to the straight line containing RS is $\sqrt{\dfrac{15}{2}}$

Q.13 Let P be the image of the point $(3,1,7)$ with respect to the plane $x - y + x = 3$. Then equation of the plane passing through P and containing the straight line $\dfrac{x}{1} = \dfrac{y}{2} = \dfrac{z}{1}$ is **(2016)**

(A) $x + y - 3z = 0$ (B) $3x + z = 0$

(C) $x - 4y + 7z = 0$ (D) $2x - y = 0$

Q.14 In R^3, consider the planes $P_1 : y=0$ and $P_2 : x+z= 1$. Let P_3 be a plane, different from P_1 and P_2, which passes through the intersection of P_1 and P_2. If the distance of the distance of the point (0, 1, 0) from P_3 is 1 and the distance a point (α, β, γ) from p_3 is 2, then which of the following relations is (are) true? **(2015)**

(A) $2\alpha + \beta + 2\gamma + 2 = 0$ (B) $2\alpha - \beta + 2\gamma + 4 = 0$

(C) $2\alpha + \beta - 2\gamma - 10 = 0$ (D) $2\alpha - \beta + 2\gamma - 8 = 0$

Q.15 In R^3 let L be a straight line passing through the origin. Suppose that all the points on L are at a constant distance from the two planes $P_1 : x + 2y - z + 1 = 0$ and $P_2 : 2x - y + z - 1 = 0$. Let M be the locus of the feet of the perpendiculars drawn from the points on L to the plane P_1. Which of the following points lie (s) on M? **(2015)**

(A) $\left(0, -\dfrac{5}{6}, -\dfrac{2}{3}\right)$ (B) $\left(-\dfrac{1}{6}, -\dfrac{1}{3}, \dfrac{1}{6}\right)$

(C) $\left(-\dfrac{5}{6}, 0, \dfrac{1}{6}\right)$ (D) $\left(-\dfrac{1}{3}, 0, \dfrac{2}{3}\right)$

Q.16 From a point $p(\lambda, \lambda, \lambda)$ perpendiculars PQ and PR are drawn respectively on the lines $y = x, z = 1$ and $y = -x, z = -1$. If p is such that $\angle QPR$ is a right angle, then the possible value(s) of λ is(are) **(2014)**

(A) $\sqrt{2}$ (B) 1   (C) -1   (D) $-\sqrt{2}$

Q.17 Perpendiculars are drawn from points on the line $\dfrac{x+2}{2} = \dfrac{y+1}{-1} = \dfrac{z}{3}$ to the plane $x + y + z = 3$. The feet of perpendiculars lie on the line **(2013)**

(A) $\dfrac{x}{5} = \dfrac{y-1}{8} = \dfrac{z-2}{-13}$ (B) $\dfrac{x}{2} = \dfrac{y-1}{3} = \dfrac{z-2}{-5}$

(C) $\dfrac{x}{4} = \dfrac{y-1}{3} = \dfrac{z-2}{-7}$ (D) $\dfrac{x}{2} = \dfrac{y-1}{-7} = \dfrac{z-2}{5}$

Q.18 Two lines $L_1 : x = 5, \dfrac{y}{3-\alpha} = \dfrac{z}{-2}$ and

$L_2 : x = \alpha, \dfrac{y}{-1} = \dfrac{z}{2-\alpha}$ are coplanar.

The α can take value(s) **(2013)**

(A) 1   (B) 2   (C) 3   (D) 4

Q.19 Consider the lines $L_1 : \dfrac{x-1}{2} = \dfrac{y}{-1} = \dfrac{z+3}{1}$,

$L_2 : \dfrac{x-4}{1} = \dfrac{y+3}{1} = \dfrac{z+3}{2}$ and the planes

$P_1 : 7x + y + 2z = 3, P_2 : 3x + 5y - 6z = 4.$

Let $ax + by + cz = d$ be the equation of the plane passing through the point of intersection of lines L_1 and L_2 and perpendicular to planes P_1 and P_2.

Match List I with List II and select the correct answer using the code given below the list: **(2013)**

List I	List II
p. a=	1. 13
q. b=	2. -3
r. c=	3. 1
s. d=	4. -2

Codes:

	p	q	r	s
(A)	3	2	4	1
(B)	1	3	4	2
(C)	3	2	1	4
(D)	2	4	1	3

Q.20 The point P is the intersection of the straight line joining the point $Q(2,3,5)$ and $R(1,-1,4)$ with the plane $5x - 4y - z = 1$. If S is the foot of the perpendicular drawn from the point $T(2,1,4)$ to QR, then the length of the line segment PS is **(2012)**

(A) $\dfrac{1}{\sqrt{2}}$ (B) $\sqrt{2}$ (C) 2   (D) $2\sqrt{2}$

Q.21 The equation of a plane passing through the line of intersection of the planes $x + 2y + 3z = 2$ and $x - y + z = 3$ and at a distance $\dfrac{2}{\sqrt{3}}$ from the point $(3,1,-1)$ is **(2012)**

(A) $5x - 11y + z = 17$ (B) $\sqrt{2}x + y = 3\sqrt{2} - 1$

(C) $x + y + z = \sqrt{3}$ (D) $x - \sqrt{2}y = 1 - \sqrt{2}$

Q.22 If $f(x) = \displaystyle\int_0^x e^{t^2}(t-2)(t-3)dt$ for all $x \in (0, \infty)$ then **(2012)**

(A) f has a local maximum at $x = 2$

(B) f is decreasing on (2, 3)

(C) There exists some $c \in (0, \infty)$ such that $f'(c) = 0$

(D) f has local minimum at $x = 3$

Q.23 If the distance between the plane $Ax - 2y + z = d$ and the plane containing the lines

$\dfrac{x-1}{2} = \dfrac{y-2}{3} = \dfrac{z-3}{4}$ and $\dfrac{x-2}{3} = \dfrac{y-3}{4} = \dfrac{z-4}{5}$ is $\sqrt{6}$,

then $|d|$ is

(2010)

Q.24 If the distance of the point $P(1,-2,1)$ from the plane $x + 2y - 2z = \alpha$, where $\alpha > 0$, is 5, then the foot of the perpendicular from P to the plane is **(2010)**

(A) $\left(\dfrac{8}{3}, \dfrac{4}{3}, -\dfrac{7}{3}\right)$ (B) $\left(\dfrac{4}{3}, -\dfrac{4}{3}, \dfrac{1}{3}\right)$

(C) $\left(\dfrac{1}{3}, \dfrac{2}{3}, \dfrac{10}{3}\right)$ (D) $\left(\dfrac{2}{3}, -\dfrac{1}{3}, \dfrac{5}{2}\right)$

Q.25 Two adjacent sides of a parallelogram ABCD are given by $\overrightarrow{AB} = 2\hat{i} + 10\hat{j} + 11\hat{k}$ and $\overrightarrow{AD} = -\hat{i} + 2\hat{j} + 2\hat{k}$ The side AD is rotated by an acute angle α in the plane of the parallelogram so that AD becomes AD'. If AD' makes a right angle with the side AB, then the cosine of the angle a is given by **(2010)**

(A) $\dfrac{8}{9}$ (B) $\dfrac{\sqrt{17}}{9}$ (C) $\dfrac{1}{9}$ (D) $\dfrac{4\sqrt{5}}{9}$

Important Questions

JEE Main/Boards

Exercise 1

Q.5	Q.10	Q.23
Q.29	Q.36	Q.40
Q.42	Q.47	Q.49
Q.50		

Exercise 2

Q.2	Q.8	Q.12
Q.13	Q.14	

Previous Years' Questions

Q.3	Q.6

JEE Advanced/Boards

Exercise 1

Q.2	Q.5	Q.8
Q.10	Q.13	

Exercise 2

Q.2	Q.5	Q.6
Q.7	Q.9	

Previous Years' Questions

Q.3	Q.5	Q.6
Q.9	Q.11	

Answer Key

JEE Main/Boards

Exercise 1

Q.1 $\langle 3, -2, 6 \rangle$

Q.2 $\langle 1, 3, -3 \rangle$

Q.3 $\cos^{-1}\left(\dfrac{7}{\sqrt{70}}\right)$

Q.4 $r = (-\hat{i} + \hat{j} + \hat{k}) + \lambda(3\hat{i} - \hat{j} - 3\hat{k})$

Q.5 $r = (-3\hat{i} + \hat{j} - \hat{k}) + \lambda(2\hat{i} + 4\hat{j} + 5\hat{k})$

Q.6 $\dfrac{x-3}{-2} = \dfrac{y-2}{1} = \dfrac{z+5}{3}$

Q.7 -3

Q.8 $3x + 2y + 5z = 7$

Q.9 $r(3\hat{i} - \hat{j} - 4\hat{k}) + 7 = 0$

Q.10 $3\hat{i} - 7\hat{j}$

Q.11 $\langle 7, 1, -2 \rangle$

Q.12 $\sin^{-1}\left(\dfrac{-2}{\sqrt{90}\sqrt{13}}\right)$

Q.13 $\dfrac{7}{\sqrt{11}}$

Q.14 $\dfrac{1}{\sqrt{3}}$

Q.15 $4, -6, 3$

Q.16 $\left(\dfrac{1}{\sqrt{14}}, \dfrac{3}{\sqrt{14}}, \dfrac{-2}{\sqrt{14}}\right)$

Q.17 $\langle 0, 1, 0 \rangle$

Q.18 $\langle 2, 3, -10 \rangle$

Q.19 $\cos^{-1}\left(\dfrac{11}{21}\right)$

Q.20 $\sin^{-1}\left(\dfrac{-1}{7\sqrt{3}}\right)$

Q.21 $\left(\dfrac{1}{\sqrt{14}}, \dfrac{2}{\sqrt{14}}, \dfrac{3}{\sqrt{14}}\right), \left(\dfrac{-1}{\sqrt{6}}, \dfrac{1}{\sqrt{6}}, \dfrac{2}{\sqrt{6}}\right)$

Q.23 $(3, 4, 5)$

Q.26 $\dfrac{x-3}{-2} = \dfrac{y}{1} = \dfrac{z-1}{3}$

Q.27 $x - 2y - 2x - 3 = 0; \ 2x + y - 2z + 3 = 0$

Q.28 $2x + 2y - 3z + 3 = 0$

Q.29 13 units

Q.30 $9x - 8y + z + 11 = 0$

Q.33 $5y - 5z - 6 = 0, \dfrac{3\sqrt{2}}{5}$

Q.34 $2x + 5y + 7z = 78$

Q.35 $5x + 9y + 11z - 8 = 0$

Q.36 $\dfrac{\pi}{4}$

Q.37 $\dfrac{x-5}{3} = \dfrac{y+7}{1} = \dfrac{z+3}{9}$

Q.39 $\left(-\dfrac{1}{12}, \dfrac{25}{12}, -\dfrac{1}{6}\right), \dfrac{13\sqrt{6}}{12}$

Q.40 $r(4\hat{i} - \hat{j} - 2\hat{k}) = 6$

Q.41 $\dfrac{25\sqrt{14}}{42}$

Q.43 $x + 2y - 5z = 0$

Q.44 $(0, 2, -2)$

Q.45 $(ax + by + cz + d) - \dfrac{a l + bm + cn}{(a'1 + b'm + c'm)}(a'x + b'y + c'z + a') = 0$

Q.46 $7y + 5z = 35$

Q.48 $\lambda = 3, -6$

Q.49 $r(3\hat{i} + 5\hat{j} - 6\hat{k}) - 7\sqrt{70} = 0$

Q.50 $r(5\hat{i} + \hat{j} - 6\hat{k}) = 4$

Exercise 2

Single Correct Choice Type

Q.1 C	**Q.2** A	**Q.3** D	**Q.4** C	**Q.5** A	**Q.6** B
Q.7 B	**Q.8** D	**Q.9** C	**Q.10** A	**Q.11** D	**Q.12** B
Q.13 B	**Q.14** D	**Q.15** A			

Previous Years' Questions

Q.1 A   **Q.2** B   **Q.3** A   **Q.4** $\sqrt{13}$ sq. units   **Q.5** $\pm\dfrac{(2\hat{i} + \hat{j} + \hat{k})}{\sqrt{6}}$ **Q.6** $\dfrac{9}{2}$ cu unit

Q.7 $2x - y + z - 3 = 0$ and $62x + 29y + 19z - 105 = 0$ **Q.8** C   **Q.9** A   **Q.10** C

Q.11 B   **Q.12** B   **Q.13** A   **Q.14** A   **Q.15** B   **Q.16** A

Q.17 D   **Q.18** B

JEE Advanced/Boards

Exercise 1

Q.2 34   **Q.3** $\dfrac{\pi}{2}$ **Q.4** $\left(\dfrac{1}{\sqrt{3}}, \dfrac{2}{\sqrt{3}}\right)$ **Q.5** $\sqrt{3}$ **Q.6** $p = 2(\vec{a} + \vec{b})$

Q.7 $x + 2y - 3z = 14$ **Q.8** No, No   **Q.9** $\sqrt{110}$ **Q.10** 13   **Q.11** 13

Q.12 $\dfrac{x-1}{2} = \dfrac{y-2}{2} = \dfrac{z-3}{-3}$ **Q.13** $\dfrac{x}{1} = \dfrac{y}{2} = \dfrac{z}{-1}$ or $\dfrac{x}{-1} = \dfrac{y}{1} = \dfrac{z}{-2}$

Exercise 2

Single Correct Choice Type

Q.1 B **Q.2** A **Q.3** D **Q.4** A **Q.5** B **Q.6** A

Q.7 A **Q.8** C **Q.9** C

Assertion Reasoning Type

Q.10 C

Previous Years' Question

Q.1 D **Q.2** A **Q.3** C **Q.4** D **Q.5** D **Q.6** B

Q.7 D **Q.8** C

Q.9 A → r; B → q; C → p; D → s **Q.10** (a) x + y − 2z = 3 (b) Q(6, 5, −2) **Q.12** B, C, D

Q.13 C **Q.14** B, D **Q.15** A, B **Q.16** C **Q.17** D **Q.18** A, D

Q.19 A **Q.20** A **Q.21** A **Q.22** B, C **Q.23** 6 **Q.24** A

Q.25 B

Solutions

JEE Main/Boards

Exercise 1

Sol 1: $l = \dfrac{3}{7}$ $m = \dfrac{-2}{7}$ $n = \dfrac{6}{7}$

Direction ratios are <3, −2, 6>

Sol 2: [2, 1, 0] & [1, −2, 3]

Direction ratios = 2 − 1, 1 + 2, 0 − 3 = <1, 3, −3>

Sol 3: $\dfrac{x}{1} = \dfrac{y}{2} = \dfrac{z}{0}$ and $\dfrac{x-1}{3} = \dfrac{y+5}{2} = \dfrac{z-3}{1}$

<1, 2, 0> and <3, 2, 1>

$\cos\theta = \dfrac{1.3 + 2.2 + 0.1}{\sqrt{5}\sqrt{14}} = \dfrac{7}{\sqrt{70}} \Rightarrow \theta = \cos^{-1}\left(\dfrac{7}{\sqrt{70}}\right)$

Sol 4: $\dfrac{x+1}{3} = \dfrac{y-1}{-1} = \dfrac{z-1}{-3} = t$

$r = \hat{i} + \hat{j} + \hat{k} + t\left(3\hat{i} - \hat{j} - 3\hat{k}\right)$

Sol 5: $r = -3\hat{i} + \hat{j} - \hat{k} + \lambda(2\hat{i} + 4\hat{j} + 5\hat{k})$

Sol 6: $x\hat{i} + y\hat{j} + z\hat{k} = (3 - 2\lambda)\hat{i} + (2 + \lambda)\hat{j} + (-5 + 3\lambda)\hat{k}$

$\dfrac{x-3}{-2} = \dfrac{y-2}{1} = \dfrac{z+5}{3}$

Sol 7: $\cos\theta = 0 = -3.1 + 3.5 + 2p.2$

$\Rightarrow 12 + 4p = 0 \Rightarrow p = -3$

Sol 8: $(xi + yj + zk) \cdot (3i + 2j + 5k) = 7$

$3x + 2y + 5z = 7$

Sol 9: $3x - y - 4z = -7$; $r(3i - j - 4k) = -7$

$r(3\hat{i} - \hat{j} - 4\hat{k}) + 7 = 0$

Sol 10: $3x - 7y = -5$

Direction ratios of normal to plane are (3, −7, 0) the vector along that normal is $3\hat{i} - 7\hat{j}$.

Sol 11: $7x + y - 2z = 1$

Direction ratios of vector normal to the plane are

$7i + j - 2k = 0$

$(7, 1, -2)$

Sol 12: Direction ratios of line $<4, -5, 7>$

Direction ratio of line perpendicular to plane $<3, 0, -2>$

$$\sin\theta = \frac{4 \times 3 + (-5) \times (0) + 7 \times (-2)}{\sqrt{16 + 25 + 49}\sqrt{9 + 0 + 4}} = \frac{-2}{\sqrt{90}\sqrt{13}}$$

Sol 13: $x + y + 3z + 7 = 0$

Distance from origin is $\dfrac{0 + 0 + 3(0) + 7}{\sqrt{1 + 1 + 9}} = \dfrac{7}{\sqrt{11}}$

Sol 14: $3x - 3y + 3z = 0$

Distance from $(1, 1, 1)$ is $\dfrac{3(1) - 3(1) + 3(1)}{\sqrt{9 + 9 + 9}} = \dfrac{3}{3\sqrt{3}} = \dfrac{1}{\sqrt{3}}$

Sol 15: $3x - 2y + 4z = 12$

Intercept on x-axis $(y, z = 0, 0)x = 4$

Intercept on y-axis $(z, z = 0, 0)y = -6$

Intercept on z-axis $(x, y = 0, 0)z = 3$

Sol 16: $<a, b, c> = <1, 3, -2>$

$$<l, m, n> = \left(\frac{1}{\sqrt{1 + 9 + 4}}, \frac{3}{\sqrt{1 + 9 + 4}}, \frac{-2}{\sqrt{1 + 9 + 4}}\right)$$

$$= \left(\frac{1}{\sqrt{14}}, \frac{3}{\sqrt{14}}, \frac{-2}{\sqrt{14}}\right)$$

Sol 17: Direction cosines of y-axis $= <0, 1, 0>$

Sol 18: $\dfrac{x + 2}{1} = \dfrac{2y - 1}{3} = \dfrac{3 - z}{5}$

$$\Rightarrow \frac{x + 2}{1} = \frac{y - \frac{1}{2}}{\frac{3}{2}} = \frac{z - 3}{-5}$$

Direction ratio are $\left\langle 1, \dfrac{3}{2}, -5 \right\rangle$ or $\langle 2, 3, -10 \rangle$

Sol 19: $\vec{r} . (i - 2j - 2k) = 1;$ $\vec{r} . (3i - 6j + 2k) = 0$

$$\cos\theta = \frac{3.1 + (-6).(-2) + (2)(-2)}{\sqrt{1 + 4 + 4}\sqrt{9 + 36 + 4}} = \frac{3 + 12 - 4}{3.7} = \frac{11}{21}$$

$\theta = \cos^{-1}\left(\dfrac{11}{21}\right)$

Sol 20: $\vec{r} = 2i - j + 3k + \lambda(3i - 6j + 2k)$ and Plane $\vec{r} . (i + j + k) = 3$

$$\sin\theta = \frac{3.1 - 6.1 + 2.1}{\sqrt{3}\sqrt{9 + 36 + 4}} = \frac{-1}{7\sqrt{3}}$$

Sol 21: $l - 5m + 3n = 0;$ $7l^2 + 5m^2 - 3n^2 = 0$

$l = 5m - 3n$

$\Rightarrow 7(25m^2 + 9n^2 - 30mn) + 5m^2 - 3n^2 = 0$

$\Rightarrow 180m^2 + 60n^2 - 210mn = 0$

$\Rightarrow 6m^2 - 7mn + 2n^2 = 0$

$\Rightarrow 6m^2 - 4mn - 3mn + 2n^2 = 0$

$\Rightarrow 2m(3m - 2n) - n(3m - 2n) = 0$

$\Rightarrow m = \dfrac{n}{2}$ or $m = \dfrac{2n}{3}$

If $m = \dfrac{n}{2}$, $l = -m$, if $m = \dfrac{2n}{3}$, $l = \dfrac{m}{2}$

The following ratio are

$$\left\langle \frac{-1}{\sqrt{6}}, \frac{1}{\sqrt{6}}, \frac{2}{\sqrt{6}} \right\rangle \text{ or } \left\langle \frac{+1}{\sqrt{14}}, \frac{2}{\sqrt{14}}, \frac{3}{\sqrt{14}} \right\rangle$$

Sol 22: Line through the points

$$\frac{x - 1}{2} = \frac{y - 2}{4} = \frac{z - 3}{6} = \lambda$$

$$\frac{x - 3}{5} = \frac{y - 3}{2} = \frac{z - 3}{-3} = \lambda$$

$$\cos\theta = \frac{2.5 + 4.2 + 6.(-3)}{\sqrt{56}\sqrt{38}} = 0$$

$\theta = 90°$

Sol 23: $\dfrac{x - 5}{4} = \dfrac{y - 4}{0} = \dfrac{z - 4}{-2}$

equation of line $= \lambda$

Let foot of $\perp$ is (α, β, γ)

$\alpha = 5 + 4\lambda;$ $\beta = 4;$ $\gamma = 4 - 2\lambda$

$\Rightarrow (\alpha - 1).4 + (\beta - 2).0 + (\gamma - 1).(-2) = 0$

$\Rightarrow (4 + 4\lambda)4 - 2(3 - 2\lambda) = 0 \Rightarrow 20\lambda + 10 = 0 \Rightarrow \lambda = \dfrac{-1}{2}$

$\Rightarrow \alpha = 5 + 4\left(\dfrac{-1}{2}\right) = 3$ $\beta = 4 = 4$

$\Rightarrow \gamma = 4 - 2\left(\dfrac{-1}{2}\right) = 5$

$(3, 4, 5)$

Sol 24: $\cos(\delta\theta)$

$$= \frac{l\cdot(l+\delta l)+m\cdot(m+\delta m)+n\cdot(n+\delta n)}{\sqrt{l^2+m^2+n^2}\sqrt{(l+\delta l)^2+(m+\delta m)^2+(n+\delta n)^2}}$$

$$\left[\begin{array}{l}\text{neglecting}\\ \delta l^2,\,\delta m^2,\,\delta n^2\end{array}\right]$$

$$= \frac{l^2+m^2+n^2+l\delta l+m\delta m+n\delta n}{\sqrt{(l^2+m^2+n^2)}\sqrt{(l^2+2l\delta l+m^2+2l\delta m+n^2+2l\delta n)}}$$

$$\frac{1-(\delta\theta)^2}{2} = \frac{1+l\delta l+m\delta m+n\delta n}{1}$$

$$(\delta\theta)^2 = -2(l\delta l+m\delta m+n\delta n) \qquad \ldots \text{(i)}$$

$$l^2+m^2+n^2 = (l+\delta l)^2+(m+\delta m)^2+(n+\delta n)^2$$

$$\Rightarrow (\delta l)^2+(\delta m)^2+(\delta n)^2 = -2l\delta l-2m\delta m-2n\delta n \quad \ldots \text{(ii)}$$

$$(\delta\theta)^2 = (\delta l)^2+(\delta m)^2+(\delta m)^2$$

Sol 25:

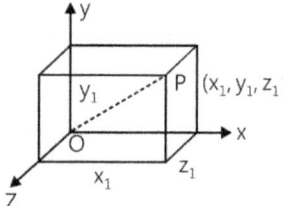

$$\left(\frac{l_1+l_2+l_3}{\sqrt{3}}\right)l_1 + \left(\frac{m_1+m_2+m_3}{\sqrt{3}}\right)m_1 + \left(\frac{n_1+n_2+n_3}{\sqrt{3}}\right)\cdot n_1$$

$$= \frac{1+l_1l_2+l_1l_3+m_1m_2+m_3m_1+n_1m_2+n_1n_3}{\sqrt{3}} = \frac{1}{\sqrt{3}}$$

Similarly dot product with l_2 and l_3 gives $\dfrac{1}{\sqrt{3}}$ as result

i.e. it makes same angle with (l_1, m_1, n_1) (l_2, m_2, n_2) and (l_3, m_3, n_3)

Sol 26: $x + 2y = 0$ ... (i)

$3y - z = 0$

$$2y - \frac{2z}{3} = 0 \qquad \ldots \text{(ii)}$$

The line will be across $(a_1, b_1, c_1) \times (a_2, b_2, c_2)$

$(1\ 2\ 0) \times (0\ 3\ -1)$

$$\begin{vmatrix} i & j & k \\ 1 & 2 & 0 \\ 0 & 3 & -1 \end{vmatrix} = i(-2) - j(-1) + k(3) = -2i + j + 3k$$

Equation of line will be $\dfrac{x-3}{-2} = \dfrac{y-0}{1} = \dfrac{z-1}{3}$

Sol 27: $x + 3y + 6 = 0$, $3x - y - 4z = 0$

$x + 3y + 6 + \lambda(3x - y - 4z) = 0$

$x(1 + 3\lambda) + y(3 - \lambda) + z(-4\lambda) + 6 = 0$

$$\text{Distance from origin} = \frac{6}{\sqrt{(1+3\lambda)^2+(4\lambda)^2+(3-\lambda)^2}} = 1$$

$36 = 1 + 9\lambda^2 + 6\lambda + 16\lambda^2 + 9 + \lambda^2 - 6\lambda$

$36 = 26\lambda^2 + 10$

$\lambda = \pm 1$

Planes are $\Rightarrow 4x + 2y - 4z + 6 = 0$ $(\lambda = 1)$

$-2x + 4y + 4z + 6 = 0$ $(\lambda = -1)$

Sol 28: $ax + by + cz = 1$ ... (i)

$(-1, 1, 1)$ lies on (1)

$\Rightarrow -a + b + c = 1$

$(1, -1, 1)$ lies on (1)

$\Rightarrow +a - b + c = 1 \Rightarrow c = 1$

If $\perp$ to $x + 2y + 2z = 5$

$a \cdot 1 + b \cdot 2 + 2 \cdot c = 0$

$a + 2b = -2$

$a - b = 0$

$a = b = \dfrac{-2}{3}$

Equation of plane is $-2x - 2y + 3z = 3$.

Sol 29: $P(-1 + r\cos\alpha, -5 + r\cos\beta, -10 + r\cos\gamma)$

are coordinates of point at distance r from $(-1, -5, -10)$ along $\langle\alpha, \beta, \gamma\rangle$

Point P lies on the given plane

$x - y + z = 5$

$-1 + r\cos\alpha + 5 - r\cos\beta + r\cos\gamma - 10 = 5$

$r\cos\alpha - r\cos\beta + r\cos\gamma = 11$

$$r = \frac{11}{\dfrac{3-4+12}{13}} = \frac{11\cdot 13}{11} = 13 \text{ units}$$

Sol 30: $ax + by + cz = 1$

$(1, 2, -4)$

$a + 2b - 4c = 1$ ... (i)

This plane is parallel

$r_1 = i + 2j + 4k + \lambda(2i + 3j + 6k)$

$r_2 = i - 3j + 5k + \lambda(i + j - k)$

$\Rightarrow 2a + 3b + 6c = 0$

$\Rightarrow a + b - c = 0$

$\Rightarrow b = -8c$

$\Rightarrow a = 9c$

$\Rightarrow 9c - 16c - 4c = 1$

$\Rightarrow c = \dfrac{-1}{11}$, $b = \dfrac{+8}{11}$, $a = \dfrac{-9}{11}$

Equation of plane is $-9x + 8y - z = 11$ or

$\Rightarrow \vec{r} \cdot (-9i + 8j - k) = 11$

Sol 31: $al + bm + cn = 0$ ... (i)

and $fmn + gnl + hlm = 0$ ... (ii)

$\Rightarrow \dfrac{f}{l} + \dfrac{g}{m} + \dfrac{h}{n} = 0$ ... (iii)

Comparing (i) and (iii)

$\dfrac{a}{f} l^2 = \dfrac{b}{g} m^2 = \dfrac{c}{h} n^2 = \lambda$

$\Rightarrow l^2 = \dfrac{f}{a} \lambda \Rightarrow l = \pm \sqrt{\dfrac{f}{a} \lambda}$

Similarly

$m^2 = \dfrac{g}{b} \lambda \Rightarrow m = \pm \sqrt{\dfrac{g}{h} \lambda}$

$n^2 = \dfrac{h}{c} \lambda \Rightarrow n = \pm \sqrt{\dfrac{h}{c} \lambda}$

Since, lines are $\perp$

$\cos \theta = l_1 l_2 + m_1 m_2 + n_1 n_2 = 0$

$-\dfrac{f}{a} \lambda - \dfrac{g}{b} \lambda - \dfrac{h}{c} \lambda = 0$

$\Rightarrow \lambda \left(\dfrac{f}{a} + \dfrac{g}{b} + \dfrac{h}{c} \right) = 0 \Rightarrow \dfrac{f}{a} + \dfrac{g}{b} + \dfrac{h}{c} = 0$

Sol 32: $\dfrac{x - b}{a} = y = \dfrac{z - d}{c}$ ... (i)

$\dfrac{x - b'}{a'} = y = \dfrac{z - d'}{c'}$ ... (ii)

These 2 are perpendicular if $aa' + cc' + 1 = 0$

Sol 33: $4x - y + z - 10 + \lambda (x + y - z - 4) = 0$

$\Rightarrow x(4 + \lambda) + y(-1 + \lambda) + z(1 - \lambda) = 10 + 4\lambda$

$\Rightarrow (4 + \lambda) \cdot 2 + (\lambda - 1) \cdot 1 + (1 - \lambda) \cdot 1 = 0$

$\Rightarrow 8 - 1 + 1 + 2\lambda = 0 \Rightarrow \lambda = -4$

$\Rightarrow -5y + 5z = -6$, equation of plane

Distance from $(1, 1, 1)$

$= \dfrac{-5 + 5 + 6}{\sqrt{25 + 25}} = \dfrac{6}{5\sqrt{2}} = \dfrac{3\sqrt{2}}{5}$

Sol 34: Ratios of line perpendicular to plane is $\{(2 - 0), (5 - 0), (7 - 0)\}$

Equation of plane is $2x + 5y + 7z = k$

$(2, 5, 7)$ lies on the plane

$2.2 + 5.5 + 7.7 = k = 78$

$2x + 5y + 7z = 78$

Sol 35: Direction ratios of line $\perp$ to the given planes

$3x + 2y - 3y = 1;$ $5x - 4y + z = 5$

$\begin{vmatrix} i & j & k \\ 3 & 2 & -3 \\ 5 & -4 & 1 \end{vmatrix} = i(2 - 12) - j(3 + 15) + k(-12 - 10)$

$= -10i - 18j - 22k$

Plane will be $10x + 18y + 22z = k$

Passes through $(-1, -1, 2)$

$2 \cdot (22) - 28 = k$ $\therefore K = +16$

$10x + 18y + 22z - 16 = 0$

$\Rightarrow 5x + 9y + 11z - 8 = 0$

Sol 36: $l + m + n = 0$ and $l^2 + m^2 = n^2$

$\Rightarrow n = -(l + m)$

$\Rightarrow l^2 + m^2 = (l + m)^2 = l^2 + m^2 = l^2 + m^2 + 2m \cdot n$

$\Rightarrow m \cdot n = 0$

$\Rightarrow m = 0$ or $n = 0$

$\Rightarrow (l, m, n) \equiv \left(-\dfrac{1}{\sqrt{2}}, 0, \dfrac{1}{\sqrt{2}} \right)$ or $\left(\dfrac{1}{\sqrt{2}}, -\dfrac{1}{\sqrt{2}}, 0 \right)$

$\Rightarrow$ Angle $= \dfrac{\pi}{4}$

Sol 37:

$\begin{vmatrix} i & j & k \\ 1 & -3 & 0 \\ 0 & -5 & 9 \end{vmatrix} = i(3) - j(-1) + k(9) = 3i + j + 9k$

$\dfrac{x - 5}{3} = \dfrac{y + 7}{1} = \dfrac{z + 3}{9}$

Sol 38: $ax + by + cz = 1$

$a + b + c = 1 \Rightarrow a + c = 1 - b$

$a - b + c = 1 \Rightarrow b = 0$

$-7a + 3b - 5 + 5a = 1$

$b = 6 + 2a/3, a = -3, c = 4$

$-3x + 4z = 1 \rightarrow$ ratio $\rightarrow$ [-3, 0, 4]

xz plane $\rightarrow$ ratio $\rightarrow$ [0, 1, 0]

$-3.0 + 0.1 + 4.0 = 0$

Hence given plane is perpendicular to xz plane.

Sol 39: $\dfrac{\alpha - 1}{2} = \dfrac{\beta - 1}{-2} = \dfrac{\gamma - 2}{4}$

$= \dfrac{-(2 - 2 + 8 + 5)}{4 + 4 + 16} = \dfrac{-13}{24}$

$\alpha = 1 - \dfrac{13}{12} = -\dfrac{1}{12}$, $\beta = 1 + \dfrac{13}{12} = \dfrac{25}{12}$

$\gamma = 2 - \dfrac{13}{6} = \dfrac{-1}{6}$

Length $= \dfrac{2 - 2 + 8 + 5}{\sqrt{24}} = \dfrac{13}{\sqrt{24}}$

Sol 40: $ax + by + cz = 1$

$(1, 0, -1) \Rightarrow a - c = 1$

$(3, 2, 2) \Rightarrow 3a + 2b + 2c = 1$

It is parallel to $\langle 1, -2, 3 \rangle$

$\Rightarrow a - 2b + 3c = 0$

$\Rightarrow 4a + 5c = 1$

$\Rightarrow 4 + 4c + 5c = 1$

$\Rightarrow c = \dfrac{-1}{3}$

$\Rightarrow a = \dfrac{2}{3}$

$\Rightarrow \dfrac{2}{3} - \dfrac{3}{3} = +2b$

$b = \dfrac{-1}{6}$

Eq. of plane $2x - \dfrac{y}{2} - z = 3$

$4x - y - 2z = 6$

$\Rightarrow \vec{r} \cdot (4\hat{i} - \hat{j} - 2\hat{k}) = 6$

Sol 41: distance between $2x - y + 3z = 4$

$2x - y + 3z = \dfrac{-13}{3}$

Distance, $d = \dfrac{4 + \dfrac{13}{3}}{\sqrt{4 + 1 + 9}} = \dfrac{25}{3\sqrt{14}} = \dfrac{25\sqrt{14}}{42}$

Sol 42: $ax + by + cz = 1$

$A\left(\dfrac{1}{a}, 0, 0\right), B\left(0, \dfrac{1}{b}, 0\right), C\left(0, 0, \dfrac{1}{c}\right)$

$\dfrac{1}{a} = 3\alpha, \dfrac{1}{b} = 3\beta, \dfrac{1}{c} = 3\gamma$

$\dfrac{x}{\alpha} + \dfrac{y}{\beta} + \dfrac{z}{\gamma} = 3$

Sol 43: $x - 2y + 3z + 4 + \lambda(x - y + z + 3) = 0$

Through origin $3\lambda + 4 = 0$; $\lambda = \dfrac{-4}{3}$

$\Rightarrow x\left(1 - \dfrac{4}{3}\right) + y\left(-2 + \dfrac{4}{3}\right) + z\left(3 - \dfrac{4}{3}\right) = 0$

$\Rightarrow \vec{r}\, \dfrac{-x}{3} - \dfrac{2y}{3} + \dfrac{5z}{3} = 0$

$\Rightarrow x + 2y - 5z = 0$

Sol 44: $2x + 2y - z - 6 + \lambda(2x + 3y - z - 8) = 0$

$x(2 + 2\lambda) + y(2 + 3\lambda) + z(-1 - \lambda) - 6 - 8\lambda = 0$ equation of plane

xz plane $\langle 0, 1, 0 \rangle$ any point on the line is $(\alpha, 2, 2\alpha - 2)$

Direction ratios of line

$\begin{vmatrix} i & j & k \\ 2 & 2 & -1 \\ 2 & 3 & -1 \end{vmatrix} = i(-2 + 3) - j(-2 + 2) + k(6 - 4)$

$= i + 2k = \langle 1, 0, 2 \rangle$

This is parallel to plane y = 0 as

$(1, 0, 2) \cdot (0, 1, 0) = 0$

$\alpha = 0$ i.e. $(0, 2, -2)$

Sol 45: The equation of Plane

$ax + by + cz + d + \lambda(a'x + b'y + c'z + d') = 0$ ... (i)

$\Rightarrow (a + \lambda a')x + (b + \lambda b)y + (c + \lambda c')z + d + \lambda d' = 0$

Which parallel to line $\dfrac{x}{l} = \dfrac{y}{m} = \dfrac{z}{n}$

$\Rightarrow (a + \lambda a')l + (b + \lambda b')m + (c + \lambda c')n = 0$

$\Rightarrow -\dfrac{al + bm + cn}{a'l + b'm + c'n} = \lambda$

Substituting in (i)

$(ax + by + cz + d) - \dfrac{al + bm + cn}{a'l + b'm + c'n}(a'x + b'y + c'z + d) = 0$

Sol 46: $ax + by + cz = 1$

$\dfrac{1}{b} = 5, \dfrac{1}{c} = 7$ {given intercepts}

$<a, b, c> \bullet <1, 0, 0> = 0$

$a = 0$

$\dfrac{y}{5} + \dfrac{z}{7} = 1;\qquad 7y + 5z = 35$

Sol 47: $ax + by + cz = 1$

$\dfrac{1}{\sqrt{a^2 + b^2 + c^2}} = P$ ......(i)

$A\left(\dfrac{1}{a}, 0, 0\right);\ B\left(0, \dfrac{1}{b}, 0\right);\ C\left(0, 0, \dfrac{1}{c}\right)$

$x = \dfrac{1}{a},\ y = \dfrac{1}{b},\ c = \dfrac{1}{z}$

$\dfrac{1}{P^2} = a^2 + b^2 + c^2 = \dfrac{1}{x^2} + \dfrac{1}{y^2} + \dfrac{1}{z^2}$ from (i)

Sol 48: $i - j + 3k$ from $5x + 2y - 7z + 9 = 0$

$\Rightarrow \left|\dfrac{5 - 2 - 21 + 9}{\sqrt{49 + 4 + 25}}\right| = \dfrac{9}{\sqrt{78}}$

$\Rightarrow (3i + \lambda j + 3k)$ from $5x + 2y - 7z + 9 = 0$

$\Rightarrow \left|\dfrac{15 + 2\lambda - 21 + 9}{\sqrt{49 + 4 + 25}}\right| = \left|\dfrac{3 + 2\lambda}{\sqrt{78}}\right| \Rightarrow |3 + 2\lambda| = 9$

$\Rightarrow \lambda = 3$ or -6

Sol 49: Normal to vector $3i + 5j - 6k$

$3x + 5y - 6z = k$

at 7 units from origin

$\left|\dfrac{k}{\sqrt{36 + 25 + 9}}\right| = 7;\qquad k = \pm 7\sqrt{70}$

$\vec{r} \cdot (3\hat{i} - 5\hat{j} - 6\hat{k}) = \pm 7\sqrt{70}$

Sol 50: $r = i - j + \lambda(i + j + k) + \mu(4i - 2j + 3k)$

$B = \begin{vmatrix} i & j & k \\ 1 & 1 & 1 \\ 4 & -2 & 3 \end{vmatrix} = i(5) - j(-1) + k(-2 - 4) = 5i + j - 6k$

Plane pass through $(1, -1, 0)$

Equation of plane $\vec{r} \bullet (5i + j - 6k) = z$

$5(1) + 1(-1) - 6(0) - z = 4$

The equation of plane $\Rightarrow \vec{r} \bullet (5i + j - 6k) = 4$

Exercise 2

Single Correct Choice Type

Sol 1: (C) $l^2 + m^2 + n^2 = \cos^2\alpha + \cos^2\beta + \cos^2\gamma = 1$

Sol 2: (A) $ax + by + cz + d = 0$ to intersect x and y axis at equal angle

$|\tan\alpha| = |\tan\beta| \Rightarrow |a| = |b|$

Sol 3: (D) Parallel to x-axis i.e. $<1, 0, 0>$

$\dfrac{x - a}{1} = \dfrac{y - b}{0} = \dfrac{z - c}{0}$

Sol 4: (C) $\cos\alpha = \dfrac{1}{\sqrt{2}}\quad \cos\gamma = \cos 60° = \dfrac{1}{2}$

$l^2 + m^2 + n^2 = 1$

$\Rightarrow \dfrac{1}{2} + \dfrac{1}{4} + m^2 = 1 \Rightarrow m^2 = \dfrac{1}{4}$

$\Rightarrow m = \dfrac{1}{2} = \cos\beta$

$\Rightarrow \beta = 60°$

Sol 5: (A) $\dfrac{x + 1}{1} = \dfrac{y + 3}{3} = \dfrac{z + 2}{-2} = \lambda$

$\Rightarrow (-1 + \lambda, -3 + 3\lambda, -2 - 2\lambda)$

$\Rightarrow 3(-1 + \lambda) + 4(3\lambda - 3) + 5(-2 - 2\lambda) = 5$

$\Rightarrow 5\lambda - 3 - 12 - 10 = 5 \Rightarrow 5\lambda = 30$

$\Rightarrow x = 6$

$(5, 15, -14)$

Sol 6: (B)

$\dfrac{x - 0}{2} = \dfrac{y - 3}{-2} = \dfrac{z - 4}{1} = -\dfrac{(0 - 6 + 4 - 10)}{9}$

$\Rightarrow \dfrac{x}{2} = \dfrac{y-3}{-2} = z-4 = \dfrac{12}{3\times 3} = \dfrac{4}{3}$

$\Rightarrow x = \dfrac{8}{3}, y = 3 - \dfrac{8}{3}, z = 4 + \dfrac{4}{3}$

$\Rightarrow \dfrac{8}{3}, \dfrac{1}{3}, \dfrac{16}{3}$

Sol 7: (B) $2x + y - z - 4 + \lambda (3x + 5z - 4) = 4$

$2 + 3\lambda = 1 \Rightarrow \lambda = \dfrac{-1}{3}$

$\Rightarrow 2x - x + y - z - \dfrac{5}{3}z - 4 + \dfrac{4}{3} = 0$

$\Rightarrow 3x + 3y - 8z - 8 = 0$

Sol 8: (D) $\begin{vmatrix} i & j & k \\ 1 & 1 & -1 \\ 2 & -3 & 1 \end{vmatrix} = i(-2) - j(3) + k(-3 -2)$

$= -2i - 3j - 5k$

It passes through (1, 0, 0)

Equation of line is $\dfrac{x-1}{2} = \dfrac{y}{3} = \dfrac{z}{5}$

Sol 9: (C) Line $\dfrac{x-1}{1} = \dfrac{y-3}{2} = \dfrac{z-4}{3}$ is parallel to plane

$ax + by + cz = 1$

If $a + 2b + 3c = 0$

Only C satisfies the condition

Sol 10: (A) $a = \sqrt{49+1+1} = \sqrt{51}$;

$b = \sqrt{1+0+1} = \sqrt{2}$; $c = \sqrt{36+1+0} = \sqrt{37}$

$s = \dfrac{\sqrt{2} + \sqrt{51} + \sqrt{37}}{2}$

$s(s - a)\,(s - b)\,(s - c)$

$\Rightarrow \dfrac{\sqrt{51} + \sqrt{37} + \sqrt{2}}{2} \left[\dfrac{\sqrt{2} + \sqrt{37} - \sqrt{51}}{2} \right]$

$\left[\dfrac{\sqrt{2} + \sqrt{51} - \sqrt{37}}{2} \right] \left[\dfrac{\sqrt{37} + \sqrt{51} - \sqrt{2}}{2} \right]$

$= \dfrac{\left[37 + 2 + 2\sqrt{74} - 51 \right]\left[31 - (37 + 2 - 2\sqrt{74}) \right]}{16}$

$= \dfrac{\left[2\sqrt{74} - 12 \right]\left[12 + 2\sqrt{74} \right]}{16}$

$= \dfrac{4\times 74 - 144}{16} = \dfrac{296 - 144}{16} = \dfrac{152}{16} = \dfrac{38}{4}$

$\Delta = \dfrac{\sqrt{38}}{2}$

Sol 11: (D) $\dfrac{x+2}{3} = \dfrac{y-1}{-1} = \dfrac{z+4}{1}$

or $\dfrac{x-1}{3} = \dfrac{y}{-1} = \dfrac{z+3}{1}$

Sol 12: (B) Let P(x, y, z) be any point on the locus, then the distances from the six faces are

$|x + 1|, |x- 1|, |y + 1|, |y- 1|, |z + 1|, |z- 1|$

According to the given condition

$|x + 1|^2 + |x - 1| + |y + 1|^2 + |y - 1|^2 + |z + 1|^2 + |z- 1|^2 = 10$

$\Rightarrow 2(x^2 + y^2 + z^2) = 10 - 6 = 4$

$\Rightarrow x^2 + y^2 + z^2 = 2$

Sol 13: (B) If $(x_1, y_1, z_1), (x_2, y_2, z_2), (x_3, y_3, z_3)$ and (x_4, y_4, z_4) are coplanar, then

$\begin{vmatrix} x_2 - x_1 & y_2 - y_1 & z_2 - z_1 \\ x_3 - x_1 & y_3 - y_1 & z_3 - z_1 \\ x_4 - x_1 & y_4 - y_1 & z_4 - z_1 \end{vmatrix} = 0$

$\begin{vmatrix} -4-0 & 4+1 & 4+1 \\ 4-0 & 5+1 & 1+1 \\ 3-0 & 9+1 & 4+1 \end{vmatrix} \Rightarrow = \begin{vmatrix} -4 & 5 & 5 \\ 4 & 6 & 2 \\ 3 & 10 & 5 \end{vmatrix} = 0$

$= -4(30-20) - 5(20-6) + 5(40-18) = -40 - 70 + 110 = 0$

Sol 14: (D) The plane $y + z + 1 = 0$

Since the plane does not have any intercepts on x-axis, therefore it is parallel to x-axis.

Then normal to plane can not be parallel to x-axis.

Sol 15: (A) Using the fact that reflection of a' x + b' y + c'z + d' = 0 in the plane ax+ by + cz + d = 0 is given by
$2 (aa' + bb' + cc') (ax+ by + cz + d)$

$= (a^2 + b^2 + c^2) (a' x + b' y + c' z + d')$

We get the required equation as

$2 (2 + 3 + 4) (x-y + z-3) = (1 + 1 + 1)(2x-3y + 4z-3)$

$6 (x- y + z - 3) = 2x- 3y + 4z -3$

$4x- 3y + 2z- 15 = 0$

Previous Years' Questions

Sol 1: (A) Given equation of straight line

$$\frac{x-4}{1} = \frac{y-2}{1} = \frac{z-k}{2}$$

Since, the line lies in the plane $2x - 4y + z = 7$

∴ Point $(4, 2, k)$ must satisfy the plane.

$$\Rightarrow 8 - 8 + k = 7 \Rightarrow k = 7$$

Sol 2: (B) Since, the lines intersect they must have a point in common

i.e., $\frac{x-1}{2} = \frac{y+1}{3} = \frac{z-1}{4} = \lambda$

and $\frac{x-3}{1} = \frac{y-k}{2} = \frac{z}{1} = \mu$

$$\Rightarrow x = 2\lambda + 1, y = 3\lambda - 1, z = 4\lambda + 1$$

and $x = \mu + 3, y = 2\mu + k, z = \mu$ are same

$$\Rightarrow 2\lambda + 1 = \mu + 3, 3\lambda - 1 = 2\mu + k, 4\lambda + 1 = \mu$$

On solving Ist and IIIrd terms, we get,

$$\lambda = -\frac{3}{2} \text{ and } \mu = -5$$

$$\therefore k = 3\lambda - 2\mu - 1 \Rightarrow k = 3\left(-\frac{3}{2}\right) - 2(-5) - 1 = \frac{9}{2}$$

$$\therefore k = \frac{9}{2}$$

Sol 3: (A) Since, $\frac{x}{a} + \frac{y}{b} + \frac{z}{c} = 1$

cuts the coordinate axes at

A(a, 0, 0), B(0, b, 0), C(0, 0, c)

and its distance from origin = 1

$$\therefore \frac{1}{\sqrt{\frac{1}{a^2} + \frac{1}{b^2} + \frac{1}{c^2}}} = 1$$

or $\frac{1}{a^2} + \frac{1}{b^2} + \frac{1}{c^2} = 1$ ... (i)

where P is centroid of triangle

$$\therefore x = \frac{a}{3}, y = \frac{b}{3}, z = \frac{c}{3}$$ ... (ii)

∴ From Eqs. (i) and (ii), we get

$$\frac{1}{9x^2} + \frac{1}{9y^2} + \frac{1}{9z^2} = 1 \quad \text{or} \quad \frac{1}{x^2} + \frac{1}{y^2} + \frac{1}{z^2} = 9 = K$$

$$\therefore K = 9$$

Sol 4: Area of $\triangle ABC = \frac{1}{2}(\vec{AB} \times \vec{AC})$, where

$$\vec{AB} = \hat{i} + 2\hat{j} - 3\hat{k} \text{ and } \vec{AC} = 2\hat{i} + 0\hat{j} + 0\hat{k}$$

$$\therefore \vec{AB} \times \vec{AC} = \begin{vmatrix} \hat{i} & \hat{j} & \hat{k} \\ 1 & 2 & -3 \\ 2 & 0 & 0 \end{vmatrix} = 2(-3\hat{j} - 2\hat{k})$$

$$\Rightarrow \text{Area of triangle} = \frac{1}{2}(\vec{AB} \times \vec{AC})$$

$$= \frac{1}{2} \cdot 2 \cdot \sqrt{9 + 4} = \sqrt{13} \text{ sq. units}$$

Sol 5: A unit vector perpendicular to the plane

determined by P, Q, R $= \pm \frac{(\vec{PQ} \times \vec{PR})}{|\vec{PQ} \times \vec{PR}|}$

where $\vec{PQ} = [\hat{i} + \hat{j}] - 3\hat{k}$ and $\vec{PR} = -\hat{i} + 3\hat{j} - \hat{k}$

$$\therefore \vec{PQ} \times \vec{PR} = \begin{vmatrix} \hat{i} & \hat{j} & \hat{k} \\ 1 & 1 & -3 \\ -1 & 3 & -1 \end{vmatrix}$$

$$= \hat{i}(-1 + 9) - \hat{j}(-1 - 3) + \hat{k}(3 + 1) = 8\hat{i} + 4\hat{j} + 4\hat{k}$$

$$\Rightarrow |\vec{PQ} \times \vec{PR}| = 4\sqrt{4 + 1 + 1} = 4\sqrt{6}$$

$$\therefore \text{Unit vector} = \pm \frac{(\vec{PQ} \times \vec{PR})}{|\vec{PQ} \times \vec{PR}|} = \pm \frac{4(2\hat{i} + \hat{j} + \hat{k})}{4\sqrt{6}} = \pm \frac{(2\hat{i} + \hat{j} + \hat{k})}{\sqrt{6}}$$

Sol 6: Let the equation of plane through (1, 1, 1) having a, b, c as DR's of normal to plane, $a(x - 1) + b(y - 1) + c(z - 1) = 0$ and plane is parallel to straight line having DR's.

(1, 0, –1) and (–1, 1, 0)

$$\Rightarrow a - c = 0 \text{ and } -a + b = 0$$

$$\Rightarrow a = b = c$$

∴ Equation of plane is $x - 1 + y - 1 + z - 1 = 0$

or $\frac{x}{3} + \frac{y}{3} + \frac{z}{3} = 1$. Its intercept on coordinate axes are

A(3, 0, 0), B(0, 3, 0), C(0, 0, 3)

Hence, the volume of tetrahedron OABC

$$= \frac{1}{6}[\vec{a}\vec{b}\vec{c}] = \frac{1}{6}\begin{vmatrix} 3 & 0 & 0 \\ 0 & 3 & 0 \\ 0 & 0 & 3 \end{vmatrix} = \frac{27}{6} = \frac{9}{2} \text{ cu units}$$

Sol 7: Equation of plane containing the lines

$2x - y + z - 3 = 0$ and $3x + y + z = 5$ is

$(2x - y + z - 3) + \lambda(3x + y + z - 5) = 0$

$\Rightarrow (2 + 3\lambda)x + (\lambda - 1)y + (\lambda + 1)z - 3 - 5\lambda = 0$

Since, distance of plane from (2, 1, –1) to above plane is $1/\sqrt{6}$.

$\therefore \left| \dfrac{6\lambda + 4 + \lambda - 1 - \lambda - 1 - 3 - 5\lambda}{\sqrt{(3\lambda + 2)^2 + (\lambda - 1)^2 + (\lambda + 1)^2}} \right| = \dfrac{1}{\sqrt{6}}$

$\Rightarrow 6(\lambda - 1)^2 = 11\lambda^2 + 12\lambda + 6$

$\Rightarrow \lambda = 0, -\dfrac{24}{5}$

$\therefore$ Equations of planes are $2x - y + z - 3 = 0$

and $62x + 29y + 19z - 105 = 0$

Sol 8: (C) The line $\dfrac{x - 3}{2} = \dfrac{y + 2}{-1} = \dfrac{z + 4}{3}$ lies in the plane,

then point $(3, -2, -4)$ lies on the plane

$\Rightarrow 3\ell - 2m = 5$ ...(i)

And line is $\perp$ to normal of plane

$\Rightarrow 2\ell - m = 3$ ...(ii)

From (i) and (ii)

$\ell = 1$ and $m = -1$

$\Rightarrow \ell^2 + m^2 = 1^2 + (-1)^2 = 2$

Sol 9: (A) The eq of line passes through $(1, -5, 9)$

along $x = y = z$ is

$\dfrac{x - 1}{1} = \dfrac{y + 5}{1} = \dfrac{z - 9}{1} = r$

The point on line $(r + 1, r - 5, r + 9)$

This point also lies on the given plane

$r + 1 - r + 5 + r + 9 = 5$

$r = -10$

The point in $(-9, -15, -1)$

Distance between $(1, -5, 9)$ and $(-9, -15, -1)$

$= \sqrt{10^2 + (-10)^2 + (10)^2} = 10\sqrt{3}$ unit

Sol 10: (C) $\dfrac{x - 2}{3} = \dfrac{y + 1}{4} = \dfrac{z - 2}{12} = r$

The point of interstation $(3r + 2, 4r - 1, 12r + 2)$

Lies on plane, then

$3r + 2 - 4r + 1 + 12r + 2 - 16 = 0$

$\Rightarrow 11r - 11 = 0$

$\Rightarrow r = 1$

The point in $(5, 3, 14)$

Distance $= \sqrt{(5 - 1)^2 + (3 - 0)^2 + (14 - 2)^2}$

$= \sqrt{16 + 9 + 144}$

$= \sqrt{169} = 13$

Sol 11: (B) Let the two lines in a same plane interest at $P(x, y, 0)$, then $2x - 5y = 3$ and $x + y = 5$

On solving, we get $P \equiv (4, 1, 0)$

Any plane $\parallel$ to $x + 3y + 6z = 1$ is

$x + 3y + 6z = \lambda$

$P(4, 1, 0)$ must satisfies it, then

$4 + 3 + 0 = \lambda \Rightarrow \lambda = 7$

The eq. to required plane

$\Rightarrow x + 3y + 6z = 7$

Sol 12: (B) The parallel planes $2x + y + 2z = 8$

and $4x + 2y + 4z = -5$

Distance $= \dfrac{|-8 \times 2 - 5|}{\sqrt{16 + 4 + 16}} = \dfrac{21}{\sqrt{36}} = \dfrac{21}{6} = \dfrac{7}{2}$

Sol 13: (A) Image of point $(1, 3, 4)$ is

$\dfrac{x - 1}{2} = \dfrac{y - 3}{-1} = \dfrac{z - 4}{1} = \dfrac{-2(2 - 3 + 4 + 3)}{4 + 1 + 1} = -2$

$\Rightarrow (-3, 5, 2)$

Since line is parallel to plane direction, ratio will not change

Eq. of imaged line $\dfrac{x + 3}{3} = \dfrac{y - 5}{1} = \dfrac{z - 2}{1}$

Sol 14: (A) $\ell + m + n = 0 \Rightarrow n = -(\ell + m)$

Substituting in $\ell^2 = m^2 + n^2$

$\ell^2 = m^2 + (\ell + m)^2$

$\Rightarrow \ell^2 = m^2 + \ell^2 + m^2 + 2m$

$\Rightarrow 2m^2 + 2m = 0$

$\Rightarrow 2m(m + 1) = 0$

$\Rightarrow m = 0, 1$

if $m = 0$, $\ell = \dfrac{-1}{\sqrt{2}}$, $n = \dfrac{1}{\sqrt{2}}$

if $m = 1$, $\ell = 0 = n$ (not possible)

Therefore direction cosine

$$\left(-\dfrac{1}{\sqrt{2}}, 0, \dfrac{1}{\sqrt{2}}\right) \text{ or } \left(-\dfrac{1}{\sqrt{2}}, \dfrac{1}{\sqrt{2}}, 0\right)$$

$$\cos\phi = \left(-\dfrac{1}{\sqrt{2}}\right)\left(-\dfrac{1}{\sqrt{2}}\right) + (0)\left(\dfrac{1}{\sqrt{2}}\right) + \left(\dfrac{1}{\sqrt{2}}\right)(0) = \dfrac{1}{2}$$

$$\Rightarrow \phi = \dfrac{\pi}{3}$$

Sol 15: (B) The lines $\dfrac{x-2}{1} = \dfrac{y-3}{1} = \dfrac{z-4}{-k}$ and

$\dfrac{x-1}{k} = \dfrac{y-4}{2} = \dfrac{z-5}{1}$ are coplanars, then

$$\begin{vmatrix} 1 & 1 & -k \\ k & 2 & 1 \\ 1 & -1 & 1 \end{vmatrix} = 0$$

$$\Rightarrow k(k+3) = 0$$

$$\Rightarrow K = 0, -3$$

Two values exist.

Sol 16: (A) Eq. of plane parallel to $x - 2y + 2z - 5 = 0$ is

$x - 2y + 2z = \lambda$

$\perp$ distance from origin is 1,

then $\dfrac{|0 - 0 + 0 - \lambda|}{\sqrt{1+4+4}} = 1 \Rightarrow \dfrac{|\lambda|}{3} = 1 \Rightarrow \lambda = \pm 3$

Eq. of plane $x - 2y + 2z = \pm 3$

Sol 17: (D) $\sin\theta = \dfrac{1 + 4 + 3\lambda}{\sqrt{1+4+9}\sqrt{1+4+\lambda}}$

$$= \dfrac{5 + 3\lambda}{\sqrt{14}\sqrt{5+\lambda}} \qquad \text{... (i)}$$

Given $\cos\theta = \sqrt{\dfrac{5}{14}}$

$$\Rightarrow \sin\theta = \sqrt{1 - \cos^2\theta} = \sqrt{1 - \dfrac{5}{14}} = \dfrac{3}{\sqrt{14}}$$

From (i) $\dfrac{3}{\sqrt{14}} = \dfrac{5 + 3\lambda}{\sqrt{14}\sqrt{5+\lambda}}$

$$\Rightarrow 3\sqrt{5 + \lambda^2} = (5 + 3\lambda) \Rightarrow 9(5 + \lambda^2) = 25 + 9\lambda^2 + 30\lambda$$

$$\Rightarrow 30\lambda = 20 \Rightarrow \lambda = \dfrac{2}{3}$$

Sol 18: (B) Statement-I: Since mid point of A(1, 0, 7) and B(1, 6, 3) is which lies on the line, therefore point B is image of A about line

Statement-II: Since it given that the line only bisects the line joining A and B, therefore not the correct explanation.

$$\left(-\dfrac{1}{\sqrt{2}}, 0, \dfrac{1}{\sqrt{2}}\right) \text{ or } \left(-\dfrac{1}{\sqrt{2}}, \dfrac{1}{\sqrt{2}}, 0\right)$$

$$\cos\theta = \left(-\dfrac{1}{\sqrt{2}}\right)\left(\dfrac{-1}{\sqrt{2}}\right) + (0)\left(\dfrac{1}{\sqrt{2}}\right) + (0)$$

$$= \dfrac{1}{2} \Rightarrow 9 = \dfrac{\pi}{3}$$

JEE Advanced/Boards

Exercise 1

Sol 1: Let point P be taken as origin and $\vec{q}, \vec{s}$ are the position vectors of Q and S points respectively.

$$\Rightarrow \overrightarrow{PR} = \vec{q} + \vec{s}$$

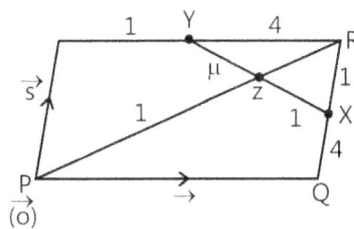

P.V. of X $= \dfrac{\vec{q} + 4(\vec{q} + \vec{s})}{5} = \dfrac{5\vec{q} + 4\vec{s}}{5}$

P.V. of Y $= \dfrac{4\vec{s} + \vec{q} + \vec{s}}{5} = \dfrac{\vec{q} + 5\vec{s}}{5}$

Let, $\dfrac{PZ}{ZR} = \dfrac{1}{\lambda}$ and $\dfrac{YZ}{ZX} = \mu$

P.V. of P $= \dfrac{\vec{q} + \vec{s}}{\lambda + 1} = \dfrac{\mu\left(\vec{q} + \dfrac{4}{5}\vec{s}\right) + \left(\dfrac{\vec{q}}{5} + \vec{s}\right)}{\mu + 1}$

$$\Rightarrow \dfrac{1}{\lambda + 1} = \dfrac{\mu + \dfrac{1}{5}}{\mu + 1} \qquad \text{... (i)}$$

$$\Rightarrow \dfrac{1}{\lambda + 1} = \dfrac{\mu + \dfrac{1}{5}}{\mu + 1} \qquad \text{... (ii)}$$

From (i) & (ii), we get

$\mu = 4$, $\lambda = \dfrac{4}{21}$

$\Rightarrow \dfrac{PZ}{ZR} = \dfrac{21}{4} \Rightarrow \dfrac{PZ}{PR} = \dfrac{21}{25}$

Sol 2: p (x, y)

$\overrightarrow{PA} \cdot \overrightarrow{PB} + 3\overrightarrow{OA} \cdot \overrightarrow{OB} = 0$

$(x-1)(x+1) + y^2 + 3(-1) = 0$

$x^2 + y^2 = 4$

$\left|\overrightarrow{PA}\right| \cdot \left|\overrightarrow{PB}\right|$

$\sqrt{(x-1)^2 + y^2} \quad \sqrt{(x+1)^2 + y^2}$

$\sqrt{(5-2x)}\sqrt{(5+2x)} = \sqrt{25 - 4x^2}$

Max is $\sqrt{25} = 5 = M$

Min $= \sqrt{9} = 3$

$M^2 + m^2 = 34$

Sol 3: $|\vec{a}| = |\vec{b}| = |\vec{c}| = |\vec{a} + \vec{b}| = 1$

$\Rightarrow |\vec{a}|^2 + |\vec{b}|^2 + 2\vec{a} \cdot \vec{b} = 1$; $\vec{a} \cdot \vec{b} = \dfrac{-1}{2}$

$\theta = 120°$

$\angle \left(2\vec{a} + \vec{b} \ \& \ \vec{b}\right)$

$\Rightarrow (2a + b)(b) = |2a + b| \, |b| \cos\theta_1$

$\Rightarrow 2a \cdot b + |b|^2 = \sqrt{4a^2 + b^2 + 4a \cdot b} \ |b| \cos\theta_1$

$\Rightarrow -1 + 1 = \cos\theta_1 \times k$

$\Rightarrow \cos\theta_1 = 0$

$\Rightarrow \theta_1 = \dfrac{\pi}{2}$

Sol 4: $c = \lambda\vec{a} + \mu\vec{b}$

$|c|^2 = \lambda^2 + \mu^2 + 2\lambda\mu\vec{a} \cdot \vec{b} = 1$

$\Rightarrow \lambda^2 + \mu^2 - \lambda\mu = 1$

$\Rightarrow \vec{c} \cdot \vec{a} = \lambda + \mu(\vec{a} \cdot \vec{b}) = 0$

$\Rightarrow \lambda - \dfrac{\mu}{2} = 0$

$\Rightarrow \lambda = \dfrac{\mu}{2} \Rightarrow u = 2\lambda$

$\Rightarrow \lambda^2 + 4\lambda^2 - 2\lambda^2 = 1$

$\Rightarrow \lambda = \dfrac{1}{\sqrt{3}}$, $\mu = \dfrac{2}{\sqrt{3}} \Rightarrow \left(\dfrac{1}{\sqrt{3}}, \dfrac{2}{\sqrt{3}}\right)$

Sol 5: $\vec{P} = x\vec{a} + y\vec{c} \Rightarrow \vec{p} = y\vec{c}$

$0 \le \vec{p} \cdot \vec{a} = x \le 1 \ x \in [0,1]$

$0 \le \vec{p} \cdot \vec{b} = x\vec{a} \cdot \vec{b} \le 1 \ x \in [-2,0] \Rightarrow x = 0$

$\vec{p} \cdot \vec{c} = y$

$\vec{c} = \dfrac{\vec{a}}{\sqrt{3}} + \dfrac{2\vec{b}}{\sqrt{3}}$

$\vec{p} \cdot \vec{c} = \dfrac{\vec{p} \cdot \vec{a}}{\sqrt{3}} + \dfrac{2\vec{p} \cdot \vec{b}}{\sqrt{3}}$

$\vec{p} = x\vec{a} + y\left(\dfrac{\vec{a} + 2\vec{b}}{\sqrt{3}}\right)$

Sol 6: For max. x and y; $x + \dfrac{y}{\sqrt{3}} = \dfrac{2y}{\sqrt{3}} \Rightarrow y = \sqrt{3}x$

$x = 1$; $y = \sqrt{3}$

$\Rightarrow \vec{p} = 2\vec{a} + 2\vec{b}$

Sol 7: The coordinates of the points, O and P, are (0, 0, 0) and (1, 2, -3) respectively.

Therefore, the direction ratios of OP are (1 - 0) = 1, (2 - 0) = 2 and (-3-0) = -3

It is knows that the equation of the plane passing through the point $\left(x_1 \ y_1 \ z_1\right)$ is

$a\left(x - x_1\right) + b\left(y - y_1\right) + c\left(z - z_1\right) = 0$, where a, b and c are the direction ratio of normal.

Here, the direction ratios of normal are 1, 2 and -3 and the point P is (1, 2, -3).

Thus, the equation of the required plane is

$1\left(x - 1\right) + 2\left(y - 2\right) - 3\left(z + 3\right) = 0$

$\Rightarrow x + 2y - 3z - 14 = 0$

Sol 8: (i) $\vec{A} = [x_1 y_1 z_1]$; $\vec{B} = [x_2 y_2 z_2]$; $\vec{C} = [x_3 y_3 z_3]$

$\vec{A}\left(\vec{B} \times \vec{C}\right) = 0$ all are coplanar

$\vec{A} \times \vec{B} = 0 = \vec{B} \times \vec{C} = \vec{C} \times \vec{A}$ i.e. all are mutually $\perp$ which simultaneously is not possible.

(ii) $P = (x_1, y_2, x_3)$ $Q = (y_1 y_2 y_3)$ O (0,0,0)

In $\triangle POQ$; $OP = x_1 i + x_2 j + x_3 k$

$OQ = y_1 i + y_2 j + y_3 k$

$OP \cdot OQ = x_1 y_1 + x_2 y_2 + x_3 y_3 \ x_1 > 0 \ y_1 > 0$

$OP \cdot OQ < 0$ [i.e. it can never be zero]

Sol 9: $A = (-5, 22, 5)$; $B = (1, 2, 3)$; $C = (4, 3, 2)$

$D = (-1, 2, -3)$

and $\Delta AEF = \sqrt{S}$

$\overline{DE} = \overline{DA} + \overline{DB}$

$OE - OD = OA - OD + OB - OD$

$\overrightarrow{OE} = -5i + 22j + 5k + i + 2j + 3k - (-i + 2j - 3k)$

$\overrightarrow{OE} = -3i + 22j + 11k$

$\overrightarrow{BF} = \overline{BA} + \overline{BC}$

$\overrightarrow{OF} = \overline{OA} + \overline{OC} - \overline{OB}$

$\overrightarrow{OF} = -5i + 22j + 5k + 4i + 3j + 2k - i - 2j - 3k$

$OF = -2i + 23j + 4k$

$\text{Area} = \dfrac{1}{2}\left[\overrightarrow{AE} \times \overrightarrow{AF}\right] = \dfrac{1}{2}\left[(2i + 6k) \times (3i + j - k)\right]$

$\dfrac{1}{2}\begin{vmatrix} \hat{i} & \hat{j} & \hat{k} \\ 2 & 0 & 6 \\ 3 & 1 & -1 \end{vmatrix} = \dfrac{1}{2}\hat{i}(-6) - \hat{j}(-2 - 18) + \hat{k}(2) = -6\hat{i} + 20\hat{j} + 2\hat{k}$

$= \dfrac{1}{2}\sqrt{36 + 400 + 4} = \sqrt{\dfrac{440}{4}} = \sqrt{110} \Rightarrow S = \sqrt{110}$

Sol 10: $\left((a-2)\alpha^2 + (b-3)\alpha + c\right)x +$

$\Rightarrow \left((a-2)\beta^2 + (b-3)\beta + c\right)y +$

$\Rightarrow \left((a-2)\gamma^2 + (b-3)\gamma + c\right)(x \times y) = 0$

$\Rightarrow a - 2 = b - 3 = c = 0$

$\Rightarrow a = 2 ; b = 3 ; c = 0$

$\Rightarrow a^2 + b^2 + c^2 = 13$

Sol 11: The equation of the given line is

$\vec{r} = 2\hat{i} - \hat{j} + 2\hat{k} + \lambda(3\hat{i} + 4\hat{j} + 2\hat{k})$ ... (i)

The equation of the given plane is

$\vec{r} \cdot \left(\hat{i} - \hat{j} + \hat{k}\right) = 5$ ... (ii)

Substituting the value of from equation (i) in equation (ii), we obtain.

$\left[2\hat{i} - \hat{j} + 2\hat{k} + \lambda\left(3\hat{i} + 4\hat{j} + 2\hat{k}\right)\right]\left(\hat{i} - \hat{j} + \hat{k}\right) = 5$

$\Rightarrow \left[(3\lambda + 2)\hat{i} + (4\lambda - 1)\hat{j}(2\lambda + 2)\hat{k}\right] \cdot \left(\hat{i} - \hat{j} + \hat{k}\right) = 5$

$\Rightarrow (3\lambda + 2) - (4\lambda - 1) + (2\lambda + 2) = 5$

$\Rightarrow \lambda = 0$

Substituting this value in equation (i), we obtain the equation of the line as $\vec{r} = 2\hat{i} - \hat{j} + 2\hat{k}$

This means that the position vector of the point of intersection of the line and the plane is $\vec{r} = 2\hat{i} - \hat{j} + 2\hat{k}$

This shows that the point of intersection of the given line and plane is given by the coordinates (2, -1, 2). The point is (-1, -5, -10).

The distance d between the points, (2, -1, 2) and (-1, -5, -10), is

$d = \sqrt{\left(-1 - 2\right)^2 + \left(-5 + 1\right)^2 + \left(-10 - 2\right)^2}$

$= \sqrt{9 + 16 + 144} = \sqrt{169} = 13$

Sol 12: $\dfrac{x-1}{a} = \dfrac{y-2}{b} = \dfrac{z-3}{c}$

Parallel to the plane $x + 5y + 4z = 0$

$\Rightarrow a + 5b + 4c = 0$

$\Rightarrow \left(-1 + 2\lambda, 2 + \lambda, -4 + 2\lambda\right) = \Rightarrow \left(1 + ka, 2 + kb, 3 + kc\right)$

$\Rightarrow \dfrac{ka + 2}{2} = \dfrac{2kb}{2} = \dfrac{7 + kc}{2}$

$\Rightarrow \dfrac{2}{2b - a} = \dfrac{7}{2b - c} = \dfrac{5}{a - c}$

$\Rightarrow 10b = 7a - 2c$ ....(i)

$\Rightarrow \dfrac{7a - 10b}{2} = \dfrac{-a - 5b}{4}$

$\Rightarrow a = b ; c = \dfrac{-3a}{2}$ $a = 2b = 2c = -3$

Sol 13: $\dfrac{x}{a} = \dfrac{y}{b} = \dfrac{z}{c} = \lambda$

$\lambda_a = 3 + 2k ; \lambda_b = 3 + k ; \lambda_c = k$

$\dfrac{1}{2} = \left|\dfrac{2a + b + c}{\sqrt{6}\sqrt{a^2 + b^2 + c^2}}\right|$ $3(a^2 + b^2 + c^2) = 2(2a + b - c)^2$

$\dfrac{3 + 2k}{a} = \dfrac{3 + k}{b} = \dfrac{k}{c}$

$\dfrac{3a - 3b}{2b - a} = \dfrac{3c}{b - c} = \dfrac{3c}{a - 2c}$

$\Rightarrow a = 1, b = 2, c = -1$ or $a = -1$ $b = 1, c = -2$

Exercise 2

Single Correct Choice Type

Sol 1: (B) Direction cosines of PQ (2, 3, –6) (3, –4, 5)

Ratios = 2 – 3, 3 + 4, –6 – 5 = –1, 7, –11

Direction cosines = $\dfrac{-1}{\sqrt{171}}, \dfrac{7}{\sqrt{171}}, \dfrac{-11}{\sqrt{171}}$

or $\dfrac{1}{\sqrt{171}}, \dfrac{-7}{\sqrt{171}}, \dfrac{11}{\sqrt{171}}$

Sol 2: (A) $\alpha = \dfrac{m+3}{m+1}$, $\beta = \dfrac{4m+1}{m+1}$, $\gamma = \dfrac{-6m-5}{m+1}$

As $\alpha = 0 \Rightarrow m = -3$[A]

Sol 3: (D) 3x + 2y + z + 5 = x + y – 2z – 3

2x – y – λz = 7x + 10y – 8z are ⊥ to each other

1. $\begin{vmatrix} i & j & k \\ 3 & 2 & 1 \\ 1 & 1 & -2 \end{vmatrix}$ = i(–5) – j (–6 –1) + k(3–2) = –5i + 7j + k

2. $\begin{vmatrix} i & j & k \\ 2 & -1 & -\lambda \\ 7 & +10 & -8 \end{vmatrix}$ = i(8 +10λ) –j (–16 + 7λ) + k (+20

+ 7) –40 – 50λ + 112 – 49λ – 127 = 0 ⇒ λ = 1

Sol 4: (A)

```
     •————m————•————1————•
   (2,4,5)   (a,b,g)   (3,5,7)
```

$\alpha = \dfrac{3m+2}{m+1} = 0 \Rightarrow m = \dfrac{-2}{3}$

Sol 5: (B) $\cos\alpha = \dfrac{1}{\sqrt{3}}$

$\cos^2\alpha + \cos^2\beta + \cos^2\gamma + \cos^2\delta = \dfrac{1}{3} + \dfrac{1}{3} + \dfrac{1}{3} + \dfrac{1}{3} = \dfrac{4}{3}$

Sol 6: (A) $\alpha(x-a) + \beta(y-b) + \gamma(z-c) = 0$

$A\left(\dfrac{\beta b + \gamma c + \alpha a}{\alpha}, 0, 0\right)$

$(h, k, l) = \left(\dfrac{\alpha a + \beta b + \gamma c}{\alpha}, \dfrac{\alpha a + \beta b + \gamma c}{\beta}, \dfrac{\alpha a + \beta b + \gamma c}{\gamma}\right)$

$(h - a)\alpha = \beta b + \gamma c$

$(k - b)\beta = \alpha a + \gamma c$

$(l - c)\gamma = \alpha a + \beta b$

$\begin{vmatrix} h-a & -b & -c \\ -a & k-b & -c \\ -a & -b & l-c \end{vmatrix} = 0$

$(h - \alpha)\left[(k-b)(l-c) - bc\right] + b\left[-al + ac - ac\right] - c\,[ab + ak - ab] = 0$

$(h - \alpha)\,[kl - kc - bl] - bal - cak = 0$

$hkl - hkc - hbl - akl + akc + abl - bal - cak = 0$

$ayz + bzx + cxy = xyz$

Sol 7: (A) $\dfrac{2x - y + 2z + 3}{3} = -\dfrac{(3x - 2y + 6z + 8)}{7}$

$p_1p_2 + q_1q_2 + r_1r_2 = 6 + 2 + 12 > 0$

+ve → acute

23x – 13y + 32z + 45 = 0[C]

Sol 8: (C) $\dfrac{x - 4/3}{2} = \dfrac{y + 6/5}{3} = \dfrac{z - 3/2}{4} = \lambda$

$\left(\dfrac{4}{3} + 2\lambda, \dfrac{-6}{5} + 3\lambda, \dfrac{3}{2} + 4\lambda\right)$

$\left(\dfrac{5}{3}, \dfrac{8}{5}, \dfrac{9}{2}\right) = \left(\dfrac{4}{3} + 5\lambda, \dfrac{-6}{5} + 8\lambda, \dfrac{3}{2} + 9\lambda\right)$

Both passes through $\left(\dfrac{+4}{3}, \dfrac{-6}{5}, \dfrac{3}{2}\right)$

Minimum distance is zero.

Sol 9: (C) $\dfrac{x}{2} + \dfrac{y}{\beta} = \dfrac{z}{\gamma}$

$\Rightarrow \alpha(b + c) + \beta(a + c) + \gamma(a + b) = 0$

$\Rightarrow \alpha(b - c) + \beta(c - a) + \gamma(a - b) = 0$

$\Rightarrow \alpha b + \beta c + \gamma a = 0$; $\alpha c + \beta a + \gamma b = 0$

$\Rightarrow \alpha = a^2 - bc$

$\Rightarrow \beta = b^2 - ac$

$\Rightarrow \gamma = c^2 - ab$

$\Rightarrow \dfrac{x}{a^2 - bc} = \dfrac{y}{b^2 - ac} = \dfrac{z}{c^2 - ab}$

Assertion Reasoning Type

Sol 10: (C) y + z + 1 = 0 [0, 1, 1]

x-axis [1, 0, 0]

$\sin\theta = 0$

R is wrong.

Previous Years' Questions

Sol 1: (D) Let the equation of plane be

$a(x - 1) + b(y + 2) + c(z - 1) = 0$ which is perpendicular to $2x - 2y + z = 0$ and

$x - y + 2z = 4$.

$\Rightarrow 2a - 2b + c = 0$ and $a - b + 2c = 0$

$\Rightarrow \dfrac{a}{-3} = \dfrac{b}{-3} = \dfrac{c}{0} \Rightarrow \dfrac{a}{1} = \dfrac{b}{1} = \dfrac{c}{0}$.

So, the equation of plane is

$x - 1 + y + 2 = 0$ or $x + y + 1 = 0$

Its distance from the point (1, 2, 2) is $\dfrac{|1 + 2 + 1|}{\sqrt{2}} = 2\sqrt{2}$.

Sol 2: (A) Given $\vec{OQ} - (1 - 3\mu)\hat{i} + (\mu - 1)\hat{j} + (5\mu + 2)\hat{k}$,

$\vec{OP} = 3\hat{i} + 2\hat{j} + 6\hat{k}$ (Where O is origin)

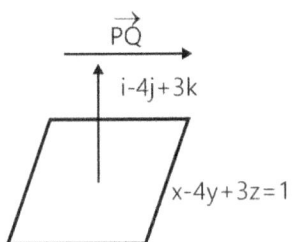

Now, $\vec{PQ} = (1 - 3\mu - 3)\hat{i} + (\mu - 1 - 2)\hat{j} + (5\mu + 2 - 6)\hat{k}$

$= (-2 - 3\mu)\hat{i} + (\mu + 3)\hat{j} + (5\mu - 4)\hat{k}$

$\because \vec{PQ}$ is parallel to the plane

$x - 4y + 3z = 1$

$\therefore -2 - 3\mu - 4\mu + 12 + 15\mu - 12 = 0$

$\Rightarrow 8\mu = 2 \Rightarrow \mu = \dfrac{1}{4}$

Sol 3: (C) Since, $l = m = n = \dfrac{1}{\sqrt{3}}$

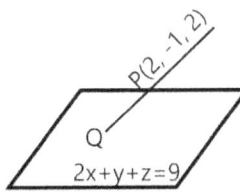

$\therefore$ Equation of line are $\dfrac{x - 2}{1/\sqrt{3}} = \dfrac{y + 1}{1/\sqrt{3}} = \dfrac{z - 2}{1/\sqrt{3}}$

$\Rightarrow x - 2 = y + 1 = z - 2 = r$

$\therefore$ Any point on the line is $Q \equiv (r + 2, r - 1, r + 2)$

$\because$ Q lies on the plane $2x + y + z = 9$

$\Rightarrow 4r + 5 = 9 \Rightarrow r = 1$

$2(x + 2) + (r - 1) + (r + 2) = 9$

$\therefore Q(3, 0, 3)$

$\therefore PQ = \sqrt{(3 - 2)^2 + (0 + 1)^2 + (3 - 2)^2} = \sqrt{3}$

Sol 4: (D) Given planes are $3x - 6y - 2z = 15$ and $2x + y - 2z = 5$ For $z = 0$, we get $x = 3$,

$y = -1$

Direction ratios of planes are $\langle 3, -6, -2 \rangle$ and $\langle 2, 1, -2 \rangle$

then the DR's of line of intersection of planes is $\langle 14, 2, 15 \rangle$ and line is

$\dfrac{x - 3}{14} = \dfrac{y + 1}{2} = \dfrac{z - 0}{15} = \lambda$ (say)

$\Rightarrow x = 14\lambda + 3, y = 2\lambda - 1, z = 15\lambda$

Hence, statement I is false.

But statement II is true.

Sol 5: (D) Given three planes are

$P_1 : x - y + z = 1$ ... (i)

$P_2 : x + y - z = -1$ ... (ii)

and $P_3 : x - 3y + 3z = 2$ ... (iii)

Solving Eqs. (i) and (ii), we have $x = 0$, $z = 1 + y$

which does not satisfy Eq. (iii)

As, $x - 3y + 3z = 0 - 3y + 3(1 + y) = 3(\neq 2)$

$\therefore$ Statement-II is true.

Next, since we know that direction ratio's of line of intersection of planes $a_1 x + b_1 y + c_1 z + d_1 = 0$

and $a_2 x + b_2 y + c_2 z + d_2 = 0$ is

$b_1 c_2 - b_2 c_1, c_1 a_2 - a_1 c_2, a_1 b_2 - a_2 b_1$

Using above result.

Direction ratio's of lines L_1, L_2 and L_3 are 0, 2, 2 ; 0; $-4, -4; 0, -2, -2$

Respectively

$\Rightarrow$ All the three lines L_1, L_2, and L_3 are parallel pairwise.

$\therefore$ Statement-I is false.

Sol 6: (B) The equation of given lines in vector form may be written as

$L_1 : \vec{r} = (-\hat{i} - 2\hat{j} - \hat{k}) + \lambda(3\hat{i} + \hat{j} + 2\hat{k})$

and $L_2 : \vec{r} = (2\hat{i} - 2\hat{j} + 3\hat{k}) + \mu(\hat{i} + 2\hat{j} + 3\hat{k})$

Since, the vector perpendicular to both L_1 and L_2.

$$\therefore \begin{vmatrix} \hat{i} & \hat{j} & \hat{k} \\ 3 & 1 & 2 \\ 1 & 2 & 3 \end{vmatrix} = -\hat{i} - 7\hat{j} + 5\hat{k}$$

$\therefore$ Required unit vector

$$= \frac{(-\hat{i} - 7\hat{j} + 5\hat{k})}{\sqrt{(-1)^2 + (-7)^2 + (5)^2}} = \frac{1}{5\sqrt{3}}(-\hat{i} - 7\hat{j} + 5\hat{k})$$

Sol 7: (D) The shortest distance between L_1 and L_2 is

$$\left| \frac{((2-(-1))\hat{i} + (2-2)\hat{j} + (3-(-1))\hat{k}) \cdot (-\hat{i} - 7\hat{j} + 5\hat{k})}{5\sqrt{3}} \right|$$

$$= \left| \frac{(3\hat{i} + 4\hat{k}) \cdot (-\hat{i} - 7\hat{j} + 5\hat{k})}{5\sqrt{3}} \right| = \frac{17}{5\sqrt{3}} \text{unit.}$$

Sol 8: (C) The equation of the plane passing through the point $(-1, -2, -1)$ and whose normal is perpendicular to both the given lines L_1 and L_2 may be written as

$(x+1) + 7(y+2) - 5(z+1) = 0$

$\Rightarrow x + 7y - 5z + 10 = 0$

$$= \left| \frac{1+7-5+10}{\sqrt{1+49+25}} \right| = \frac{13}{\sqrt{75}} \text{units.}$$

Match the Columns

Sol 9: $A \to r; B \to q; C \to p; D \to s$

$$\text{Let } \Delta = \begin{vmatrix} a & b & c \\ b & c & a \\ c & a & b \end{vmatrix}$$

$$= -\frac{1}{2}(a+b+c)[(a-b)^2 + (b-c)^2 + (c-a)^2]$$

(A) If $a + b + c \neq 0$ and $a^2 + b^2 + c^2$

$\Rightarrow \Delta = 0$ and $a = b = c \neq 0$

The equations represent identical planes.

(B) $a + b + c = 0$ and $a^2 + b^2 + c^2 \neq ab + bc + ca$

$\Rightarrow \Delta = 0$

$\Rightarrow$ the equations have infinitely many solutions.

$ax + by = (a+b)z$

$bx + cy = (b+c)z$

$\Rightarrow (b^2 - ac)y = (b^2 - ac)z \Rightarrow y = z$

$\Rightarrow ax + by + cy = 0$

$\Rightarrow ax = ay$

$\Rightarrow x = y = z.$

(C) $a + b + c \neq 0$ and $a^2 + b^2 + c^2 \neq ab + bc + ca$

$\Rightarrow \Delta \neq 0$

$\Rightarrow$ The equation represent planes meeting at only one point.

(D) $a + b + c = 0$ and $a^2 + b^2 + c^2 = ab + bc + ca$

$\Rightarrow a = b = c = 0$

$\Rightarrow$ The equations represent whole of the three dimensional space.

Sol 10: (i) Equations of a plane passing through $(2, 1, 0)$ is

$a(x-2) + b(y-1) + c(z) = 0$

It also passes through $(5, 0, 1)$ and $(4, 1, 1)$

$3a - b + c = 0$ and $2a + 0b + c = 0$

On solving, we get $\dfrac{a}{-1} = \dfrac{b}{-1} = \dfrac{c}{2}$

$\therefore$ Equation of plane is

$-(x-2) - (y-1) + 2(z-0) = 0$

$\Rightarrow -x + 2 - y + 1 + 2z = 0$

$\Rightarrow x + y - 2z = 3$

(ii) Let the coordinate of Q (α, β, γ)

Equation of line $PQ = \dfrac{x-2}{1} = \dfrac{y-1}{1} = \dfrac{z-6}{-2}$

Since, mid point of P and Q is $\left(\dfrac{\alpha+2}{2}, \dfrac{\beta+1}{2}, \dfrac{\gamma+6}{2} \right)$.

Which lies in a line P

$$\Rightarrow \frac{\dfrac{\alpha+2}{2} - 2}{1} = \frac{\dfrac{\beta+1}{2} - 1}{1} = \frac{\dfrac{\gamma+6}{2} - 6}{-2}$$

$$= \frac{1\left(\dfrac{\alpha+2}{2} - 2 \right) + 1\left(\dfrac{\beta+1}{2} - 1 \right) - 2\left(\dfrac{\gamma+6}{2} - 6 \right)}{1 \cdot 1 + 1 \cdot 1 + (-2)(-2)} = 2$$

$$\left\{ \text{since,} \left(\dfrac{\alpha+2}{2} \right) + 1\left(\dfrac{\beta+1}{2} \right) - 2\left(\dfrac{\gamma+6}{2} \right) = 3 \right\}$$

$\Rightarrow \alpha = 6, \beta = 5, \gamma = -2$

$\Rightarrow$ Q$(6, 5, -2)$

Sol 11: Let the equation of the plane ABCD be ax + by + cz + d = 0, the point A″ be (α, β, γ) and the height of the parallelepiped ABCD be h.

$$\Rightarrow \frac{|a\alpha + b\beta + c\gamma + d|}{\sqrt{a^2 + b^2 + c^2}} = 90\% h$$

$$\Rightarrow a\alpha + b\beta + c\gamma + d = \pm 0.9h\sqrt{a^2 + b^2 + c^2}$$

∴ Locus is, ax + by + cz + d = $\pm 0.9h\sqrt{a^2 + b^2 + c^2}$

∴ Locus of A″ is a plane parallel to the plane ABCD.

Sol 12: (B, C, D) According to given data, we have

$$P(3,0,0), Q(3,3,0), R(0,3,0), S\left(\frac{3}{2}, \frac{3}{2}, 3\right)$$

$$\overrightarrow{OQ} = 3\hat{i} + 3\hat{j}$$

$$\overrightarrow{OS} = \frac{3}{2}\hat{i} + \frac{3}{2}\hat{j} + 3\hat{k}$$

$$\overrightarrow{OQ}.\overrightarrow{OS} = |\overrightarrow{OQ}||\overrightarrow{OS}|\cos\phi$$

$$\frac{9}{2} + \frac{9}{2} = 9\sqrt{2} \times \frac{3\sqrt{3}}{\sqrt{2}}\cos\phi \Rightarrow 9 = 9\sqrt{3}\cos\phi$$

$$\Rightarrow \phi = \cos^{-1}\left(\frac{1}{\sqrt{3}}\right)$$

The equation of plane containing ∆OQR is x − y = 0

The ⊥ distance of point (3, 0, 0) from the plane x − y = 0 is given by

$$= \left|\frac{3-0}{\sqrt{2}}\right| = \frac{3}{\sqrt{2}}$$

The equation of RS
Direction ratios of RS $< \frac{-3}{2}, \frac{3}{2}, -3 >$ or $< 1, -1, 2 >$

Equation of line RS $\frac{x}{1} = \frac{y-3}{-1} = \frac{z}{2} = r$

$\Rightarrow$ point on line $(r, 3-r, 2r)$

$r + (3-r)(-1) + 2(2r) = 0 \Rightarrow r - 3 + r + 4r = 0$

$\Rightarrow r = \frac{1}{2} \Rightarrow$ point $\left(\frac{1}{2}, \frac{5}{2}, 1\right)$

Perpendicular distance

$$= \sqrt{\left(\frac{1}{2}\right)^2 + \left(\frac{5}{2}\right)^2 + 1} = \sqrt{\frac{1}{4} + \frac{25}{4} + 1} = \sqrt{\frac{30}{4}} = \sqrt{\frac{15}{2}}$$

Sol 13: (C) Let $P^1(3, 1, 7)$

The image of P′ given by

$$\frac{x-3}{1} = \frac{y-1}{-1} = \frac{z-7}{1} = -\frac{2(3-1+7-3)}{3} = -4$$

$$\Rightarrow P(x, y, z) \equiv (-1, 5, 3)$$

Any plane passing through $P(-1,5,3)$ and containing line $\frac{x}{1} = \frac{y}{2} = \frac{z}{1}$

$$\begin{vmatrix} x & y & z \\ -1 & 5 & 3 \\ 1 & 2 & 1 \end{vmatrix} = 0$$

$$x(5-6) - y(-1-3) + z(-2-5) = 0$$

$$\Rightarrow x - 4y + 72 = 0$$

Sol 14: (B, D) Any plane passes through point of intersection of plane P_1 and P_2 is $x + z - 1 + \lambda y = 0$

Given:

$$\left|\frac{0+0-1+\lambda}{\sqrt{1+1+\lambda^2}}\right| = 1 \Rightarrow |\lambda - 1| = \sqrt{\lambda^2 + 2} \Rightarrow \lambda = -\frac{1}{2}$$

$\Rightarrow P_3$ is $2x - y + 2z = 2$

Now, distance of P_3 from (α, β, γ) is 2.

$$\Rightarrow \left|\frac{2\alpha - \beta + 2\gamma - 2}{\sqrt{4+4+1}}\right| = 2$$

$\Rightarrow 2\alpha - \beta + 2\gamma = 8$ and $2\alpha - \beta + 2\gamma = -4$

Sol 15: (A, B) Since all the points on L are at same distance from planes P_1 and P_2 implies that line L is parallel to line of intersection of P_1 and P_2.

Let direction ratio of line L be α, β, γ then

$\alpha + 2\beta - \gamma = 0$ and $2\alpha - \beta + \gamma = 0$

$\Rightarrow \alpha : \beta : \gamma \equiv 1 : -3 : -5$

Eq. of line L passes through origin

$$\frac{x-0}{1} = \frac{y-0}{-3} = \frac{z-0}{-5} = r$$

Foot of perpendicular from origin to the plane $P_1 \equiv x + 2y - z + 1 = 0$ can be obtained as

$$\frac{x-0}{1} = \frac{y-0}{2} = \frac{z-0}{-1} = \frac{-(0+0-0+1)}{1+4+1} = \frac{-1}{6}$$

$$\Rightarrow \left(\frac{-1}{6}, \frac{-1}{3}, \frac{1}{6} \right)$$

Now equation of perpendicular from any point on L is

$$\frac{x+\frac{1}{6}}{1} = \frac{y+\frac{1}{3}}{-3} = \frac{z-\frac{1}{6}}{-5} = \lambda$$

Any point on line $\left(\lambda - \frac{1}{6}, -3\lambda - \frac{1}{3}, -5\lambda + \frac{1}{6} \right)$

Point $\left(0, \frac{-5}{6}, \frac{-2}{3} \right)$ and $\left(\frac{-1}{6}, \frac{-1}{3}, \frac{1}{6} \right)$ satisfy the line.

Sol 16: (C) Given: $P \equiv (\lambda, \lambda, \lambda)$

$$L_1 \equiv \frac{x-0}{1} = \frac{y-0}{1} = \frac{z-1}{0} = m$$

$$L_2 \equiv \frac{x-0}{1} = \frac{y-0}{1} = \frac{z-1}{0} = n$$

$$\Rightarrow Q \equiv (m, m, 1)$$

$$\Rightarrow R \equiv (n, -n, -1)$$

$$\overrightarrow{PQ} = (\lambda-m)\hat{i} + (\lambda-m)\hat{j} + (\lambda-1)\hat{k}$$

Since $\overrightarrow{PQ}$ is perpendicular to L_1

$$\Rightarrow \lambda - m + \lambda - m + 0 = 0 \quad \Rightarrow \lambda = m \quad \Rightarrow Q(\lambda, \lambda, 1)$$

Similarly, $R = (0, 0, -1)$

Now, $PQ \perp PR$

$$\Rightarrow (\lambda-m).(\lambda-n) + (\lambda-m).(\lambda+n) + (\lambda-1).(\lambda+1) = 0$$

$$\Rightarrow 0 + 0 + (\lambda-1).(\lambda+1) = 0 \quad \Rightarrow \lambda = \pm 1$$

Negotiating $\lambda = 1$ became points p and Q will coincide. $\lambda = -1$

Sol 17: (D) Let any point P on the line $\frac{x+2}{2} = \frac{y-1}{-1} = \frac{z}{3}$ be $(2\gamma-2, -\gamma-1, 3\gamma)$

P lies on the plane $x + y + 2 = 3$

$$\Rightarrow 2\gamma - 2(-\gamma-1) + 3\gamma = 3 \Rightarrow 4\gamma = 6 \Rightarrow \gamma = \frac{3}{2}$$

$$P \equiv \left(1, \frac{-5}{2}, \frac{9}{2} \right)$$

Point $(-2, -1, 0)$ lies on the line, the feet of perpendicular Q is given by

$$\frac{x+2}{1} = \frac{y+1}{1} = \frac{z-0}{1} = -\frac{(-2-1+0-3)}{1^2+1^2+1^2}$$

$$\Rightarrow Q \equiv (0, 1, 2)$$

Direction ratio of line PQ joining feets of perpendicular are $\left(1, \frac{-7}{2}, \frac{5}{2} \right)$

Equation of PQ $\frac{x}{2} = \frac{y-1}{-7} = \frac{z-2}{5}$

Sol 18: (A, D) Given lines

$$L_1 \equiv \frac{x-5}{0} = \frac{y-0}{3-\alpha} = \frac{z-0}{-2}$$

$$L_2 \equiv \frac{x-\alpha}{0} = \frac{y}{-1} = \frac{z}{2-\alpha}$$

L_1 and L_2 will be co-planar, then

$$\begin{vmatrix} 0 & 3-\alpha & 2 \\ 0 & -1 & 2-\alpha \\ 5-\alpha & 0 & 0 \end{vmatrix} = 0$$

$$\Rightarrow (5-\alpha)\left[(3-\alpha)(2-\alpha) + 2 \right] = 0$$

$$\Rightarrow (5-\alpha)(\alpha-1)(\alpha-4) = 0$$

$$\Rightarrow \alpha = 1, 4, 5$$

Sol 19: (A) $L_1 \equiv \frac{x-1}{2} = \frac{y}{-1} = \frac{z+3}{1} = r_1$

$$L_2 \equiv \frac{x-y}{1} = \frac{y+3}{1} = \frac{z+3}{2} = r_2$$

For point of intersection of L_1 and L_2

$$2r_1 + 1 = r_2 + 4 \Rightarrow 2r_1 - r_1 = 3 \qquad \dots \text{(i)}$$

$$-r_1 = r_2 - 3 \qquad \dots \text{(ii)}$$

and $r_1 - 3 = 2r_3 - 3$ \qquad \dots \text{(iii)}

Form (i), (ii), (iii), we get $r_1 = 2$, $r_2 = 1$

The point of intersection $(5, -2, -1)$

Now, direction ratio of plane $\perp$ to P_1 and P_2 given by

$$\begin{vmatrix} i & j & k \\ 3 & 5 & -6 \\ 7 & 1 & 2 \end{vmatrix}$$

$$= i(10+6) - j(6+42) + k(3-35)$$

$$= 16i - 48j - 32k$$

Any plane passes through $(5, -2, -1)$ and having direction ratio of normal

$$16(x-5) - 48(y+2) - 32(z+1) = 0$$

$$\Rightarrow (x-5) - 3(y+2) - 2(z+1) = 0$$

$$\Rightarrow x - 3y - 2z = 13$$

$$\Rightarrow a = 1, b = -3, c = -2 \text{ and } d = 13$$

Sol 20: (A) Given: $Q(2,3,5)$ $R(1,-1,4)$

Direction ratio of line QR is $(1,4,1)$

The eq. of QR

$$\frac{x-2}{1} = \frac{y-3}{4} = \frac{z-5}{1} = r$$

Any point on it $P(r+2, 4r+3, r+5)$

P lies on the plane $5x - 4y - z = 1$

$$5(r+2) - 4(4r+3) - (r+5) = 1$$

$$\Rightarrow 5r + 10 - 16r - 12 - r - 5 = 1$$

$$\Rightarrow r = -\frac{8}{12} = -\frac{2}{3}$$

$$\Rightarrow P \equiv \left(\frac{4}{3}, \frac{1}{3}, \frac{13}{3}\right)$$

$$PT = \sqrt{\left(\frac{4}{3} - 2\right)^2 \left(\frac{1}{3} - 1\right)^2 + \left(\frac{13}{3} - 4\right)^2}$$

Now, direction ration of PT is $(2, 2, -1)$

Angle between PT and QR

$$\cos\theta = \frac{1\times2 + 4\times2 + 1\times-1}{\sqrt{1+16+1}\sqrt{4+4+1}}$$

$$= \frac{9}{3\sqrt{2}\times3} = \frac{1}{\sqrt{2}}$$

$$\Rightarrow \theta = 45^0$$

$$TS = PS = \frac{1}{\sqrt{2}}$$

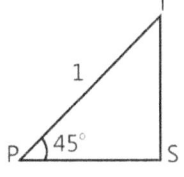

Sol 21: (A) Eq of plane

$$x + 2y + 3z - 2 + k(x - y + z - 3) = 0$$

$$\Rightarrow x(1+k) + y(2-k) + z(3+k) - 2 - 3k = 0$$

Distend from point $(3,1,-1)$ is $\frac{2}{\sqrt{3}}$

$$\left|\frac{3(1+k) + 1(2-k) - 1(3+k) - (2+3k)}{\sqrt{(1+k)^2 + (2-k)^2 + (3+k)^2}}\right|$$

On solving, we get $k = \frac{-7}{3} = \frac{2}{\sqrt{3}}$

Eq. of plane is $5x - 11y + z = 17$

Sol 22: (B, C) The lines $\frac{x-1}{2} = \frac{y+1}{k} = \frac{z}{2}$

and $\frac{x+1}{5} = \frac{y+1}{2} = \frac{z}{k}$

are coplanar, then $\begin{vmatrix} 2 & 0 & 0 \\ 2 & k & 2 \\ 5 & 2 & k \end{vmatrix} = 0$

$$2(k^2 - 4) = 0 \Rightarrow k = \pm 2$$

For $k = 2$, the lines are

$$\frac{x-1}{2} = \frac{y+1}{2} = \frac{z}{2} \text{ and } \frac{x+1}{5} = \frac{y+1}{2} = \frac{z}{2}$$

Clearly plane $y + 1 = z$ contains both the lines

For $k = -2$, the lines are

$$\frac{x-1}{2} = \frac{y+1}{2} = \frac{z}{2} \text{ and } \frac{x+1}{5} = \frac{y+1}{2} = \frac{z}{-2}$$

From options plane $y + z + 1 = 0$ also contains both the lines.

Sol 23: Let the direction ratio of plane containing both the given lines are a, b, c then

$$2a + 3b + 4c = 0 \text{ and } 3a + 4b + 5c = 0$$

$$\Rightarrow \frac{a}{-1} = \frac{b}{2} = \frac{c}{-1}$$

Now, the equation of plane is

$$a(x - 2) + b(y - 3) + c(z - 4) = 0$$

$$\Rightarrow -(x - 2) + 2(y - 3) - (z - 4) = 0$$

$$\Rightarrow -x + 2 + 2y - 6 - z + 4 = 0$$

$$\Rightarrow -x + 2y - z = 0$$

$$\Rightarrow x - 2y + z = 0$$

Distance between planes

$$\frac{|d - 0|}{\sqrt{1+4+1}} = \sqrt{6} \Rightarrow |d| = 6$$

Sol 24: (A) Distance of point $p(1, -2, 1)$

From plane $x + 2y - 2z = \alpha$ is 5, then

$$\left|\frac{1 + 2x - 2 - 2\times1 - \alpha}{\sqrt{1+4+4}}\right| = 5$$

$$|-5 - \alpha| = 5$$

$$\Rightarrow \alpha = 10$$

For foot of perpendicular is M, then

$$\frac{x-1}{1} = \frac{y+2}{2} = \frac{z-1}{-2} = \lambda$$

$M \equiv \left(\lambda+1,\ 2\lambda-2,\ -2\lambda+1\right)$ lies on plane

$$\Rightarrow \lambda+1+2(2\lambda-2)-2(2\lambda+1)-10 = 0$$
$$\Rightarrow \lambda+1+4\lambda-4+4\lambda-2-10 = 0$$
$$\Rightarrow 9\lambda = 15$$
$$\Rightarrow \lambda = \frac{5}{3} \quad \Rightarrow Q\left(\frac{8}{3}, \frac{4}{3}, \frac{-7}{3}\right)$$

Sol 25: (B) Angle between $\overrightarrow{AB}$ and $\overrightarrow{AD}$

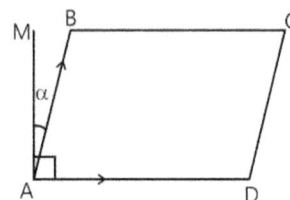

$$\cos\theta \frac{(2i+10j+11k).(-i+2j+2k)}{\sqrt{4+100+121}\sqrt{1+4+4}} = \frac{40}{15\times3} = \frac{8}{9}$$

$$\Rightarrow \alpha = \frac{\pi}{2} - \cos^{-1}\left(\frac{8}{9}\right) = \sin^{-1}\left(\frac{8}{9}\right) = \cos^{-1}\left(\frac{\sqrt{17}}{9}\right)$$

$$\Rightarrow \cos\alpha = \left(\frac{\sqrt{17}}{9}\right)$$

3. PROBABILITY

1. INTRODUCTION

There are various phenomena in nature leading to an outcome, which cannot be predicted beforehand. For example, tossing a coin may result into two outcomes- a head or a tail. Probability theory aims to measure the uncertainties of such outcomes. Consequently, probability is the measure of uncertainty of random experiments.

2. RANDOM EXPERIMENT

An experiment is said to be random if it has more than one possible known outcomes which cannot be predicted in advance. For example - Throwing of a die is a random experiment.

Sample Space: The set of all possible outcomes of a trial (random experiment) is called its sample space. It is generally denoted by S and each outcome of the trial is said to be a sample point.

For example - In throwing of a die, the sample space for the number that shows up on the top face would be:

S = {1, 2, 3, 4, 5, 6}

EVENT: Every subset of a sample space is called an event.

For example, in throwing a dice, the sample space

S = {1, 2, 3, 4, 5, 6} and n(S) = 6

E_1 = {1, 3, 5} ⊂ S. So E_1 is an event and n (E_1) = 3.

The event E_1 = {1, 3, 5} can also be expressed as the event of getting an odd number in throwing a dice.

(a) **Simple event:** A simple event or an elementary event is an event containing only a single sample point.

(b) **Compound events:** Compound events or decomposable events are those events that are obtained by combining together two or more elementary events.

For instance, the event of drawing a heart from a deck of cards is the subset A = {heart} of the sample space S = {heart, spade, club, diamond}. Therefore, A is a simple event. The event B of drawing a red card is a compound event since B = {heart U diamond} = {heart and diamond}.

(c) **Mutually exclusive or disjoint events:** Events are said to be mutually exclusive or disjoint or incompatible if the occurrence of any one of them prevents the occurrence of all the other events.

(d) **Mutually non-exclusive events:** The events which are not mutually exclusive are known as compatible events or mutually nonexclusive events.

(e) **Independent events:** Events are said to be independent, if the happening(or non-happening) of one event is not affected by the happening (or non-happening) of other events.

(f) Dependent events: Two or more events are said to be dependent, if the happening of one event affects (partially or totally) the other event.

The relationship between mutually exclusive and independent events

Mutually exclusive events can't happen at the same time. Mathematically put,

Independent events: $P(A \text{ and } B) = P(A) \times P(B)$

Mutually exclusive: $P(A \text{ and } B) = 0$, where A and B are two events.

$\rightarrow$ On comparing two definitions, we see that **the events can't be independent and Mutually exclusive at the same time.**

Equally likely events: Events which have the same chance of occurring are said to be equally likely events.

For example, in the experiment of tossing a coin,

where,

A: The event of getting a "HEAD" and

B: The event of getting a "TAIL"

Events "A" and "B" are said to be equally likely events.

[Both the events have the same chance of occurrence]

In the experiment of throwing a die,

where,

A: The event of getting 1

B: The event of getting 2

...

...

F: The event of getting 6

Events "A", "B", "C", "D", "E" and "F" are said to be equally likely events.

[All these events have the same chance of occurrence.]

3.COMPLEMENT OF EVENTS

The complement of an event 'A' with respect to a sample space S is the set of all elements of 'S' which are not in A. It is usually denoted by A', $\bar{A}$ or A^c

(a) The union $E_1 \cup E_2$ of events E_1 and E_2 is the event of at least one of the events E_1, E_2 happening.

(b) The intersection $E_1 \cap E_2$ of events E_1 and E_2 is the event of both the events E_1, E_2 happening.

e.g. Tossing of coin sample space s = {H, T}. Event of getting head in tossing of coin A = {H} $\Rightarrow$ A^C = {T}

NOMORECLASS CONCEPTS

Terminology: Being closely familiar with the terminology of probability helps a lot in thinking with clearly. In particular, always think of outcomes as the most elementary results of an experiment, events as a set of outcomes, the sample space as the set of all possible outcomes and events as subsets of the sample space.

4. ALGEBRA OF EVENTS

Verbal description of the event	Equivalent set theoretic notation
Not A	$\bar{A}$
A or B (at least one of A or B)	$A \cup B$
A and B	$A \cap B$
A but not B	$A \cap \bar{B}$
Neither A nor B	$\bar{A} \cap \bar{B}$
At least one of A, B or C	$A \cup B \cup C$
Exactly one of A and B	$(A \cap \bar{B}) \cup (\bar{A} \cap B)$
All three of A, B and C	$A \cap B \cap C$
Exactly two of A, B and C	$(A \cap B \cap \bar{C}) \cup (A \cap \bar{B} \cap C) \cup (\bar{A} \cap B \cap C)$

5.PROBABILITY

If a random experiment results in n mutually exclusive, equally likely and exhaustive outcomes out of which m are favorable to the occurrence of an event A, then the probability of occurrence of A is given by

$$P(A) = \frac{m}{n} = \frac{\text{Number of outcomes favourable to A}}{\text{Number of total outcomes}}$$

It is obvious that $0 \leq m \leq n$. If an event A is certain to happen, then m = n, thus P (A) = 1.

If A is impossible to happen, then m = 0 and so P (A) = 0.

Hence we conclude that $0 £P(A) \leq 1$

NOMORECLASS CONCEPTS

Working Rule to find probability

Step 1. For the given experiment, find out all possible outcomes n(S) of the sample space.

Step 2. Identify the outcomes n (A), which are favorable to the event A, whose probability is required.

Step 3. Apply the formula to find P (A),

$$P(A) = \frac{n(A)}{n(S)}$$

Equal Likelihood, the formula that we apply to calculate probability,

$\dfrac{\text{Number of favorable outcome}}{\text{Number of total outcome}}$, is valid only when all the cases have equal likelihood of occurrence.

This important point is overlooked a lot of times. For example, if a rolling die is not fair, then you cannot assign a probability of $\frac{1}{6}$ for each face showing up. Sometimes, the way you count the total cases and favorable cases can lead to a mistake. Consider a random experiment involving the rolling of two dice simultaneously. Suppose you have to evaluate the probability of getting a total of less than 6. The following argument has a mistake: "There are a total of 11 possible cases, namely {2,3,4,5,6,7,8,9,10,11,12}, out of which 4 are favorable, namely {2,3,4,5}, and thus the required probability is $\frac{4}{11}$."

The mistake is that, the different cases do not have equal likelihood of occurrence. For example, the sum 6 is more likely to occur than the sum 2 (why?). The correct way to solve this problem would be to consider the 36 equally likely outcomes (x, y) where x and y can take integer values from 1 to 6, and then consider those outcomes from this set of 36 outcomes, which leads to a sum of less than 6. You can verify that there will be 10 such favorable outcomes. And now, it would be correct to apply the formula

$\dfrac{\text{Number of favorable outcome}}{\text{Number of total outcome}}$ to obtain the required probability as $\dfrac{10}{36}$.

Illustration 1: In a single case with two fair dice, find the chance of getting

(A) Two 4's (B) A doublet (C) Five-six (D) A sum of 7 **(JEE MAIN)**

Sol: Write all the possible outcomes and the favorable events in each case.

(A) There are 6 × 6 equally likely cases (as any face of any die may turn up)

$\Rightarrow$ 36 possible outcomes. For this event, only one outcome (4 – 4) is favourable

$\therefore$ Probability = 1/36.

(B) A doublet can occur in six ways {(1, 1), (2, 2), (3, 3), (4, 4) (5, 5), (6, 6)}.

Therefore, probability of doublet = 6/36 = 1/6.

(C) Two favorable outcomes {(5, 6), (6, 5)}.Therefore, probability = 2/36 = 1/18.

Sample space = (1, 1), (1, 2), (1, 3), (1, 4), (1, 5), (1, 6)

(2, 1), (2, 2), (2, 3), (2, 4), (2, 5), (2, 6)

(3, 1), (3, 2), (3, 3), (3, 4), (3, 5), (3, 6)

(4, 1), (4, 2), (4, 3), (4, 4), (4, 5), (4, 6)

(5, 1), (5, 2), (5, 3), (5, 4), (5, 5), (5, 6)

(6, 1), (6, 2), (6, 3), (6, 4), (6, 5), (6, 6)

(D) A sum of 7 can occur in the following cases {(1, 6), (2, 5), (3, 4), (4, 3), (5, 2), (6, 1)} which are 6 in number. Therefore, probability = 6/36 = 1/6.

Illustration 2: Seven accidents occur in a week. What is the probability that they take place on the same day? **(JEE MAIN)**

Sol: Find the total number of ways accidents can happen. And clearly, all the accidents can take place on the same day in 7 ways.

Total no. of cases = Total no. of ways in which 7 accidents can happen in a week (or be distributed= 7^7

Favorable No. of cases out of these = number of those in which all 7 happen on one day (any the week)= 7

$\therefore$ Required probability = $\dfrac{7}{7^7} = \dfrac{1}{7^6}$

6. MUTUAL INDEPENDENCE AND PAIRWISE INDEPENDENCE

Three events A, B, C are said to be mutually independent if, $P(A \cap B) = P(A).P(B)$, $P(A \cap C) = P(A).P(C)$, $P(B \cap C) = P(B).P(C)$ and $P(A \cap B \cap C) = P(A).P(B).P(C)$

These events would be said to be pairwise independent if, $P(A \cap B \cap C) = \Rightarrow P(A \cap B) = P(A).P(B)$, $P(B \cap C) = P(B).P(C)$ and $P(A \cap C) = P(A).P(C)$

Thus, mutually independent events are pairwise independent but the converse may not be true.

Illustration 3: From a bag containing 5 white, 7 red and 4 black balls, a man draws 3 balls at random. Find the probability of them being all white. **(JEE MAIN)**

Sol: Use the principle of restricted combination.

Total number of balls in the bag = 5 + 7 + 4 = 16

Total number of ways in which 3 balls can be drawn is $^{16}C_3 = \dfrac{16 \times 15 \times 14}{3 \times 2 \times 1} = 560$

Thus, the sample space S for this experiment has 560 outcomes i.e. n(S) = 560

Let E be the event of all the three balls being white. Total number of white balls is 5. So, the number of ways in which

3 white balls can be drawn = $^5C_3 = \dfrac{5 \times 4 \times 3}{3 \times 2 \times 1} = 10$

Thus, E has 10 element of S, $\therefore$ n(E) = 10

$\therefore$ Probability of E, $P(E) = \dfrac{n(E)}{n(S)} = \dfrac{10}{560} = \dfrac{1}{56}$

NOMORECLASS CONCEPTS

Exhaustive Event: A Set of events is said to be exhaustive if the performance of random experiments always results in the occurrence of at least one of them. For instance, consider an ordinary pack of cards. The events 'drawn card is heart', drawn card is diamond', 'drawn card is club' and 'drawn card is spade' is a set of events that is exhaustive. In other words all sample points put together (i.e. sample space itself) would give us an exhaustive event.

If 'E' is an exhaustive event then P (E) = 1.

7. ODDS IN FAVOUR, ODDS AGAINST

(a) The odds in favor of the event E = $\dfrac{P(E)}{P(E')}$.

(b) The odds against the event E = $\dfrac{P(E')}{P(E)}$

(c) If odds in favor of the event E = a: b then $P(E) = \dfrac{a}{a+b}$

(d) If odds against the event E = a: b then $P(E) = \dfrac{b}{a+b}$

8. A FEW THEOREMS ON PROBABILITY

(a) If A and B are two mutually exclusive events, then $P(A \cup B) = P(A) + P(B)$.

(b) If A is any event, then $P(A') = 1 - P(A)$

(c) If A and B are two events, then $P(A \cap B') = P(A) - P(A \cap B)$

(d) If A and B are two events, then $P(A \cup B) = P(A) + P(B) - P(A \cap B)$

(e) If A and B are two events, then

P(exactly one of A, B occurs)

$= P[(A \cap B') \cup (A' \cap B)] = P(A) - P(A \cap B) + P(B) - P(A \cap B)$

$= P(A) + P(B) - 2P(A \cap B) = P(A \cup B) - P(A \cap B)$

Also, P(exactly one of A, B occurs)

$= P(A \cap B') + P(A' \cap B) = P(B') - P(A' \cap B') + P(A') - P(A' \cap B') = P(A') + P(B') - 2P(A' \cap B')$

$= P(A' \cup B') - P(A' \cap B')$

(f) If A and B are two events, $P(A' \cup B') = 1 - P(A \cap B)$ and $P(A' \cap B') = 1 - P(A \cup B)$

(g) If $A_1, A_2, ..., A_n$ are n events, then $P(A_1 \cup A_2 \cup ... \cup A_n)$

$$= \sum_{i=1}^{n} P(A_i) - \sum_{1 \le i < j \le n} p(A_i \cap A_j) + \sum_{1 \le i < j < k \le n} P(A_i \cap A_j \cap Ak) - ... + (-1)^{n-1} P(A_1 \cap A_2) ... \cap A_n)$$

(h) If A, B and C are three events, then

$P(A \cup B \cup C) = P(A) + P(B) + P(C) - P(B \cap C) - P(C \cap A) - P(A \cap B) + (A \cap B \cap C)$

(i) P(at least two of A, B, C occur) $= P(B \cap C) + P(C \cap A) + P(A \cap B) - 2P(A \cap B \cap C)$

(ii) P(exactly two of A, B, C occur) $= P(B \cap C) + P(C \cap A) + P(A \cap B) - 3P(A \cap B \cap C)$

(iii) P(exactly one of A, B, C occurs)

$= P(A) + P(B) + P(C) - 2P(B \cap C) - 2P(C \cap A) - 2P(A \cap B) + 3P(A \cap B \cap C)$

(i) If $A_1, A_2, ..., A_n$ are n events, then

(i) $P(A_1 \cup A_2 \cup ... \cup A_n) \le P(A_1) + P(A_2) + ... + P(A_n)$

(ii) $P(A_1 \cap A_2 \cap ... \cap A_n) \ge 1 - P(A'_1) - P(A'_2) - ... - P(A'_n)$

(j) If $A_1, A_2, ..., A_n$ are n events, then $P(A_1 \cap A_2 \cap ... \cap A_n)^3 P(A_1) + P(A_2) + ... + P(A_n) - (n-1)$

(k) If A and B are two events, such that $A \subseteq B$, then $P(A) \le P(B)$

9. BOOLE'S INEQUALITY

(a) For any two events A and B

(i) $P(A \cup B) = P(A) + P(B) - P(A \cap B)$

(ii) $\therefore P(A \cup B) \le P(A) + P(B)$ $\{\because P(A \cap B) \ge 0\}$

(b) In general for any n events $A_1 A_2 ... A_n$

(i) $P(A_1 \cup A_2 \cup ... \cup A_n) \le P(A_1) + P(A_2) + ... + P(A_n)$

Illustration 4: Let A, B, C be three events. If the probability of the occurrence of one event out of A and B is $1-a$, out of B and C is $1-2a$, out of C and A is $1-a$ and that of occurrence of three events simultaneously is a^2, then prove that the probability that at least one event out of A, B, C will occur is greater than or equal to 0.5. **(JEE ADVANCED)**

Sol: Apply Boole's Inequality.

Probability that exactly one event out of A and B occur is $P(A) + P(B)=2P(A \cap B)$ and probability that exactly one event out of B and C occur is $P(B) + P(C) - 2P(B \cap C)$ and so on.

Now, $P(A \cup B \cup C) = P(A) + P(B) + P(C) - P(A \cap B) - P(A \cap C) - P(B \cap C) + P(A \cap B \cap C)$

$= \dfrac{1}{2}[P(A)+P(B) -2P(A \cap B) + P(B) + P(C)-2P(B \cap C) + P(C) + P(A) -2P(A \cap C)] + P(A \cap B \cap C)$

$\Rightarrow \dfrac{1-a+1-2a+1-a}{2} + a^2 = a^2 - 2a + \dfrac{3}{2}$. Let, $a^2 - 2a + \dfrac{3}{2} = y \Rightarrow a^2 - 2a + \dfrac{3}{2} - y = 0$

Since a is real, so $4 - 4\left(\dfrac{3}{2} - y\right) \geq 0 \Rightarrow y \geq \dfrac{1}{2}$

Illustration 5: From a pack of 52 cards, two cards are drawn at random. Find the probability of the following events:

(A) Both cards are of spade.

(B) One card is of spade and one card is of diamond. **(JEE MAIN)**

Sol: Use combination to calculate the number of favorable ways and the total number of ways in both cases.

The total number of ways in which 2 cards can be drawn $= {}^{52}C_2 = \dfrac{52 \times 51}{1 \times 2} = 26 \times 51 = 1326$

$\therefore$ Number of elements in the space S are $n(S) = 1326$

(A) Let the event that both cards are of spade be denoted by E_1. Then, $n(E_1)$ = Number of elements in E_1 = Number of ways in which 2 cards can be selected out of 13 cards of spade $= {}^{13}C_2 = \dfrac{13 \times 12}{1 \times 2} = 78$.

$\therefore$ Probability of $E_1 = P(E_1) = \dfrac{n(E_1)}{n(S)} = \dfrac{78}{1326} = \dfrac{1}{17}$.

(B) Let E_2 be the event that one card is of spade and one is of diamond. Then, $n(E_2)$ = number of elements in E_2 = number of ways in which one card of spade can be selected out of 13 spade cards and one card of diamond can be selected out of 13 diamond cards. $= {}^{13}C_1 \times {}^{13}C_1 = 13 \times 13 = 169$

$\therefore P(E_2) = \dfrac{n(E_2)}{n(S)} = \dfrac{169}{1326} = \dfrac{13}{102}$.

Illustration 6: Two numbers x and y are chosen at random from the set $\{1, 2, 3,..., 3n\}$. Find the probability that $x^2 - y^2$ is divisible by 3. **(JEE ADVANCED)**

Sol: Divide the above given set in three subsets such that the difference of any two elements in any of these three sets is divisible by 3. Use this partition of set to find the answer.

$x^2 - y^2 = (x + y)(x - y)$ and 3 is a prime number.

$\therefore x^2 - y^2$ is divisible by 3 if $x + y$ or $x - y$ is divisible by 3.

Now, $\{1, 2, 3,..., 3n\} = \{3, 6, 9,..., 3n\} \cup \{1, 4, 7,..., 3n - 2\} \cup \{2, 5, 8,..., 3n - 1\} = A \cup B \cup C$ (say).

Clearly, if x, y are selected from A or B or C then $x + y$ or $x - y$ are divisible by 3; and, if x, y are selected one from B and the other from C then $x + y$ is divisible by 3.

$\therefore$ The probability of $x^2 - y^2$ is divisible by 3

= Probability of selecting both x, y from A or B or C + probability of selecting x, y one from B and the other from C

$$= \frac{{}^{n}C_2}{{}^{3n}C_2} \times 3 + \frac{{}^{n}C_1 \times {}^{n}C_1}{{}^{3n}C_2} = \frac{3n(n-1)}{3n(3n-1)} + \frac{2n^2}{3n(3n-1)} = \frac{3n-3}{3(3n-1)} + \frac{2n}{3(3n-1)} = \frac{5n-3}{3(3n-1)}.$$

10. CONDITIONAL PROBABILITY

The probability of occurrence of an event A, given that B has already occurred is called the conditional probability of occurrence of A. It is denoted by $P(A \mid B)$. If the event B has already occurred, then the sample space reduces to B. Not the outcome favorable to the occurrence of A (given that B has already occurred) are those that are common to both A and B, that is, those which belong to $A \cap B$.

Thus, $P(A \mid B) = \dfrac{N_{A \cap B}}{N_B}$

where $N_{A \cap B}$ is the number of elements in $A \cap B$ and $N_B \neq 0$ is the number of

elements in B and N the total number of elements in

$$\text{S.} \Rightarrow P(A \mid B) = \frac{N_{A \cap B}/N}{N_B/N} = \frac{P(A \cap B)}{P(B)}$$

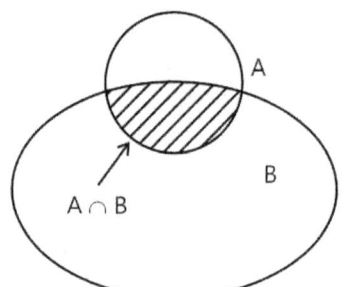

$A \cap B$

Hence, $P(A \cap B) = \begin{cases} P(B)\,P(A \mid B) & \text{if } P(B) \neq 0 \\ P(A)\,P(B \mid A) & \text{if } P(A) \neq 0 \end{cases}$

NOMORECLASS CONCEPTS

Trick to solve conditional probability

The trick is to identify when a probability is a conditional probability (in a word problem)

When dealing with conditional probability, the difference is that, we know for certain that something else has already happened.

This means that in our definition of probability that says

$$P(E) = \frac{\text{Number of ways for something to Happen}}{\text{Total Number of Ways}}$$

$$P(A/(B) = \frac{P(A \cap B)}{P(B)}$$

Consider events A and B.

If A and B are independent events, then $P(A/B) = P(A)$ and $P(B/A) = P(B)$. Therefore, $P(A \cap B) = P(A)\,P(B)$

Also, $P(A \cup B) = P(A) + P(B) - P(A \cap B)$

$= P(A) + P(B) - P(A)P(B) = 1 - (1 - P(A) - P(B) + P(A)\,P(B)] = 1 - [(1 - P(\bar{A}))(1 - P(\bar{B}))] = 1 - P(\bar{A})\,P(\bar{B})$

Illustration 7: If m is a natural such that $m \leq 5$, then the probability that the quadratic equation $x^2 + mx + \dfrac{1}{2} + \dfrac{m}{2} = 0$ has real roots is **(JEE MAIN)**

(A) 1/5   (B) 2/3   (C) 3/5   (D) 2/5

Sol: Apply Discriminant ≥ 0.

Discriminant D of the quadratic equation $x^2 + mx + \dfrac{1}{2} + \dfrac{m}{2} = 0$

$D = m^2 - 4\left(\dfrac{1}{2} + \dfrac{m}{2}\right) = m^2 - 2m - 2 = (m-1)^2 - 3$

Now, $D \geq 0$ Û $(m-1)^2 \geq 3$

This is possible for m = 3, 4 and 5. Also, the total number of ways of choosing m is 5.

$\therefore$ Probability of the required event = 3/5.

11. PROBABILITY OF AT LEAST ONE OF THE N INDEPENDENT EVENTS

If $P_1, P_2, ... P_n$ are the probabilities of n independent events $A_1, A_2, A_3 ... A_n$ then the probability that at least one of these events will happen is $1 - [(1 - P_1)(1 - P_2)....(1 - P_n)]$

$P(A_1 \cup A_2 \cup A_3 \cup ... \cup A_n) = 1 - P(\bar{A}_1) P(\bar{A}_2) P(\bar{A}_3) ...P(\bar{A}_n)$

Illustration 8: A mathematics problem is given to three students A, B and C whose chances of solving it are 1/2, 1/3, 1/4 respectively. Then the probability that the problem is solved is   **(JEE MAIN)**

Sol: Apply the principle of probability for independent events.

Obviously, the events of solving the problem by A, B and C are independent. Therefore required probability

$$= 1 - \left[\left(1 - \dfrac{1}{2}\right)\left(1 - \dfrac{1}{3}\right)\left(1 - \dfrac{1}{4}\right)\right] = 1 - \dfrac{1}{2} \cdot \dfrac{2}{3} \cdot \dfrac{3}{4} = \dfrac{3}{4}$$

(1) Multiplication theorems on probability

(i) If A and B are two events associated with a random experiment:

then $P(A \cap B) = P(A) .P(B/A)$, If $P(A) \neq 0$ or $P(A \cap B) = P(B). P(A/B)$, if $P(B) \neq 0$.

(ii) Extension of multiplication theorem: If $A_1, A_2,..., A_n$ are n events related to a random experiment, then

$P(A_1 \cap A_2 \cap A_3 \cap ... \cap A_n) = P(A_1) P(A_2 / A_1) P(A_3 / A_1 \cap A_2)$

$... P(A_n / A_1 \cap A_2 \cap ... \cap A_{n-1})$ where $P(A_i / A_1 \cap A_2 \cap ... \cap A_{i-1})$ represents the conditional probability of the event A_i, given that the events $A_1, A_2, ..., A_{i-1}$ have already happened.

(iii) Multiplication theorem for independent events: If A and B are independent events associated with a random experiment, then $P(A \cap B) = P(A) . P(B)$ i.e., the probability of simultaneous occurrence of two independent events is equal to the product of their probabilities. By multiplication theorem, we have $P(A \cap B) = P(A). P(B/A)$. Since A and B are independent events, therefore $P(B/A) = P(B)$. Hence, $P(A \cap B) = P(A).P(B)$.

(iv) Extension of multiplication theorem for independent events: If $A_1, A_2, ... A_n$ are independent events associated with a random experiment, then $P(A_1 \cap A_2 \cap A_3 \cap ... \cap A_n) = P(A_1) P(A_2) ... P(A_n)$.

By multiplication theorem, we have

$P(A_1 \cap A_2 \cap A_3 \cap ... \cap A_n) = P(A_1) P(A_2 / A_1) P(A_3 / A_1 \cap A_2)... P(A_n / A_1 \cap A_2 \cap ... \cap A_{n-1})$

Since $A_1, A_2,...,A_{n-1}, A_n$ are independent events, therefore

$P(A_2 / A_1) = P(A_2), P(A_3 / A_1 \cap A_2) = P(A_3),.........,$

$P(A_n / A_1 \cap A_2 \cap ....\cap A_{n-1}) = P(A_n)$

Hence, $P(A_1 \cap A_2 \cap ....\cap A_n) = P(A_1)P(A_2).....P(A_n)$

Illustration 9: The probability that a married man watches a certain T.V. show is 0.4 and the probability that a married woman watches the show is 0.5. The probability that a man watches the show, given that his wife does, is 0.7. Find **(JEE MAIN)**

(A) the probability that married couples watch the show

(B) the probability that a wife watches the show given that her husband does.

(C) the probability that at least one person of a married couple will watch the show.

Sol: Refer to Multiplication theorems on probability.

Let 'H' be the event that a married man watches the show and 'W' be the probability that a married woman watches the show,

$\Rightarrow$ P(H) $= 0.4$, P(W) $= 0.5$, P(H/W) $= 0.7$

(A) P(H $\cap$ W) $=$ P(W).P(H/W) $= 0.5 \times 0.7 = 0.35$

(B) P(W/H) $= \dfrac{P(H \cap W)}{P(H)} = \dfrac{0.35}{0.4} = \dfrac{7}{8}$

(C) P(H $\cup$ W) $=$ P(H) + P(W) $-$ P(H $\cap$ W) $= 0.4 + 0.5 - 0.35 = 0.55$

Illustration 10: Consider the sample space 'S' representing the adults in a small town who have completed the requirements for a college degree. They have been categorized according to sex and employment as under: **(JEE MAIN)**

	Employed	Unemployed
Male	460	40
Female	140	260

An employed person is selected at random. Find the probability that the chosen person is male.

Sol: Same as previous illustration.

Let M be the event that a man is chosen and E be the event that the chosen one is employed.

From the concept of reduced sample space we immediately get, P(M/E) $= \dfrac{460}{600} = \dfrac{23}{30}$

Also, P(E) $= \dfrac{600}{900} = \dfrac{2}{3}$; P(E $\cap$ M) $= \dfrac{460}{900} = \dfrac{23}{45} \Rightarrow$ P(M/E) $= \dfrac{23/45}{2/3} = \dfrac{23}{30}$

Illustration 11: A bag contains 3 white balls and 2 black balls, another contains 5 white and 3 blackballs. If a bag is chosen at random and a ball is drawn from it, what is the probability that it is white? **(JEE MAIN)**

Sol: Consider two cases. Case I – When the ball is chosen from the first bag and Case II – When the ball is chosen from the second bag.

The probability that the first bag is chosen is 1/2 and the chance of drawing a white ball from it is 3/5.

∴ Chance of choosing the first bag and drawing a white ball is 1/2, 3/5 respectively

Similarly the chance that the second bag is chosen and a white ball is drawn is 1/2, 5/8 respectively

∴ The chance of randomly choosing a bag and drawing a white ball is

$= \dfrac{1}{2}.\dfrac{3}{5} + \dfrac{1}{2}.\dfrac{5}{8}$ (Mutually exclusive cases) $= 49/80$.

Illustration 12: Find the probability that a year chosen at random has 53 Sundays. **(JEE MAIN)**

Sol: Divide the solution in two parts, when the year is a leap year and otherwise.

Let $P(L) \rightarrow$ be the probability that a year chosen at random is a leap year $P(L) = 1/4$.

$\therefore P(\bar{L}) = 3/4$

Let $P(S) \rightarrow$ be the probability that a year chosen at random has 53 Sundays.

$\therefore P(S) = P(L) \cdot P(S/L) + P(\bar{L}) \cdot P(S/\bar{L})$

Now, $P(S/L)$ is the probability that a leap year has 53 Sundays.

A leap year has 366 days, 52 weeks + the remaining 2 days may be Sunday-Monday, M-T, T-W, W-Th, Th-F, F-Sat or Sat –Sunday,

Out of the 7 possibilities, 2 are favorable

$\therefore P(S/L) = \dfrac{2}{7}$. Similarly $P(S/\bar{L}) = \dfrac{1}{7}$

$\therefore P(S) = \dfrac{1}{4} \cdot \dfrac{2}{7} + \dfrac{3}{4} \cdot \dfrac{1}{7} = \dfrac{5}{28}$

Theorem of total probability: If $E_1, E_2,...E_n$ are mutually exclusive and exhaustive events such that $P(E_i) \neq 0$ for each i and A is an event, then $P(A) = P(E_1) P(A|E_1) + P(E_2) P(A|E_2) +...+ P(E_n) P(A|E_n)$

Bayes' Theorem: If $E_1, E_2,...,E_n$ are n mutually exclusive and exhaustive events such that $P(E_i) > 0$ $(1 \leq i \leq n)$ and A is an event, then for $1 \leq k \leq n$,

$$P(E_k|A) = \frac{P(E_k)P(A \mid E_k)}{P(E_i)P(A \mid E_i) + P(E_2)P(A \mid E_2) + ... + P(E_n)P(A \mid E_n)}$$

The probabilities $P(E_j)$ $(1 \leq j \leq n)$ are called '*a priori probabilities*' and conditional probabilities

$P(E_j|A)$ are known as '*posteriori probabilities*'. Two events are said to be independent if occurrence (non-occurrence) of one does not affect the probability of occurrence (non occurrence) of the other i.e. $P(B|A) = P(B)$

Bayes' Theorem: The probability of event A, given that event B has subsequently occurred, is

$$P(A|B) = \frac{P(A).P(B \mid A)}{[P(A).P(B \mid A)] + [P(\bar{A}).P(B \mid \bar{A})]}$$

This is a direct result from condition probability and theorem of total probability. In general we

can write Bayes' theorem as $P(A_i \mid B) = \dfrac{P(B \mid A_i) \times P(A_i)}{P(B \mid A_1)P(A_1) + P(B \mid A_2)P(A_2) + ... + P(B \mid A_n)P(A_n)}$

NOMORECLASS CONCEPTS

Generally, Bayes' theorem is remembered as a formula, and whenever students encounter an inverse probability problem, they try to apply that formula without in-depth analysis of that problem. In our opinion, in any such problem, you should always draw a probability tree corresponding to the situation described. This will always give you more insight into the problem than a direct application of the formula and it may even prevent you from obtaining wrong results.

The method of probability tree diagrams has been discussed later.

Illustration 13: One bag contains four white balls and three black balls. The second bag contains three white balls and five black balls. One ball is drawn from the first bag and placed unseen in the second bag. What is the probability that a ball now drawn from the second bag is black? **(JEE ADVANCED)**

Sol: Use Total Probability theorem. Consider two events

A_1 – When white ball is transferred from bag I to II, and

A_2 – When black ball is transferred from bag I to II and proceed.

Bag-I	Bag-II
4W	3W
3B	5B

Let A_1 be the event that a white ball is transferred from bag-I to bag-II and A_2 be the event that a black ball is transferred from bag-I to bag-II.

$P(A_1) = \dfrac{4}{7}$, $P(A_2) = \dfrac{3}{7}$

Let 'A' be the probability that finally a black ball is drawn from the second bag

$P(A/A_1) = \dfrac{5}{9}$, $P(A/A_2) = \dfrac{6}{9}$

Now from total probability theorem we get, $P(A) = P(A_1).P(A/A_1) + P(A_2).P(A/A_2) = \dfrac{4}{7} \cdot \dfrac{5}{9} + \dfrac{3}{7} \cdot \dfrac{6}{9} = \dfrac{38}{63}$

Illustration 14: A real estate man has eight master keys to open several new homes. Only one master key will open any given house. If 40% of these homes are usually left unlocked, what is the probability that the real estate man can get into a specific home if he selects three master keys at random before leaving the office? **(JEE ADVANCED)**

Sol: Use Total Probability theorem. Let A_1 and A_2 be the events that the specific home is left unlocked and is left locked respectively

$\Rightarrow P(A_1) = 0.4$, $P(A_2) = 0.6$

Let 'A' be the event that the real estate man get into the specific home $P(A/A_1) = 1$,

$P(A/A_2) = \dfrac{^7C_2}{^8C_3} = \dfrac{3}{8} \Rightarrow P(A) = P(A_1) P(A/A_1) + P(A_2) P(A/A_2) = (0.4)(1) + (0.6)(3/8) = \dfrac{4}{10} + \dfrac{18}{80} = \dfrac{5}{8}$.

Illustration 15: A bag 'A' contains 2 white balls and 3 red balls, a bag 'B' contains 4 white and 5 blackballs. A bag is selected at random and a ball is drawn from it. Drawn ball is observed to be white. Find the probability that bag 'B' was selected. **(JEE ADVANCED)**

Sol: Take two cases, when bag A is selected and another when bag B is selected.

Bag A	Bag B
2W, 3R	4W, 5B

Let A_1 be the event that bag 'A' is selected and A be the event that bag B is selected

$P(A_1) = P(A_2) = 1/2$

Let 'A' be the event that a white ball is drawn from the selected bag.

$\Rightarrow P(A/A_1) = 2/5$, $P(A/A_2) = \dfrac{4}{9}$

$P(A) = P(A_1) . P(A/A_1) + P(A_2) .P(A/A_2) = \dfrac{1}{2} \cdot \dfrac{2}{5} + \dfrac{1}{2} \cdot \dfrac{4}{9} = \dfrac{1}{2}\left(\dfrac{2}{5} + \dfrac{4}{9}\right) = \dfrac{38}{90}$

$$= \text{Finally, } P(A_2/A) = \frac{P(A_2).P(A/A_2)}{P(A)} = \frac{(1/2).(4/9)}{(38/90)} = \frac{90 \times 4}{18 \times 38} = \frac{10}{19}$$

Illustration 16: A card from a pack of 52 cards is lost. From the remaining cards, two cards are drawn and are found to be spades. Find the probability that the missing card is also a spade. **(JEE ADVANCED)**

Sol: Take two cases, when the missing card is a spade or a non-spade. Let A_1 be the event that missing card is spade and A_2 be event that missing card is non-spade.

$$\Rightarrow P(A_1) = \frac{1}{4}.P(A_2) = \frac{3}{4}$$

Let 'A' be the event that 2 spade cards are drawn from the remaining cards,

$$P\left(\frac{A}{A_1}\right) = \frac{^{12}C_2}{^{51}C_2} \text{ and } P\left(\frac{A}{A_2}\right) = \frac{^{13}C_2}{^{51}C_2}; \ P(A) = P(A_1).P\left(\frac{A}{A_1}\right) + P(A_2).P\left(\frac{A}{A_2}\right)$$

$$= \frac{1}{4}\frac{^{12}C_2}{^{51}C_2} + \frac{3}{4}\frac{^{13}C_2}{^{51}C_2} = \frac{1}{4.^{51}C_2}\left[^{12}C_2 + 3.^{13}C_2\right]$$

$$\text{Now, } P\left(\frac{A_1}{A}\right) = \frac{P(A_1)P\left(\frac{A}{A_1}\right)}{P(A)} = \frac{\frac{1}{4}\frac{^{12}C_2}{^{51}C_2}}{\frac{1}{4.^{51}C_2}\left[^{12}C_2 + 3.^{13}C_2\right]} = \frac{11}{50}$$

12. BINOMIAL DISTRIBUTIONS

Binomial distributions occur in relation to those experiments that are binary in nature, i.e. whose outcomes can be grouped into two classes, say, success and failure, or, say 1 and 0. For example, when you toss a coin, there are only two outcomes possible: Heads (which you may call success) and Tails (which then becomes Failure).

NOMORECLASS CONCEPTS

Note that an experiment need not have only two outcomes for it to be called binary. For example, consider the experiment of rolling a die.If youmake the following definitions-

Success: Numbers 1, 2 or 3

Failure: Numbers 4, 5 and 6

Then, with respect to this definition, the experiment is binary. Thus, an experiment needs to have two **classes of outcomes** for it to be called binary.

Let us consider a binomial experiment which has been repeated 'n' times. Let the probability of success and failure in any trial be p and q respectively. We are interested in the probability of occurrence of exactly 'r' successes in these n trials. Now, number of ways of choosing 'r' success in 'n' trials = ^nC_r.

Probability of 'r' successes and (n-r) failures is $p^r.q^{n-r}$. Thus probability of having exactly r successes = $^nC_r.$ $p^r.q^{n-r}$

Let 'X' be a random variable representing the number of successes, then

$P(X = r) = {}^nC_r . p^r.q^{n-r} (r = 0, 1, 2, ..., n)$

$1 = (p + q)^n = {}^nC_0 p^0q^n + {}^nC_1 p^1 q^{n-1} + {}^nC_2 p^2q^{n-2} + ... + {}^nC_r p^r q^{n-r} + ... + {}^nC_r p^nq^0$

$X \rightarrow$ Number of successes 0, 1, 2, ... r, ... N

- Probability of at most 'r' successes in n trials = $\sum_{r=0}^{r} {}^nC_r p^r q^{n-r}$

- Probability of at least 'r' successes in n trials = $\sum_{r=r}^{n} {}^nC_r p^r . q^{n-r}$

- Probability of having 1st success at the rth trial = $p.q^{r-1}$

13. BINOMIAL PROBABILITY DISTRIBUTION

(a) A probability distribution spells out how a total probability is distributed over several values of a random variable.

(b) Mean of any probability distribution of a random variable is given by $\mu = \dfrac{\sum p_i x_i}{\sum p_i} = \sum p_i x_i$ (Since $\Sigma p_i = 1$)

(c) Variance of a random variable is given by, $\sigma^2 = \Sigma(x_i - \mu)^2 p_i$

$\therefore \sigma^2 = \Sigma p_i x_i^2 - \mu^2$ (Note that SD = $+\overline{\sigma^2}$)

(d) The probability distribution for a binomial variable 'X' is given by $P(X = r) = {}^nC_r\, p^r\, q^{n-r}$ where $p(X = r)$ is the probability of r successes.

The recurrence formula $\dfrac{P(r+1)}{P(r)} = \dfrac{n-r}{r+1} . \dfrac{p}{q}$ is very helpful for computing P(1). P(2) . P(3) etc. quickly, if P(0) is known.

(e) Mean of BPD = np ; Variance of BPD = npq.

(f) If P represents a person's chance of success in any venture and 'M' represents the sum of only what he will receive in case of success, then his expectations of probable value = PM.

14. MODE AND MEDIAN

Usually the mode of a binomial B(n, p) distribution is equal to $\left[(n+1)p\right]$, where[.] is the greater integer function. However when $(n + 1)p$ is an integer and p is neither 0 nor 1, then the distribution has two modes: $(n + 1)p$ and $(n + 1)p - 1$. When p is equal to 0 or 1, the mode will be 0 and n correspondingly. These cases can be summarized as follows:

$$mode = \begin{cases} \left[(n+1)p\right] & \text{if } (n+1)p \text{ is 0 or a non integer,} \\ (n+1)p \text{ and } (n+1)p-1 & \text{if } (n+1)p \in \{1,...,n\}, \\ n & \text{if}(n+1)p = n+1 \end{cases}$$

In general, there is no single formula to find the median for a binomial distribution and it may even be non-unique. However several special results have been established:

(a) If np is an integer, then the mean, median, and mode coincide and equal np.

(b) When p = 1/2 and n is odd, any number m in the interval $\dfrac{1}{2}(n - 1) \le m \le 1/2(n + 1)$ is a median of the binomial distribution. If p = 1/2 and n is even, then m = n/2 is the unique median.

PROBLEM SOLVING TACTICS

Following are some extra methods which may be useful to solve probability questions:

Venn Diagrams: It is a diagram in which the sample space is represented by a rectangle and the element of the sample space by points within it. Subsets (or events) of the sample space are represented by the region within the rectangle, usually using circles.

For example, consider the following events when a die is thrown,

A = {odd numbers} = {1, 3, 5}

B = {even numbers} = {2, 4, 6}

C = {prime numbers} = {2, 3, 5}

Let us see how Venn diagrams are to be applied by using them to prove some results as follows:

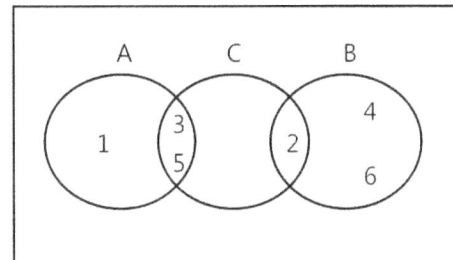

Theorem 1: For any two events A and B, $A \subseteq B \Rightarrow P(A) \leq P(B)$.

Proof. From the adjoining diagram, we have

$A \cup (B - A) = B$ and $A \cap (B - A) = f$

$\therefore P(B) = P[A \cup (B - A)]$ $[\because A \cap (B - A) = f]$

$\Rightarrow P(B) = P(A) + P(B - A)$ $[\because P(B - A) \geq 0]$

$\Rightarrow P(A) \leq P(B)$ **Proved**

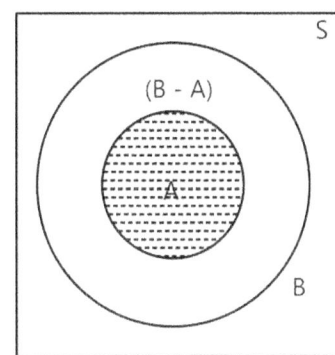

Theorem 2: For any two events A and B, $P(A - B) = P(A) - P(A \cap B)$

Proof: Let A and B be two compatible events. Then $A \cap B \neq \phi$. From the adjoining Venn diagram. it is clear that:

$(A - B) \cap (A \cap B) = f$ and $(A - B) \cup (A \cap B) = A$

$\Rightarrow P(A - B) + P(A \cap B) = P(A)$

$\Rightarrow P(A - B) = P(A) - P(A \cap B)$ **Proved**

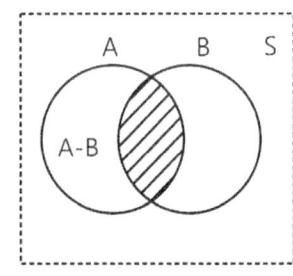

Remarks: This result may be expressed as

$P(A \cap \bar{B}) = P(A) - P(A \cap B)$

Also $P(\bar{A} \cap B) = P(B) - P(A \cap B)$

Theorem 3: For any three events A, B, C

$P(A \cup B \cup C) = P(A) + P(B) + P(C) - P(A \cap B) - P(B \cap C) - P(C \cap A) + P(A \cap B \cap C)$

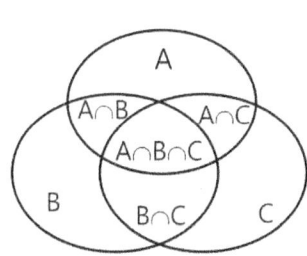

Proof: We have $\quad , P(A \cup B \cup C) = P[(A \cup B) \cup C]$

$= P(A \cup B) + P(C) - P[(A \cup B) \cap C]$

$= P(A \cup B) + P(C) - P[(A \cap C) \cup (B \cap C)]$ (Distributive Law)

$= [P(A) + P(B) - P(A \cap B)] + P(C) - P[(A \cap B) \cup (C \cap C)]$ [Addition law]

$= P(A) + P(B) - P(A \cap B) + P(C) - P[(A \cap C) \cup (B \cap C)]$

$= P(A) + P(B) + P(C) - P(A \cap B) - [P(A \cap C) + P(B \cap C) - P[(A \cap C) \cap (B \cap C)]$

$= P(A) + P(B) + P(C) - P(A \cap B) - P(A \cap C) + P(B \cap C) + P[A \cap B \cap C]$ **Proved**

Probability Tree Diagrams: Calculating probabilities can be hard. Sometimes you add them, sometimes you multiply them and often, it is hard to figure out what to do. That's when **tree diagrams come to the rescue!**

Here is a tree diagram for two tosses of a coin:

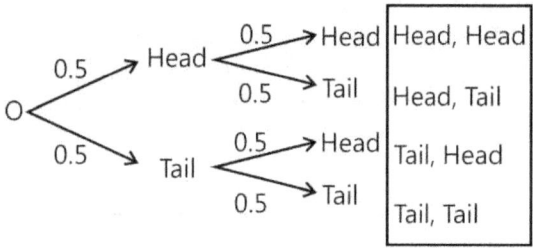

How do you calculate the overall probabilities?

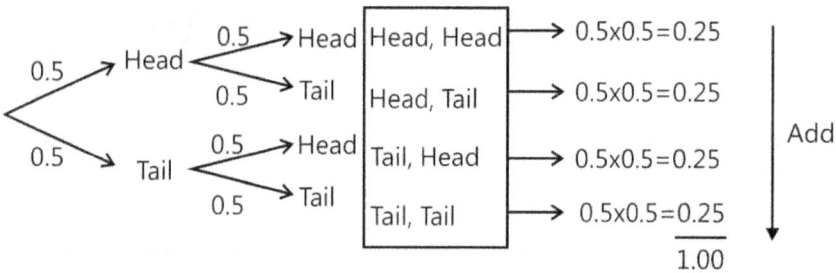

So, there you go. When in doubt, draw a tree diagram, multiply along the branches and add the columns. Make sure all probabilities add to 1 and you are good to go!

FORMULAE SHEET

(a) Mathematical definition of probability:

$$\text{Probability of an event} = \frac{\text{Number of favorable cases to event A}}{\text{Total Number of cases}}$$

Note: (i) $0 \leq P(A) \leq 1$

(ii) Probability of an impossible event is zero

(iii) Probability of a sure event is one.

(iv) $P(A) + P(\text{Not A}) = 1$ i.e. $P(A) + P(\overline{A}) = 1$

(b) Odd for an event: If $P(A) = \dfrac{m}{n}$ and $P(\overline{A}) = \dfrac{n-m}{n}$

Then odds in favor of A $= \dfrac{P(A)}{P(\overline{A})} = \dfrac{m}{n-m}$ and odd in against of A $= \dfrac{p(\overline{A})}{P(A)} = \dfrac{n-m}{m}$

(c) Set theoretical notation of probability and some important results:

(i) $P(A \cup B) = 1 - P(\overline{A} \cap \overline{B})$

(ii) $P(A/B) = \dfrac{P(A \cap B)}{P(B)}$

(iii) $P(A \cup B) = P(A \cap B) + P(\bar{A} \cap B) + P(A \cap \bar{B})$

(iv) $A \subseteq B \Rightarrow P(A) \leq P(B)$

(v) $P(\bar{A} \cap B) = P(B) - P(A \cap B)$

(vi) $P(A \cap B) \leq P(A) \, P(B) \leq P(A \cup B) \leq P(A) + P(B)$

(vii) P(Exactly one event) $= P(A \cap \bar{B}) + P(\bar{A} \cap B)$

(viii) $P(\bar{A} \cup \bar{B}) = 1 - P(A \cap B) = P(A) + P(B) - 2P(A \cap B) = P(A + B) - P(A \cap B)$

(ix) P(neither A nor B) $= P(\bar{A} \cap \bar{B}) = 1 - P(A \cup B)$

(x) When a coin is tossed n times or n coins are tossed once, the probability of each simple event is $\dfrac{1}{2^n}$

(xi) When a dice is rolled n times or n dice are rolled once, the probability of each simple event is $\dfrac{1}{6^n}$

(xii) When n cards are drawn $(1 \leq n \leq 52)$ from well shuffled deck of 52 cards, the probability of each simple event is $\dfrac{1}{{}^{52}C_n}$.

(xiii) If n cards are drawn one after the other with replacement, the probability of each simple event is $\dfrac{1}{(52)^n}$

(xiv) P(none) $= 1 - P$ (at least one)

(xv) Playing cards

- Total cards: 52 (26 red, 26 black)
- Four suits: Heart, diamond, spade, club (13 cards each)
- Court (face) cards: 12 (4 kings, 4 queens, 4 jacks)
- Honor cards: 16 (4 Aces, 4 kings, 4 queens, 4 Jacks)

(xvi) Probability regarding n letters and their envelopes:

If n letters corresponding to n envelopes are placed in the envelopes at random, then

- Probability that all letters are in the right envelopes $= \dfrac{1}{n!}$

- Probability that all letters are not in the right envelopes $= 1 - \dfrac{1}{n!}$

- Probability that no letter is in the right envelope $= \dfrac{1}{2!} - \dfrac{1}{3!} + \dfrac{1}{4!} + ... + (-1)^n \dfrac{1}{n!}$

- Probability that r letters are in the right envelope $= \dfrac{1}{r!}\left[\dfrac{1}{2!} - \dfrac{1}{3!} + \dfrac{1}{4!} + ... + (-1)^{n-r}\dfrac{1}{(n-r)!}\right]$

(d) Addition Theorem of Probability:

(i) When events are mutually exclusive

i.e. $n(A \cap B) = 0 \qquad \Rightarrow P(A \cap B) = 0$

$\therefore P(A \cup B) = P(A) + P(B)$

(ii) When events are not mutually exclusive i.e. $P(A \cap B) \neq 0$

$\therefore P(A \cup B) = P(A) + P(B) - P(A \cap B)$ or $P(A + B) = P(A) + P(B) - P(AB)$

(iii) When events are independent i.e. $P(A \cap B) = P(A) \, P(B)$

$\therefore P(A + B) = P(A) + P(B) - P(A) \, P(B)$

(e) Conditional probability:

P(A/B) = Probability of occurrence of A, given that B has already happened = $\dfrac{P(A \cap B)}{P(B)}$

P(B/A) = Probability of occurrence of B, given that A has already happened = $\dfrac{P(A \cap B)}{P(A)}$

Note: If the outcomes of the experiment are equally likely, then

$P(A/B) = \dfrac{\text{Number of sample points in } A \cap B}{\text{Number of points in B}}$

(i) If A and B are independent events, then P(A/B) = P(A) and P(B/A) = P(B)

(ii) Multiplication Theorem:

P(A ∩ B) = P(A/B). P(B), P(B) ≠ 0 or P(A ∩ B) = P(B/A) P(A), P(A) ≠ 0

Generalized: $P(E_1 \cap E_2 \cap E_3 \cap ... \cap E_n)$

$= P(E_1) P(E_2/E_1) P(E_3/E_1 \cap E_2) P(E_4/E_1 \cap E_2 \cap E_3) ...$ If events are independent, then

$P(E_1 \cap E_2 \cap E_3 ... \cap E_n) = P(E_1) P(E_2) ... P(E_n)$

(f) Probability of at least one of the n Independent events: If $P_1, P_2, ... P_n$ are the probabilities of n independent events $A_1, A_2, ... A_n$ then the probability that at least one of these events will happen is $1 - [(1 - P_1) (1 - P_2) ... (1 - P_n)]$

or $P(A_1 + A_2 + ... + A_n) = 1 - P(\bar{A}_1) P(\bar{A}_2) ... P(\bar{A}_n)$

(g) Total probability: Let $A_1, A_2, ... A_n$ be n mutually exclusive & set of exhaustive events. If event A can occur through any one of these events, then the probability of occurrence of A

$P(A) = P(A \cap A_1) + P(A \cap A_2) + ... + P(A \cap A_n) = \displaystyle\sum_{r=1}^{n} P(A_r)P(A/A_r)$

(h) Bayes' Rule: Let A_1, A_2, A_3 be any three mutually exclusive & exhaustive events (i.e. $A_1 \cup A_2 \cup A_3$ = sample space & $A_1 \cap A_2 \cap A_3 = \phi$) of a sample space S and B is any other event on sample space then,

$P(A_i/B) = \dfrac{P(B/A_i)(P(A_i)}{P(B/A_1) P(A_1) + P(B/A_2)P(A_2) + P(B/A_3)P(A_3)}, i = 1, 2, 3$

(i) Probability distribution:

(i) If a random variable x assumes values $x_1, x_2, ... x_n$ with probabilities $P_1, P_2, ... P_n$ respectively then

- $P_1 + P_2 + P_3 + ... + P_n = 1$
- Mean E(x) = $\Sigma P_i x_i$
- Variance = $\sum x^2 P_i - (\text{mean})^2 = \sum (x^2) - (E(x))^2$

(ii) Binomial distribution: If an experiment is repeated n times, the successive trials being independent of one another, then the probability of r success is $^nC_r \, P^r \, q^n$ 'at least r success is $\displaystyle\sum_{k=r}^{n} {}^nC_k P^k q^{n-k}$ where p is probability of success in a single trial, q = 1 – p

- Mean E(x) = np
- E(x2) = npq + n2p2
- Variance E(x2) – (E(x))2 = npq
- Standard deviation = $\sqrt{npq}$

(j) **Truth of the statement:**

(i) If two persons A and B speak the truth with probabilities P_1 & P_2 respectively and if they agree on a statement, then the probability that they are speaking the truth will be given by $\dfrac{P_1 P_2}{P_1 P_2 + (1-P_1)(1-P_2)}$.

(ii) If A and B both assert that an event has occurred, the probability of occurrence of which is α, then the probability that the event has occurred $\dfrac{\alpha P_1 P_2}{\alpha P_1 P_2 + (1-\alpha)(1-P_1)(1-P_2)}$ given that the probability of A & B speaking truth is $p_1\, p_2$ respectively.

(iii) If in the second part, the probability that their lies coincide is β, then from the above case, the required probability will be $\dfrac{\alpha P_1 P_2}{\alpha P_1 P_2 + (1-\alpha)(1-P_1)(1-P_2)\beta}$

Solved Examples

JEE Main/Boards

Example 1: If there are two events A and B such that $P(A') = 0.3$ $P(B) = 0.5$ and $P(A \cap B) = 0.3$, then $P(B|A \cup B')$ is:

(A) 3/8   (B) 2/3   (C) 5/6   (D) 1/4

Sol: Use set theory and probability of complimentary events to calculate $P(A \cup B')$

We have $P(A \cup B')$

$= P(A) + P(B') - P(A \cap B')$

$= [1 - P(A')] + [1 - P(B)] - [P(A) - P(A \cap B)]$

$= (1 - 0.3) + (1 - 0.5) - (0.7 - 0.3) = 0.8$

Now, $P(B|A \cup B') = \dfrac{P[B \cap (A \cup B')]}{P(A \cup B')}$

$= \dfrac{P[(B \cap A) \cup (B \cap B')]}{P(A \cup B')} = \dfrac{P(A \cap B)}{P(A \cup B')} = \dfrac{0.3}{0.8} = \dfrac{3}{8}$

Example 2: Seven white balls and three black balls are randomly placed in a row. The probability that no two black balls are placed adjacently equals:

(A) 1/2   (B) 7/15   (C) 2/15   (D) 1/3

Sol: Each black balls can be arranged in between any two white balls. Use this idea to find the number of ways in which no two black balls are together.

The number of ways of placing 3 black balls at 10 places is $^{10}C_3$. The number of ways in which two black balls are

not together is equal to the number of ways of choosing 3 places marked with X out of eight places

X W X W X W X W X W X W X

This can be done in $^8C_3$ ways. Thus, probability of the required event is $\dfrac{^8C_3}{^{10}C_3} = \dfrac{8 \times 7 \times 6}{10 \times 9 \times 8} = \dfrac{7}{15}$

Example 3: A group of 2n boys and 2n girls is randomly divided into two equal groups. The probability that catch contains the same number of boys and girls is:

(A) 1/2   (B) 1/n

(C) 1/2n   (D) None of these

Sol: If one group is selected the second group automatically gets created. Hence, select n boys and n girls from the given group.

Total number of ways of choosing a group is $^{4n}C_{2n}$ The number of ways in which each group contains equal number of boys and girls is $(^{2n}C_n)(^{2n}C_n)$

$\therefore$ Required probability $= \dfrac{(^{2n}C_n)^2}{^{4n}C_{2n}}$.

Example 4: Let A and B be two events such that $P(A) = 0.3$ and $P(A \cup B) = 0.8$. If A and B are independent events, then $P(B)$ is:

(A) 3/7   (B) 4/7   (C) 5/7   (D) 6/7

Sol: If say A and B are two independent events then $P(A \cap B) = P(A) \times P(B)$

We have $0.8 = P(A \cup B)$

$= P(A) + P(B) - P(A \cap B) = P(A) + P(B) - P(A) P(B)$

[$\because$ A and B are independent]

$= 0.3 + P(B) - (0.3) P(B)$

$\Rightarrow 0.5 = (0.7) P(B) \Rightarrow P(B) = \dfrac{5}{7}$

Example 5: A natural number x is chosen at random from the first one hundred natural numbers.

The probability that $\dfrac{(1-20)(x-40)}{x-30} < 0$ is:

(A) 1/50   (B) 3/50   (C) 3/25   (D) 7/25

Sol: Find the range of values the variable x can take and then find the required probability.

Let $E = \dfrac{(x-20)(x-40)}{x-30} = \dfrac{(x-20)(x-30)(x-40)}{(x-30)^2}$

Sign of E is same as that of sign of1

$(x-20)(x-30)(x-40) = F(\text{say})$

Note that $F < 0$ if and only if

$0 < x < 20$ or $30 < x < 40$

$\therefore E < 0$ in $(0, 20) \cup (30, 40)$

Thus E is negative for $x = 1, 2, \dots, 19, 31, 32, \dots, 39$ that is E, < 0 for 28 natural numbers

$\therefore$ Required probability $= \dfrac{28}{100} = \dfrac{7}{25}$

Example 6: Let E and F be two independent events such that $P(E) < P(F)$. The probability that both E and F happen is $\dfrac{1}{12}$ and the probability that neither E nor F happen is $\dfrac{1}{2}$. Then,

(A) P(E) = 1/3, P(F) = $\dfrac{1}{2}$ (B) P(E) = 1/2, P(F) = $\dfrac{2}{3}$

(C) P(E) = 2/3, P(F) = $\dfrac{3}{4}$ (D) P(E) = 1/4, P(F) = $\dfrac{1}{3}$

Sol: Use the concept of Probability for independent events.

We are given $P(E \cap F) = \dfrac{1}{12}$ and $P(E' \cap F') = \dfrac{1}{2}$

As E and F are independent, we get P(E) P(F)

$= \dfrac{1}{12}$ and $P(E') P(F') = \dfrac{1}{2}$

$\Rightarrow (1 - P(E) (1 - P(F)) = \dfrac{1}{2}$

$\Rightarrow 1 - (P(E) + P(F) - P(E) P(F)) = \dfrac{1}{2}$

$\Rightarrow P(E) + P(F) = 1 + \dfrac{1}{12} - \dfrac{1}{2} = \dfrac{7}{12}$

$\therefore$ Equation whose roots are P(E) and P(F) is

$x^2 - (P(E) + P(F))x + P(E) P(F) = 0$

or $x^2 - \dfrac{1}{12}x + \dfrac{1}{12}$

$\Rightarrow 12x^2 - 7x + 1 = 0$

$\Rightarrow (3x - 1)(4x - 1) = 0$

$\Rightarrow x = \dfrac{1}{3}, \dfrac{1}{4}$

As $P(E) < P(F)$, we take $P(E) = \dfrac{1}{4}$ and $P(F) = \dfrac{1}{3}$

Example 7: Fifteen coupons are numbered 1, 2, ……. , 15 respectively. Seven coupons are selected at random one at a time with replacement. The probability that the largest number on a selected coupon as 9 is:

(A) $\left(\dfrac{9}{15}\right)^6$ (B) $\left(\dfrac{8}{15}\right)^7$

(C) $\left(\dfrac{3}{5}\right)^7$ (D) None of these

Sol: Calculate the probability for getting highest number as 9 and 8. Subtract the two to get the desired probability.

Let p = the probability that a selected coupon bears number ≤ 9.

$\Rightarrow p = \dfrac{9}{15} = \dfrac{3}{5}$ and

n = Number of coupons drawn with replacement

X = The number of coupons bearing number ≤ 9

Note that $X - B(n, p)$

Probability that the largest number on the selected coupons does not exceed 9

= probability that all the coupons bear number ≤ 9

$= P(X = 7) = {}^7C_7 p^7 = \left(\dfrac{3}{5}\right)^7$

Similarly, probability that largest number on the selected coupon is ≤ 8 is $\left(\dfrac{8}{15}\right)^7$.

Hence, probability of the required event $= \left(\dfrac{3}{5}\right)^7 - \left(\dfrac{8}{15}\right)^7$.

Example 8: A four digit number (numbered from 0000 to 9999) is said to be lucky if the sum of its first two digits is equal to the sum of its last two digits. If a four digit number is picked up at random, then the probability that it is lucky is:

(A) 0.065   (B) 0.064   (C) 0.066   (D) 0.067

Sol: The sum of the first two digits can be any number from 0 to 18. Use the formula for the number of non-negative integral solutions of x+y=m to proceed further.

The total number of ways of choosing the ticket is 10000.

Let the four digits number on the ticket be x_1 $x_2 x_3 x_4$. Note that $0 \le x_1 + x_2 \le 18$ and $0 \le x_3 + x_4 \le 18$.

Also, the number of non-negative integral solutions of $x + y = m$ (with $0 \le x, y \le 9$) is

$m + 1$ if $0 \le m \le 9$ and is $19 - m$ if $10 \le m £18$.

Thus, the number of favorable ways

$= 1 \times 1 + 2 \times 2 + \ldots\ldots + 10 \times 10 + 9 \times 9 +$

$8 \times 8 + \ldots\ldots + 1 \times 1$

$= 2\left\{\dfrac{9 \times 10 \times 19}{6}\right\} + 100 = 670$

$\therefore$ Probability of required event $= \dfrac{670}{10000} = 0.067$

Example 9: Three numbers are chosen at random without replacement from {1, 2, 3, ...10).The probability that minimum of the chosen number is 3 or their maximum is 7, is:

(A) $\dfrac{11}{30}$ (B) $\dfrac{11}{40}$ (C) $\dfrac{1}{7}$ (D) $\dfrac{1}{8}$

Sol: Find the probability for getting 3 as the minimum and 7 as the maximum number among the three numbers selected. Then use the formula $P(A \cup B) = P(A) + P(B) - P(A \cap B)$

Let A and B denote the following events

A: minimum of the chosen number is 3

B: maximum of the chosen number is 7

We have, P(A) = P(choosing 3 and two other numbers

from 4 to 10)$= \dfrac{^7C_2}{^{10}C_3} = \dfrac{7 \times 6}{2} \times \dfrac{3 \times 2}{10 \times 9 \times 8} = \dfrac{7}{40}$

P(B) = P(Choosing 7 and two other numbers

from 1 to 6)$= \dfrac{^6C_2}{^{10}C_3} = \dfrac{6 \times 5}{2} \times \dfrac{3 \times 2}{10 \times 9 \times 8} = \dfrac{1}{8}$

$P(A \cap B)$ = P (choosing 3 and 7 and one other

number from 4 to 6)$= \dfrac{3}{^{10}C_3} = \dfrac{3 \times 3 \times 2}{10 \times 9 \times 8} = \dfrac{1}{40}$

Now, $P(A \cup B) = P(A) + P(B) - P(A \cap B) = \dfrac{7}{40} + \dfrac{1}{8} - \dfrac{1}{40} = \dfrac{11}{40}$

Example 10: A signal which can be green or red with probability $\dfrac{4}{5}$ and $\dfrac{1}{5}$ respectively, is received by station A and then transmitted to station B. The probability of each station receiving the signal correctly is $\dfrac{3}{4}$. If the signal received at station B is green, then the probability that the original signal was green is:

(A) $\dfrac{3}{5}$ (B) $\dfrac{6}{7}$ (C) $\dfrac{20}{23}$ (D) $\dfrac{9}{20}$

Sol: Draw a tree diagram for all the possibilities and calculate the probability for all the different cases.

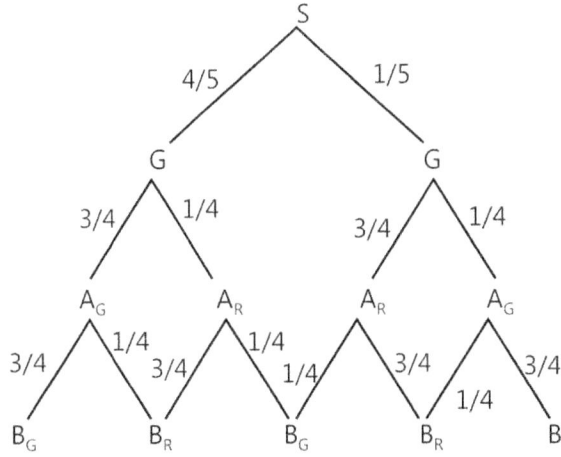

Let G, E_1, E_2 and E denote the following events:

G: Original signal is green

E_1: A receives the signal correctly

E_2:B receives the signal correctly

E = B receives the green signal

We have,

$E = GE_1E_2 \cap GE'_1E'_2 \cap G'E_1E'_2 \cap G'E'_1E_2$

$\Rightarrow P(E) = P(GE_1E_2) + P(G'E'_1E'_2) + P(G'E_1E'_2) + P(G'E'_1E_2)$

$= \left(\dfrac{4}{5}\right)\left(\dfrac{3}{4}\right)\left(\dfrac{3}{4}\right) + \left(\dfrac{4}{5}\right)\left(\dfrac{1}{4}\right)\left(\dfrac{1}{4}\right) +$

$\left(\dfrac{1}{5}\right)\left(\dfrac{3}{4}\right)\left(\dfrac{1}{4}\right) + \left(\dfrac{1}{5}\right)\left(\dfrac{1}{4}\right)\left(\dfrac{3}{4}\right)$

$= \dfrac{36 + 4 + 3 + 3}{80} = \dfrac{23}{40}$

Also, $P(G \cap E) = P(GE_1E_2) + P(GE'_1E'_2) = \dfrac{40}{80} = \dfrac{1}{2}$

$\therefore P(G/E) = \dfrac{P(G \cap E)}{P(E)} = \dfrac{1/2}{23/40} = \dfrac{20}{23}$

JEE Advanced/Boards

Example 1: Let A, B, C, be three mutually independent events. Consider the two statement S_1 and S_2

S_1: A and $B \cup C$ are independent

S_2: A and $B \cap C$ are independent

Then,

(A) Both S_1 and S_2 are true   (B) Only S_1 is true

(C) Only S_2 is true   (D) Neither S_1 nor S_2 is true

Sol: Use the basic understanding of sets and probability of union and intersection of two sets to find the answer.

We are given that

$P(A \cap B) = P(A) \, P(B)$

$P(B \cap C) = P(B) \, P(C), P(C \cap A) = P(C) \, P(A),$

and $P(A \cap B \cap C) = P(A) \, P(B) \, P(C)$

We have

$P(A \cap (B \cap C)) = P(A \cap B \cap C)$

$= P(A) \, P(B) \, P(C) = P(A) \, P(B \cap C)$

$\Rightarrow$ A and $B \cap C$ are independent. Therefore, S_2 is true.

Also $P[(A \cap (B \cup C)] = P[(A \cap B) \cup (A \cap C)]$

$= P(A \cap B) + P(A \cap C) - P[(A \cap B) \cap (A \cap C)]$

$= P(A \cap B) + P(A \cap C) - P(A \cap B \cap C)$

$= P(A) \, P(B) + P(A) \, P(C) - P(A) \, P(B) \, P(C)$

$= P(A) \, [P(B) + P(C) - P(B) \, P(C)]$

$\therefore$ A and $B \cup C$ are independent.

Example 2: A bag contains some white and some black balls, all combinations of balls being equally likely. The total number of balls in the bag is 10. If three balls are drawn at random without replacement and all of them are found to be black, the probability that the bag contains 1 white and 9 black balls is:

(A) $\dfrac{14}{55}$ (B) $\dfrac{12}{55}$ (C) $\dfrac{2}{11}$ (D) $\dfrac{8}{55}$

Sol: In this case, the number of black balls can be anything between 3 and 10. Apply Baye's theorem to find the required probability.

Let E_i denote the event that the bag contains i black and (10 – i) white balls (i = 0, 1, 2, ..., 10). Let A denote the event that the three balls drawn at random from the bag are black. We have,

$P(E_i) = \dfrac{1}{11}$ (i = 0, 1, 2, ... , 10)

$P(A|E_i) = 0$ for i = 0, 1, 2

and $P(A|E_i) = \dfrac{{}^iC_3}{{}^{10}C_3}$ for $i \geq 3$

Now, by the total probability rule,

$P(A) = \sum_{i=0}^{10} P(E_i)P(A \mid E_i)$

$= \dfrac{1}{11} \times \dfrac{1}{{}^{10}C_3} [{}^3C_3 + {}^4C_3 + ... + {}^{10}C_3]$

But ${}^3C_3 + {}^4C_3 + {}^5C_3 + ... + {}^{10}C_3$

$= {}^4C_4 + {}^4C_3 + {}^5C_3 + ... + {}^{10}C_3$

$= {}^5C_4 + {}^5C_3 + {}^6C_3 + ... + {}^{10}C_3$

$= {}^6C_4 + {}^6C_3 + ... + {}^{10}C_3 = ... = {}^{11}C_4$

Thus, $P(A) = \dfrac{{}^{11}C_4}{11 \times {}^{10}C_3} = \dfrac{1}{4}$

By the Bayes' rule

$P(E_9|A) = \dfrac{P(E_9)P(A \mid E_9)}{P(A)} = \dfrac{\dfrac{1}{11}\dfrac{({}^9C_3)}{{}^{10}C_3}}{\dfrac{1}{4}} = \dfrac{14}{55}$

Example 3: A pair of biased dice is rolled together till a sum of either 5 or 7 is obtained. The probability that 5 comes before 7 is:

(A) 2/5   (B) 3/5

(C) 4/5   (D) None of these

Sol: The possible outcomes could be 5, X5, XX5, XXX5, XXXX5 and so on, where X denotes a sum of neither 5 nor 7. Also it can be easily understood that this sequence goes on till infinity.

Let A denote the event that a sum of 5 occurs, B denote the event that a sum of 7 occurs and C the event that neither a sum of 5 nor a sum of 7 occurs. We have,

$$P(A) = \frac{4}{36} = \frac{1}{9}$$

$$P(B) = \frac{6}{36} = \frac{1}{6}$$

and $P(C) = \frac{26}{36} = \frac{13}{18}$

Thus,

P(A occurs before B) = P[A or (C ∩ A) or (C ∩ C ∩ A) or ....]

$= P(A) + P(C \cap A) + P(C \cap C \cap A) + ...$

$= P(A) + P(C)\, P(A) + P(C)^2\, P(A) + ...$

$$= \frac{1}{9} + \left(\frac{13}{18}\right)\frac{1}{9} + \left(\frac{13}{18}\right)^2 \frac{1}{9} + ...$$

$$= \frac{1/9}{1 - \dfrac{13}{18}} = \frac{2}{5} \quad \text{[sum of an infinite G.P.]}$$

Example 4: If A, B and C are three events such that $P(B) = \frac{3}{4}$, $P(A \cap B \cap C') = \frac{1}{3}$ and $P(A' \cap B \cap C') = \frac{1}{3}$, then $P(B \cap C)$ is equal to:

(A) $\frac{1}{12}$ (B) $\frac{1}{6}$ (C) $\frac{1}{15}$ (D) $\frac{1}{9}$

Sol: Apply the knowledge of set theory to write B ∩ C' in terms of A ∩ B ∩ C' and A' ∩ B ∩ C'.

We have, P(B ∩ C') = P[(A ∪ A') ∩ (B ∩ C')]

$= P(A \cap B \cap C') + P(A' \cap B \cap C') = \frac{1}{3} + \frac{1}{3} = \frac{2}{3}$

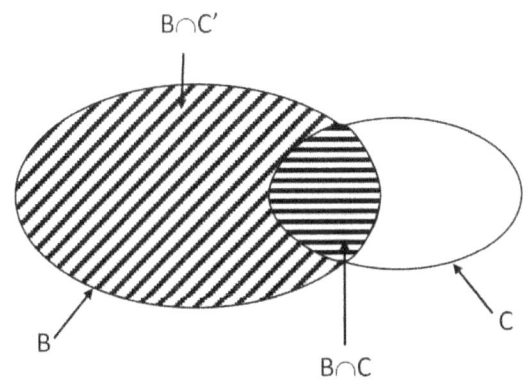

Now, $P(B \cap C) = P(B) - P(B \cap C') = \frac{3}{4} - \frac{2}{3} = \frac{1}{12}$

Example 5: A ship is fitted with three engines E_1, E_2 and E_3. The engines function independently of each other with respective probabilities $\frac{1}{2}$, $\frac{1}{4}$ and $\frac{1}{4}$. For the ship to be operative at least two of its engine must function. Let X denote the event that ship is operational and let X_1, X_2 and X_3 respectively the events that the engines E_1, E_2 and E_3 are functioning. Let,

(A) $P(X'_1 / X) = \frac{3}{8}$

(B) $P(X/X_2) = \frac{7}{8}$

(C) P(Exactly two engines are functioning) $= \frac{7}{8}$

(D) $P(X/X_1) = 7/16$

Sol: The ship can be operational in four possible cases. Calculate the probability of the ship being operational and then proceed accordingly.

We have, $X = (X_1 X_2 X'_3) \cup (X_1 X'_2 X_3) \cup (X'_1 X_2 X_3) \cup (X_1 X_2 X_3)$

and $X'_1 \cap X = X'_1 X_2 X_3$

Now, $P(X'_1 / X) = \dfrac{P(X'_1 \cap X)}{P(X)} = \dfrac{P(X'_1 X_2 X_3)}{P(X)}$

We have $P(X'_1 X_2 X_3) = P(X'_1) P(X_2) P(X_3)$

$\left(\dfrac{1}{2}\right)\left(\dfrac{1}{4}\right)\left(\dfrac{1}{4}\right) = \dfrac{1}{32}$

and $P(X) = \left(\dfrac{1}{2}\right)\left(\dfrac{1}{4}\right)\left(\dfrac{3}{4}\right) + \left(\dfrac{1}{2}\right)\left(\dfrac{3}{4}\right)\left(\dfrac{1}{4}\right) +$

$\left(\dfrac{1}{2}\right)\left(\dfrac{1}{4}\right)\left(\dfrac{1}{4}\right) + \left(\dfrac{1}{2}\right)\left(\dfrac{1}{4}\right)\left(\dfrac{1}{4}\right) = \dfrac{1}{4}$

$\therefore P(X'_1 / X) = \dfrac{1}{8}$

Next, $X \cap X_2 = X - X_1 X'_2 X_3$

$P(X \cap X_2) = P(X) - P(X_1 x'_2 X_3) = \dfrac{5}{32}$

$\therefore P(X / X_2) = \dfrac{P(X \cap X_2)}{P(X_2)} = \dfrac{5/32}{1/4} = \dfrac{5}{8}$

Example 6: A fair coin is tossed 100 times. The probability of getting tails 1, 3 ... ...49 times is:

(A) $\frac{1}{2}$ (B) $\frac{1}{4}$ (C) $\frac{1}{8}$ (D) $\frac{1}{16}$

Sol: Let p = probability of getting a tail in a single trial $= \frac{1}{2}$ and n = number of trials = 100

X = Number of trials in 100 trials

We have, $P(X = r) = {}^{100}C_r \, p^r \, q^{n-r}$

$$= {}^{100}C_r \left(\frac{1}{2}\right)^r \left(\frac{1}{2}\right)^{100-r} = {}^{100}C_r \left(\frac{1}{2}\right)^{100}$$

Now,

$P(X = 1) + P(X = 3) + ... + P(X = 49) = {}^{100}C_1 \left(\frac{1}{2}\right)^{100} +$

$${}^{100}C_3 \left(\frac{1}{2}\right)^{100} + ... + {}^{100}C_{49} \left(\frac{1}{2}\right)^{100}$$

$$= \left({}^{100}C_1 + {}^{100}C_3 + ... + {}^{100}C_{49}\right) \left(\frac{1}{2}\right)^{100}$$

But ${}^{100}C_1 + {}^{100}C_3 + ... + {}^{100}C_{99} = 2^{99}$

Also, ${}^{100}C_{99} = {}^{100}C_1,$

${}^{100}C_{97} = {}^{100}C_3, ... {}^{100}C_{51}$

${}^{100}C_{49}$

Thus, $2\left({}^{100}C_1 + {}^{100}C_3 + ... + {}^{100}C_{49}\right) = 2^{99}$

$\Rightarrow {}^{100}C_1 + {}^{100}C_3 + ... + {}^{100}C_{49} = 2^{98}$

$\therefore$ Probability of required even $\dfrac{2^{98}}{2^{100}} = \dfrac{1}{4}$

Example 7: If A, B and C are three events, then

(A) P (exactly two of A, B, C occur) $\leq P(A \cap B) + P(B \cap C) + P(C \cap A)$

(B) $P(A \cup B \cup C) \leq P(A) + P(B) + P(C)$

(C) P(exactly one of A, B, C occur $\leq P(A) + P(B) + P(C) - P(B \cap C) - P(C \cap A) - P(A \cap B)$

(D) P (A and at least one of B, C occurs)

$\leq P(A \cap B) + P(A \cap C)$

Sol: Apply Boole's Inequality.

We have, P(exactly two of A, B, C occur)

$= P(A \cup B) + P(B \cap C) + P(C \cap A) - 3P(A \cap B \cap$

$C) \leq P(A \cap B) + P(B \cap C) + P(C \cap A)$

Also, $P(A \cup B \cup C) \leq P(A \cup B) + P(C) \leq P(A) +$

$P(B) + P(C)$

Next P(exactly one of A, B, C occurs)

$= P(A) + P(B) + P(C) - 2P(A \cap B) - 2P(B \cap C)$

$- 2P(C \cap A) + 3P(A \cap B \cap C)$

$= P(A) + P(B) + P(C) - P(A \cap B) - P(C \cap A)$

$- [P(A \cap B) + P(B \cap C) + P(C \cap A) - 3P(A \cap B \cap C)]$

$= P(A) + P(B) + P(C) - P(A \cap B) - P(B \cap C)$

$P(C \cap A) - P$ (exactly two of A, B, C occur)

$\leq P(A) + P(B) + P(C) - P(A \cap B) - P(B \cap C)$

$P(C \cap A)]$

Lastly P(A and at least one of B, C occur)

$= P[A \cap (B \cup C)] = P[(A \cap B) \cup (A \cap C)]$

$= P(A \cap B) + P(A \cap C) - P[(A \cap B) \cap (A \cap C)]$

$= P(A \cap B) + P(A \cap C) - P(A \cap B \cap C)$

$\leq P(A \cap B) + P(A \cap C)$

Example 8: For two events A and B, if P(A) =

$P(A \mid B) = \dfrac{1}{4}$ and $P(B \mid A) = \dfrac{1}{2}$, then

(A) A and B are independent

(B) A and B are mutually exclusive

(C) $P(A' \mid B) = \dfrac{3}{4}$

(D) $P(B' \mid A') = \dfrac{1}{2}$

Sol: Use basic formulae.

We have, $P(A) = P(A/B) = \dfrac{P(A \cap B)}{P(B)} \Rightarrow P(A \cap B) = P(A)$ $P(B)$

Therefore, A and B are independent. Also

$$P(A \cap B) = P(A) \, P(B \mid A) = \left(\frac{1}{4}\right)\left(\frac{1}{2}\right) = \left(\frac{1}{8}\right) 10$$

$\therefore$ A and B are mutually exclusive.

As A and B are independent

$$P(A' \mid B) = P(A') = 1 - P(A) = 1 - \left(\frac{1}{4}\right) = \left(\frac{3}{4}\right)$$

Since A and B are independent.

$P(B) = P(B \mid A) = \dfrac{1}{2}$

$\Rightarrow P(B' \mid A') = P(B') = 1 - P(B) = \dfrac{1}{2}$

Example 9: Let X be a set containing n elements. If two subsets, A and B, of X are picked at random, the probability that A and B have the same number of elements is:

(A) $\dfrac{^{2n}C_n}{2^{2n}}$

(B) $\dfrac{1}{^{2n}C_n}$

(C) $\dfrac{1.3.5 \ \text{........} \ (2n-1)}{2^n(n!)}$

(D) $\dfrac{3^n}{4^n}$

Sol: The number of ways of choosing a set of k elements is ^nC_k. The total number of subsets from a set of n elements is 2^n.

We know that the number of subsets of a set containing n elements is 2^n. Therefore, the number of ways of choosing A and B is $2^n . 2^n = 2^{2n}$. We also know that the number of subsets (of X) which contain exactly r elements is ^nC_r. Therefore, the number of ways of choosing A and B so that they have the same number of elements is

$(^nC_0)^2 + (^nC_1)^2 + (^nC_2)^2 + \text{........} + (^nC_n)^2$

$= {}^{2n}C_n = \dfrac{1.2.3 \ \text{.......} \ (2n-1)(2n)}{n!n!}$

$= \dfrac{[1.3.5.......(2n-1)][2.4.6.....(2n)]}{n!n!}$

Thus, the probability of the required event is

$\dfrac{^{2n}C_n}{2^{2n}} = \dfrac{1.3.5.....(2n-1)}{2^n(n!)}$

Example 10: Two numbers are selected at random from the number 1, 2, ..., n. Let p denote the probability that the difference between the first and second is not less than m (where 0 < m < n). If n = 25 and m = 10, find 5p.

Sol: Apply Total probability theorem.

Let the first number be x and the second be y. Let A denote the event that the difference between the first and second numbers is at least m. Let E_x denote the event that the first number chosen is x. We must have $x - y \geq m$ or $y \leq x - m$. Therefore x > m and y < n – m.

Thus, $P(E_x) = 0$ for $0 < x \leq m$ and $P(E_x) = 1/n$ for $m < x \leq n$.

Also, $P(A \mid E_x) = (x - m)/(n - 1)$.

Therefore, $P(A) = \displaystyle\sum_{x=1}^{n} P(E_x) P(A \mid E_x)$

$= \displaystyle\sum_{x=m+1}^{n} P(E_x) P(A \mid E_x) = \sum_{x=m+1}^{n} \dfrac{1}{n} . \dfrac{x-m}{n-1}$

$\dfrac{1}{n(n-1)}[1 + 2 + \text{......} + (n-m)] = \dfrac{(n-m)(n-m+1)}{2n(n-1)}$

Put n = 25 and m = 10

$\Rightarrow 5P = 5\dfrac{(25-10)(25-10+1)}{2.25(25-1)} = 1$

JEE Main/Boards

Exercise 1

Q.1 Given $P(A) = \dfrac{1}{4}$, $P(B) = \dfrac{2}{3}$ and $P(A \cup B) = \dfrac{3}{4}$. Are the events independent?

Q.2 Given $P(A) = \dfrac{1}{2}$, $P(B) = \dfrac{1}{3}$ and $P(A \cap B) = \dfrac{1}{6}$. Are the events A and B independent?

Q.3 A die is thrown twice. Find the probability of getting a number 6 on the first throw and number > 4 on the second.

Q.4 Given P(A) = 0.3, P(B)= 0.2. Find P(B/A) if A and B are mutually exclusive events.

Q.5 If P(A)=0.4, P(B)=p and P(A∪B)=0.7. Find the value of p, if A and B are independent set of events.

Q.6 Does the following table represents a probability distribution? Give reasons.

X	−2	−1	0	1	2
P(X)	0.1	0.2	−0.2	0.4	0.5

Q.7 Given P(A) = 0.2, P(B) = 0.3 and P(A∩B) = 0.1 Find P(A/B).

Q.8 The parameters n and p of a binomial distribution are 12 and 1/3 respectively, Find the standard deviation.

Q.9 Given P(A) = 0.4, P(B) = 0.7 and P(B/A) = 0.6. Find P(A∪ B).

Q.10 A coin is tossed three times and the Random variable X represents "number of heads". What values X can take?

Q.11 Does the following table represents a probability distribution? Give reasons.

X	0	1	2
P(X)	$\frac{1}{3}$	$\frac{1}{3}$	$\frac{1}{6}$

Q.12 Find the value of k, such that the following distribution represents a probability distribution

X	0	1	2	3	4
P(X)	k	0	3k	2k	4k

Q.13 Two cards are drawn successively, with replacement, from a deck of 52 cards. Find the probability of getting both spades.

Q.14 Find the mean of the distribution.

X	1	2	3
P(X)	0.4	0.3	0.3

Q.15 A coin is tossed 7 times, write the Probability distribution of getting r heads.

Q.16 In two successive throws of a pair of dice, find the probability of getting a total of 8 each time.

Q.17 Events E and F are given to be independent. Find P(F) if it is given that: P(E) = 0.60 and P(E∩F) = 0.35.

Q.18 If A and B are two independent events such that P(A∪B)=0.7 P(A)=0.4. Find P(B).

Q.19 Two cards are drawn from a pack of 52 cards at random and kept out. Then one card is drawn from the remaining 50 cards. Find the probability that it is an ace.

Q.20 Three cards are drawn with replacement from a well shuffled pack of cards. Find the probability that the cards are a king, a queen and a jack.

Q.21 A policeman fires four bullets on a dacoit. The probability that the dacoit will be killed by one bullet is 0.6. What is the probability that dacoit is still alive?

Q.22 A bag contains tickets numbered 1,2,3,..... , 50 of which five tickets x_1, x_2,......
x_5 are drawn at random and arranged in ascending order of magnitude $x_1 < x_2 < x_3 < x_4 < x_5$. What is the probability that $x_3 = 30$?

Q.23 A random variable X has the following probability distribution:

X	−2	−1	0	1	2	3
P(X)	0.1	k	0.2	2k	0.3	k

Find the value k(i) P(X ≤ 1)

(ii) P(X ≥ 0)

Q.24 Two cards are drawn successfully with replacement from a well shuffled pack of 52cards.Find the probability distribution of number of aces.

Q.25 In a lottery, a person choose six different numbers at random from 1 to 20 and if these six numbers match with the six numbers already fixed by the lottery committee, he wins the prize. What is the probability of winning the prize in the game, (order of numbers is not important)?

Q.26 The probability of students A passing an examination is 3/5 and of student B is 4/5. Assuming that the two events "A passes", "B passes" as independent. Find the probability of:

(i) Both the students passing the examination (ii) only A passing the examination

(iii) Only one of them passing the examination (iv) none of them passing the examination.

Q.27 A box contains 12 items of which 3 are defective. A sample of 3 items is selected from the box. Let x denote the number of defective items in a sample, find the probability distribution of X.

Q.28 Two dice are thrown. Find the probability that the numbers appeared have a sum 8 if it is known that the second dice always exhibits 4.

Q.29 In an examination, an examinee either guesses or copies or knows the answer of multiple choice questions with four choices. The probability that he makes a guess is 1/3 and probability that he copies the answer is 1/6.The probability that his answer is correct, given that he copied it, is 1/8. Find the probability that he known the answer to the question, given that he correctly answered it.

Q.30 There are three bags which contains 2 white, 3 black; 4 white, 1 black; 3 white, 7 black balls respectively A ball is drawn at random from one of the bags and is found to be black. Find the probability that is was drawn from the bag containing

(i) Maximum number of black balls.

(ii) Maximum number of white balls.

Q.31 Two cards are drawn successively with replacement from a pack of 52 cards. Find the probability distribution of the number of aces. Find its mean and standard deviation.

Q.32 The probability that a bulb produced by a factory will fuse after 150 days of use is 0.05.Find the probability that out of 5 such bulbs.

(i) None

(ii) Not more than one

(iii) More than one will fuse after 150 days of use.

(iv) At least one

Q.33 In a hurdle race, a player has to cross 10 hurdles. The probability that he will clear each hurdle is 5/6. What is the probability that he will knock down fewer than 2 hurdles?

Q.34 If on an average 1 ship in every 10 sinks, then find the chance that out of 5 ships at least 4 will arrive safely.

Q.35 About 70% of certain kind of seeds sold in the retail market germinate when planted under normal conditions. Suppose one packed contains 10 seeds. If these are planted, then what is the probability of 2 of these germinating?

Q.36 A man takes a step forward with probability0.4 and backwards with probability 0.6. Find the probability that at the end of eleven steps, he is just one step away from the starting point.

Q.37 A bag contains 10 balls, each marked with one of the digits 0 to 9. If four balls are drawn successively with replacement from the bag, what is the probability that none is marked with digit 6?

Q.38 Six dice are thrown 729 times. How many times do you expect at least three dice to show five or six?

Q.39 A survey of200 families each having 4 children was conducted. In how many families do you expect 3 boys and 1 girl if boys and girls are equal probable?

Q.40 Past experience shows that 80% of Operations performed by a doctor are successful. If he performs 4 operations in a day, what is the probability that at least three operations will be successful?

Q.41 The probability that a student entering a collage will graduate is 0.6. Find the probability that out of a group of 6 students,

(i) None   (ii) Atleast one

(iii) At most 3 will graduate.

Q.42 The probability of a bomb hitting a target is $\frac{1}{3}$. Two bombs are enough to destroy a bridge. If five bombs are dropped at the bridge, find the probability that the bridge is destroyed.

Q.43 In a binomial distribution, the sum of mean and variance is 42 and their product is 360.Findthe distribution.

Q.44 A bag contains 3 red and 7 black balls. Two balls are selected are selected at random without replacement. If the second selected is given to be red, what is the probability that the first selected is also red?

Q.45 Five dice are thrown simultaneously. If the occurrence of an odd number in a single die is considered a success, find the probability that there are odd number of successes.

Q.46 A die is thrown 10 times. If getting an even number is a success, find the probability of getting at least 9 successes.

Q.47 There are three urns A, B and C. Urn A contain4 red balls and 4 green balls. Urn B contains red ball and 5 green balls. Urn c contains 5 red balls and 2 green balls. One ball is drawn from each of the three urns. What is the probability out of these three drawn, two are green ball one is a red ball?

Q.48 A bag contains 4 red, 3 black and 3 white ball two balls are drawn from the bag. What is the probability that none of the balls drawn is white ball?

Q.49 A and B appear for an interview for two post, the probability of A's selection is 1/3 and the of B's selection is 2/5. Find the probability the only of them will be selected.

Q.50 A coin is tossed thrice and all eight out come are assumed equally likely Let the event E"the first throw results in head" and event F"the last throw results in tail". Find whether events E and F are independent.

Exercise 2

Single Correct Choice Type

Q.1 If cards are drawn at random from a pack of 52 playing cards without replacement then the probability that a particular card is drawn at the nth draw is:

(A) 1/(53-n)   (B) 1/52

(C) n/52   (D) n/(53 – n)

Q.2 4 persons are asked the same question by an interviewer. If each has independent probability 1/6 of answering the question correctly. The probability that at least one answers correctly is:

(A) 2/3   (B) $(1/6)^4$

(C) $1 – (5/6)^4$ (D) $1 – (1/6)^4$

Q.3 A person draws a card from a pack of 52 cards, replaces it & shuffles the pack. He continues doing this till he draws a spade. The probability that he will fail exactly the first two times is:

(A) 1/64   (B) 9/64   (C) 36/64   (D) 60/64

Q.4 A committee of 5 is to be chosen from a group of 9 people. The probability that a certain married couple will either serve together or not at all is:

(A) 1/2   (B) 5/9   (C) 4/9   (D) 2/3

Q.5 For a biased die the probabilities for the different faces to turn up are given below:

Faces: 1 2 3 4 5 6

Prob.: 0.10 0.32 0.21 0.15 0.05 0.17

The die is tossed & you are told that either face one or face two has turned up. Then the probability that it is face one is:

(A) 1/6   (B) 1/10   (C) 5/49   (D) 5/21

Q.6 For any 2 events A&B, the probabilities P(A), P(A+B), P(AB) & P(A) + P(B) when arranged in the increasing order of their magnitudes is:

(A) P(AB)≤ P(A) ≤ P(A+B)≤ P(A)+P(B)

(B) P(A)+P(B)≤ P(A+B) ≤ P(AB)≤P(A)

(C) P(A+B) ≤ P(AB) ≤ P(A)+P(B)≤ P(A)

(D) P(AB) ≤ P(A) ≤ P(A)+P(B) ≤ P(A+B)

Q.7 An integer x is chosen from the first 50 positive integers. The probability that, $x + \dfrac{100}{x} > 50$, is:

(A) $\dfrac{1}{25}$ (B) $\dfrac{2}{25}$ (C) $\dfrac{1}{10}$ (D) None of these

Q.8 The probability of India winning a test match against West Indies is 1/2. Assuming independent from match to match the probability that in a 5 match series, India's second win occurs at the 3rd test is:

(A) 1/4   (B) 1/8   (C) 1/2   (D) 2/3

Previous Years' Questions

Q.1 The probability that an event A happens in one trial of an experiment is 0.4. Three independent trials of the experiments are performed. The probability that the event A happens at least ones is   *(1980)*

(A) 0.936   (B) 0.784

(C) 0.904   (D) None

Q.2 If A and B are two independent events such that P(A)> 0,and P(B) ≠ 1, then $P(\bar{A}/\bar{B})$ is equal to   *(1982)*

(A) 1 – P(A/B)   (B) 1 –P(A/$\bar{B}$)

(C) $\dfrac{1 - P(A \cup B)}{P(B)}$ (D) $\dfrac{P(\bar{A})}{P(\bar{\bar{B}})}$

Q.3 The probability that at least one of the events A and B occurs is 0.6. If A and B occur simultaneously with probability 0.2, then $P(\bar{A}) + P(\bar{B})$ is: *(1987)*

(A) 0.4   (B) 0.8   (C) 1.2   (D) 1.4

(Here $\bar{A}$ and $\bar{B}$ are complements of A and B, respectively).

Q.4 One hundred identical coins, each with probability p, of showing up heads are tossed once. If 0< p< 1 and the probability of heads showing on 50 coins is equal to that of heads showing on 51 coins, then the value of p is **(1988)**

(A) $\frac{1}{2}$ (B) 49/101   (C) 50/101   (D) 51/101

Q.5 Seven white balls and three black balls are randomly placed in a row. The probability that no two black balls are placed adjacently, equals **(1998)**

(A) $\frac{1}{2}$ (B) $\frac{7}{15}$ (C) $\frac{2}{15}$ (D) $\frac{1}{3}$

Q.6 If E and F are events with P(E)≤ P(F) and P(E∩F)>0, then: **(1998)**

(A) Occurrence of E $\Rightarrow$ occurrence of F

(B) Occurrence of F $\Rightarrow$ occurrence of E

(C) Non-occurrence of E $\Rightarrow$ non-occurrence of F

(D) None of the above implication holds

Q.7 If P(B) = $\frac{3}{4}$, P($\bar{A} \cap \bar{B} \cap \bar{C}$) = $\frac{1}{3}$ and

P($\bar{A} \cap \bar{B} \cap \bar{C}$) = $\frac{1}{3}$ then P(B ∩ C) is: **(2002)**

(A) $\frac{1}{12}$ (B) $\frac{1}{6}$ (C) $\frac{1}{15}$ (D) $\frac{1}{9}$

Q.8 If three distinct numbers are chosen randomly from the first100 natural numbers, then the probability that all three of them are divisible by both 2 and 3, is **(2004)**

(A) $\frac{4}{55}$ (B) $\frac{4}{35}$ (C) $\frac{4}{33}$ (D) $\frac{4}{1155}$

Q.9 One Indian and four American men and their wives are to be seated randomly around a circular table. Then, the conditional probability that the Indian man is seated adjacent to his wife given that each American man is seated adjacent to his wife, is **(2007)**

(A) 1/2   (B) 1/3   (C)2/5   (D)1/5

Q.10 An experiment has 10 equally likely outcomes. Let A and B be two non-empty events of the experiment. If A consists of 4 outcomes, the number of outcomes that B must have so that A and B are independent, is **(2008)**

(A) 2,4 or 8   (B) 3, 6 or 9

(C) 4 or 8   (D)5 or 10

Q.11 Let (0 be a complex cube root of unity with $\omega \neq 1$. A fair die is thrown three times. If r_1, r_2 and r_3 are the numbers obtained on the die, then the probability that $\omega^{r_1} + \omega^{r_2} + \omega^{r_3} = 0$ is: **(2010)**

(A) 1/18   (B) 1/9   (C) 2/9   (D) 1/36

Q.12 A signal which can be green or red with probability $\frac{4}{5}$ and $\frac{1}{5}$ respectively, is received by station A and then transmitted to station B. The probability of each station receiving the signal correctly is $\frac{3}{4}$. If the signal received at station B is green, then the probability that the original signal is green is **(2010)**

(A) $\frac{3}{5}$ (B) $\frac{6}{7}$ (C) $\frac{20}{23}$ (D) $\frac{9}{20}$

Q.13 It is given that the event A and B are such that $P(A) = \frac{1}{4}$, $P\left(\frac{A}{B}\right) = \frac{1}{2}$ and $P\left(\frac{B}{A}\right) = \frac{2}{3}$. Then P(B) is **(2008)**

(A) $\frac{1}{6}$ (B) $\frac{1}{3}$ (C) $\frac{2}{3}$ (D) $\frac{1}{2}$

Q.14 A die thrown. Let A be the event that the number obtained is greater than 3. Let B be the event that the number obtained is less than 5. Then P(A∪B) is **(2008)**

(A) $\frac{3}{5}$ (B) 0   (C) 1   (D) $\frac{2}{5}$

Q.15 The mean of the number a, b, 8, 5, 10 is 6 and the variance is 6.80. Then which one of the following gives possible values of a and b? **(2008)**

(A) a = 0, b = 7   (B) a = 5, b = 2

(C) a = 1, b = 6   (D) a = 3, b = 4

Q.16 In a stop there are five types of ice-creams available. A child buys six ice-creams

Statement-I: The number of different ways the child can buy the six ice-creams is $^{10}C_5$

Statement-II: The number of different ways the child can buy the six ice-creams is equal to the number of different ways of arranging 6 A/s and 4 B's in a row. **(2008)**

(A) Statement-I is false, statement-II is true.

(B) Statement-I is true, statement-II is true; statement-II

is a correct explanation for statement-I

(C) Statement-I is true, statement-II is true ; statement-II is not a correct explanation for statement-I.

(D) Statement-I is true, statement-II is false

Q.17 How many different words can be formed by jumbling the letters in the word MISSISSIPPI in which no two S are adjacent? *(2008)*

(A) $8 \cdot {}^6C_4 \cdot {}^7C_4$ (B) $6 \cdot 7 \cdot {}^8C_4$

(C) $6 \cdot 8 \cdot {}^7C_4$ (D) $7 \cdot {}^6C_4 \cdot {}^8C_4$

Q.18 If the mean deviation of number 1, 1 + d, 1 + 2d, ..., 1 + 100d from their mean is 255, then the d is equal to *(2009)*

(A) 10.0   (B) 20.0   (C) 10.1   (D) 20.2

Q.19 From 6 different novels and 3 different dictionaries, 4 movies and 1 dictionary are to be selected and arranged in a row on the shelf so that the dictionary is always in the middle. Then the number of such arrangements is *(2009)*

(A) Less than 500

(B) At least 500 but less than 750

(C) At least 750 but less than 1000

(D) At least 1000

Q.20 One ticket is selected at random from 50 tickets numbered 00, 01, 02, ..., 49. Then the probability that the sum of the digits on the selected ticket is 8, given that the product of these digits is zero, equals. *(2009)*

(A) $\frac{1}{14}$ (B) $\frac{1}{7}$ (C) $\frac{5}{14}$ (D) $\frac{1}{50}$

Q.21 Statement-I: The variance of first n even natural numbers is $\frac{n^2-1}{4}$

Statement-II: The sum of first n natural number is $\frac{n(n+1)}{2}$ and the sum of squares of first n natural numbers is $\frac{n(n+1)(2n+1)}{6}$ *(2009)*

(A) Statement-I: is true, statement is true; statement-II is a correct explanation for statement-I

(B) Statement-I is true, statement-II is true; statement-II is not a correct explanation for Statement-I

(C) Statement-I is true, statement-II is false

(D) Statement-I is false, statement-II is true

Q.22 Four numbers are chosen at random (without replacement) from the set {1, 2, 3, ..., 20} *(2010)*

Statement-I: The probability that the chosen numbers when arranged in some order will form an AP is $\frac{1}{85}$.

Statement-II: If the four chosen numbers from an AP, then the set of all possible values of common difference is $\{\pm1, \pm2, \pm3, \pm4, \pm5\}$.

(A) Statement-I is true, statement-II is true; statement-II is not the correct explanation for statement-I

(B) Statement-I is true, statement-II is false

(C) Statement-I is false, statement-II is true

(D) Statement-I is true, statement-II is true; statement-II is the correct explanation for statement-I

Q.23 Let $S_1 = \sum_{j=1}^{10} j(j-1)\,{}^{10}C_j$, $S_2 \Rightarrow \sum_{j=1}^{10} j\,{}^{10}C_j$ and $S_3 = \sum_{j=1}^{10} j^2\,{}^{20}C_j$. *(2010)*

Statement-I: $S_3 = 55 \times 2^9$

Statement-II: $S_1 = 55 \times 2^8$ and $S_2 = 10 \times 2^8$

(A) Statement-I is true, statement-II is true; statement-II is not the correct explanation for statement-I

(B) Statement-I is true, statement-II is false

(C) Statement-I is false, statement-II is true

(D) Statement-I is true, statement-II is true, Statement-II is the correct explanation for statement-I

Q.24 There are two urns. urn A has 3 distinct red balls and urn B has 9 distinct blue balls. From each urn two balls are taken out at random and then transferred to the other. The number of ways in which this can be done is *(2010)*

(A) 36   (B) 66   (C) 108   (D) 3

Q.25 An urn contains nine balls of which three are red, four are blue and two are green. Three balls are drawn at random without replacement from the urn. The probability that the three balls have different colour is *(2010)*

(A) $\frac{2}{7}$ (B) $\frac{1}{21}$ (C) $\frac{2}{23}$ (D) $\frac{1}{3}$

Q.26 For two data sets, each of size 5, the variances are given to be 4 and 5 and the corresponding means are given to be 2 and 4, respectively. The variance of the combined data set is *(2010)*

(A) $\dfrac{11}{2}$　　(B) 6　　(C) $\dfrac{13}{2}$　　(D) $\dfrac{5}{2}$

Q.27 Assuming the balls to be identical except for difference in colours, the number of ways in which one or more balls can be selected from 10 white, 9 green and 7 black balls is *(2012)*

(A) 880　　(B) 629　　(C) 630　　(D) 879

Q.28 Three numbers are chosen at random without replacement $\{1, 2, 3, ..., 8\}$. The probability that their minimum is 3, given that their maximum is 6, is *(2012)*

(A) $\dfrac{3}{8}$　　(B) $\dfrac{1}{5}$　　(C) $\dfrac{1}{4}$　　(D) $\dfrac{2}{5}$

Q.29 All the students of a class performed poorly in Mathematics. The teacher decided to give grace marks of 10 to each of the students. Which of the following statistical measures will not change even after the grace marks were given? *(2013)*

(A) Mean　　　　　　(B) Median

(C) Mode　　　　　　(D) Variance

Q.30 Let A and B be two events such that $P(\overline{A \cup B}) = \dfrac{1}{6}$, $P(A \cap B) = \dfrac{1}{4}$ and $P(\overline{A}) = \dfrac{1}{4}$, where $\overline{A}$ stands for the complement of the event A. Then the events A and B are *(2014)*

(A) Independent but not equally likely

(B) Independent and equally likely

(C) Mutually exclusive and independent

(D) Equally likely but not independent

Q.31 The variance of first 50 even natural numbers is *(2014)*

(A) 437　　(B) $\dfrac{437}{4}$　　(C) $\dfrac{833}{4}$　　(D) 833

Q.32 The number of integers greater than 6,000 that can be formed, using the digits 3, 5, 6, 7 and 8 without repetition is: *(2015)*

(A) 216　　(B) 192　　(C) 120　　(D) 72

Q.33 If 12 identical balls are to be placed in 3 identical boxes, then the probability that one of the boxes contains exactly 3 balls is *(2015)*

(A) $\dfrac{55}{3}\left(\dfrac{2}{3}\right)^{11}$　　　　(B) $55\left(\dfrac{2}{3}\right)^{10}$

(C) $220\left(\dfrac{1}{3}\right)^{12}$　　　　(D) $22\left(\dfrac{1}{3}\right)^{11}$

Q.34 The mean of the data set comprising of 16 observations is 16. If one of the observation valued 16 is deleted and three new observation valued 3, 4 and 5 are added to the data, then the mean of the resultant data, is *(2015)*

(A) 16.8　　(B) 16.0　　(C) 15.8　　(D) 14.0

Q.35 If the standard deviation of the numbers 2, 3, a and 11 is 3.5, then which of the following is true *(2016)*

(A) $3a^2 - 32a + 84 = 0$　　(B) $3a^2 - 34a + 91 = 0$

(C) $3a^2 - 23a + 44 = 0$　　(D) $3a^2 - 26a + 55 = 0$

Q.36 Let two fair six-faced dice A and B be thrown simultaneously. If E_1 is the event that die A shown up four, E_2 is the event that die B shows up two and E_3 is the event that the sum of numbers on both dice is odd, then which of the following statements is NOT true? *(2016)*

(A) E_2 and E_3 are independent

(B) E_1 and E_3 are independent

(C) E_1, E_2 and E_3 are independent

(D) E_1 and E_2 are independent

Q.37 If all the words (with or without meaning) having five letters, formed using the letter of the word SMALL and arranged as in dictionary, then the position of the work SMALL is *(2016)*

(A) 59　　(B) 52　　(C) 58　　(D) 46

Exercise 1

Q.1 There are 2 groups of subjects one of which consists of 5 science subjects and 3 engineering subjects and other consists of 3 science and 5 engineering subjects. An unbiased die is cast. If the number 3 or 5 turns up a subject is selected at random from first group, otherwise the subject is selected from 2ndgroup. Find the probability that an engineering subject is selected.

Q.2 A pair of fair dice is tossed. Find the probability that the maximum of the two numbers is greater than 4.

Q.3 In a given race, the odds in favor of four horses A, B, C & D are 1: 3, 1: 4, 1: 5 and 1: 6 respectively. Assuming that a dead heat is impossible, find the chance that one of them wins the race.

Q.4 A covered basket of flowers has some lilies and roses. In search of rose, sweety and shweta alternately pick up a flower from the basket but puts it back if it is not a rose. Sweety is 3 times more likely to be the first one to pick a rose. If Sweety begin this 'rose hunt' and if there are 60 lilies in the basket, find the number of roses in the basket.

Q.5 A hotel packed breakfast for each of the three guests. Each breakfast should have consisted of three types of rolls, one each of nut, cheese and fruit rolls. The preparer wrapped each of the nine rolls and once warped, the rolls were indistinguishable from one another. She then randomly put three rolls in a bag for each of the guests. If the probability that each guest got one roll of each type is m where m and n are relatively prime integers, find the value of(m + n).

Q.6 Players A and B alternately toss a biased coin, with A going first. A wins if A tosses a Tail before B tosses a Head; otherwise B wins. If the probability of a head is p, find the value of p for which the game is fair to both players.

Q.7 The entries in a two-by-two determinant $\begin{vmatrix} a & b \\ c & d \end{vmatrix}$ are integers that are chosen randomly and independently, and, for each entry, the probability that the entry is odd is p. If the probability that the value of the determinant is even is 1/2, then find the value of p.

Q.8 Let an ordinary fair dice is thrown for five times. If P = $\frac{a}{b}$ expressed in lowest form be the probability that the outcome of the fifth throw was already thrown, then find the value of (a + b).

Q.9 A bomber wants to destroy a bridge. Two bombs are sufficient to destroy it. If four bombs are dropped, what is the probability that it is destroyed, if the chance of a bomb hitting the target is 0.4

Q.10 The chance of one event happening is the square of the chance of a 2nd event, but odds against the first are the cubes of the odds against the 2nd. Find the chances of each, (assume that both events are neither sure nor impossible).

Q.11 A box contains 5 radio tubes of which 2 are defective. The tubes are tested one after the other until the 2 defective tubes are discovered. Find the probability that the process stopped on the (i) Second test; (ii) Third test. If the process stopped on the third test, (iii) find the probability that the first tube is non-defective.

Q.12 An aircraft gun can take a maximum of four shots at an enemy's plane moving away from it. The probability of hitting the plane at first, second, third & fourth shots are 0.4, 0.3, 0.2 & 0.1 respectively. What is the probability that the gun hits the plane?

Q.13 In a batch of 10 articles, 4 articles are defective. 6 articles are taken from the batch for inspection. If more than 2 articles in this batch are defective, the whole batch is rejected. Find the probability that the batch will be rejected.

Q.14 A game is played with a special fair cubic die which has one red side, two blue sides, and three green sides. The result is the colour of the top side after the die has been rolled. If the die is rolled repeatedly, the probability that the second blue result occurs on or before the tenth roll, can be expressed in the form $\frac{3^p - 2^q}{3^r}$ where p, q, r are positive integers, find the value of $p^2 + q^2 + r^2$.

Q.15 An author writes a good book with a probability of 1/2. If it is good it is published with a probability of 2/3. If it is not, it is published with a probability of 1/4. Find the probability that he will get atleast one book published if he writes two.

Q.16 Consider 4 independent trials in which an event A occurs with probability $\frac{1}{3}$. The event B will occur with probability if the event A occurs atleast twice, it can not occur if the event A does not occur and it occurs with a probability $\frac{1}{2}$ if the event A occurs once. If the probability P of the occurrence of event B can be expressed as $\frac{m}{n}$, find the least value of (m + n), where m, n $\in$ N.'

Q.17 A uniform unbiased die is constructed in the shape of a regular tetrahedron with faces numbered 1,2,3 and 4 and the score is taken from the face on which the die lands. If two such dice are thrown together, find the probability of scoring.

(i) Exactly 6 on each of 3 successive throws.

(ii) More than 4 on at least one of the three successive throws.

Q.18 Two cards are drawn from a well shuffled pack of 52 cards. Find the probability that one of them is a red card & the other is a queen.

Q.19 A person flips 4 fair coins and discards those which turn up tails. He again flips the remaining coin and then discards those which turn up tails. Find the probability that he discards atleast 3 coins.

Q.20 Each of the 'n' passengers sitting in a bus may get down from it at the next stop with probability p. Moreover, at the next stop either no passenger or exactly one passenger boards the bus. The probability of no passenger boarding the bus at the next stop being p_o. Find the probability that when the bus continues on its way after the stop, there will again be 'n' passengers in the bus.

Q.21 A jar contains 2n thoroughly mixed balls, n white and n black balls. n persons each of whom draw 2 balls simultaneously from the bag without replacement.

(i) If the probability that each of the n person draw both balls of different colours is 8 35, then find the value of n.

(ii) If n = 4 then find the probability that each of the 4 persons draw balls of the same colour.

(iii) If n = 7 then the find the probability that each of the 7 persons draw balls of same colour.

Q.22 Eight players P_1, P_2, P_3, ... P_8 play a knock-out tournament. It is known that whenever the players P_i and P_j play, the player P_i will win if i<j. Assuming that the players are paired at random in each round, what is the probability that the player P_4 reaches the final.

Q.23 Let A & B be two events defined on a sample space. Given P (A) = 0.4 ; P(B) = 0.80 and $P(\bar{A}/B)$ = 0.10. Then find;

(i) $P(\bar{A} \cup B)$ and $P\left[(\bar{A} \cap B) \cup (A \cap \bar{B})\right]$

Q.24 Mr. A randomly picks 3 distinct numbers from the set {1, 2, 3, 4, 5, 6, 7, 8, 9} and arranges them in the descending order to form a three digit number. Mr. B randomly picks 3 distinct numbers from the set {1, 2, 3, 4, 5, 6, 7, 8} and also arranges them in descending order to form a 3 digit number.

(i) Find the probability that A and B has the same three digit number.

(ii) Find the probability that Mr. A's number is larger than Mr. B's number.

Q.25 A pair of students is selected at random from a probability class. The probability that the pair selected will consist of one male and one female student is $\frac{10}{19}$. Find the maximum number of students the class can contain.

Exercise 2

Single Correct Choice Type

Q.1 Suppose, that it is 9 to 7 against a person A who is now 35 years of age living till he is 65 and 3 to 2 against a person B now 45 living till he is 75, then the chance that one at least of these persons will be alive 30 years hence is:

(A) 14/27   (B) 53/80

(C) I/2   (D) None of these

Q.2 An experiment results in four possible outcomes S_1, S_2, S_3 and S_4 with probabilities p_1, p_2, p_3 & p_4 respectively. Which one of the following probability assignment is possible? [Assume S_1, S_2, S_3, S_4 are mutually exclusive]

(A) $p_1 = 0.25$, $p_2 = 0.35$, $p_3 = 0.10$, $p_4 = 0.05$

(B) $p_1 = 0.40$, $p_2 = -0.20$, $p_3 = 0.60$, $p_4 = 0.20$

(C) $p_1 = 0.30$, $p_2 = 0.60$, $p_3 = 0.10$, $p_4 = 0.10$

(D) $p_1 = 0.20$, $p_2 = 0.30$, $p_3 = 0.40$, $p_4 = 0.10$

Q.3 Let P denotes the probability that in a group of 4 persons all are born on different days of the week, then P must lie in the interval:

(A) $\frac{1}{3} < P < \frac{1}{2}$ 　　　　(B) $\frac{1}{4} < P < \frac{1}{5}$

(C) $\frac{1}{6} < P < \frac{1}{3}$ 　　　　(D) None of these

Q.4 The probability that 4th power of a positive integer ends in the digit 6 is:

(A) 10% 　　(B) 20% 　　(C) 25% 　　(D) 40%

Q.5 India plays 2 matches each with West-Indies & Australia. In any match the probabilities of India getting points 0,1 & 2 are 0.45, 0.05 & 0.50 respectively Assuming that the outcomes are independent, the probability of India getting atleast 7 points is:

(A) 0.8750 　　　　(B) 0.0875

(C) 0.0625 　　　　(D) 0.0250

Q.6 A women has 'n' keys, of which one will open her door. If she tries the keys randomly, discarding those that do not work (with out using the discarded key again), the probability that she will open the door with the last key is:

(A) $\frac{1}{n-1}$ 　(B) $\frac{1}{n}$ 　(C) $\frac{1}{(n-1)!}$ 　(D) $\frac{1}{2^n}$

Q.7 If A & B are two independent events, each with probability P, ($P \neq 0$) then $P(A/A \cup B)$ is:

(A) 1/P 　　　　(B) 1/2

(C) 2/P 　　　　(D) $1/(2 - P)$

Q.8 The probability of obtaining more tails than heads in 6 tosses of a fair coins is:

(A) 2/64 　　　　(B) 22/64

(C) 21/64 　　　　(D) None of these

Q.9 A gambler has one rupee in his pocket. He tosses an unbiased normal coin unless either he is ruined or unless the coin has been tossed for a maximum of five times. If for each head he wins a rupee and for each tail he loses a rupee, then the probability that the gambler is ruined is:

(A) 1/2 　　(B) 5/8 　　(C) 3/8 　　(D) 22/32

Q.10 If x be chosen randomly from the set of first 50 natural numbers, then the probability that x^x is perfect square of a natural number is-

(A) 12/25 　(B) 1/2 　(C) 29/50 　(D) 31/50

Q.11 A and B independently solve a problem. The chance that A and B will solve the problem correctly are P & 1/2 respectively. The chance that they will make the same mistake is $\frac{1}{100}$. If the probability that their answer is correct and they get the same answer which is $\frac{300}{301}$, then P is:

(A) 1/2 　(B) 3/4 　(C) 1/4 　(D) None of these

Q.12 Two dice are thrown until a 6 appears on atleast one of them. Then the probability that for the first time, a 6 appears in the second throw is:

(A) 175/1296 　　　　(B) 275/1296

(C) 375/1296 　　　　(D) None of these

Q.13 Box A has 3 white & 2 red balls, box B has 2 white & 4 red balls. If two balls are selected at random (without replacement) from A & 2 more are selected at random from B, the probability that all the four balls are white is:

(A) 10% 　(B) 2% 　(C) 12% 　(D) 4%

Q.14 A & B are two independent events such that $P(\bar{A}) = 0.7$, $P(\bar{B}) = a$ & $P(A \cup B) = 0.8$, then, a =

(A) 5/7 　(B) 2/7 　(C) 1 　(D) None

Q.15 A writes a letter to his friend B and gives it to his son to post it. The chance that his son will post the letter is 1/2 and the chance that a letter posted will reach it's destination is 5/6. If the letter was not received by B, the chance A'sson did not post the letter is-

(A) $\frac{5}{11}$ 　(B) $\frac{6}{11}$ 　(C) $\frac{2}{3}$ 　(D) $\frac{6}{7}$

Q.16 Two numbers are randomly selected from the set of first 20 natural numbers. Find the chance that their product is even given that their sum is odd-

(A) $\frac{9}{19}$ (B) $\frac{10}{19}$ (C) $\frac{29}{38}$ (D) None of these

Previous Years' Questions

Q.1 A box contains 2 black, 4 white and 3 red balls. One ball is drawn at random from the box and kept aside. From the remaining balls in the box, another ball is drawn at random and kept beside the first. This process is repeated till all the balls are drawn from the box. Find the probability that the balls drawn are in the sequence of 2 black, 4 white and 3 red. *(1979)*

Q.2 A and B are two independent events. The probability that both A and B occur is $\frac{1}{6}$ and the probability that neither of them occurs is $\frac{1}{3}$. Find the probability of the occurrence of A. *(1984)*

Q.3 In a multiple-choice question there are four alternative answers, of which one or more are correct. A candidate will get marks in the question only if he ticks the correct answers. The candidates decide to tick the answers at random, if he is allowed up to three chances to answer the questions, find the probability that he will get marks in the question. *(1985)*

Q.4 Three players. A, B and C. toss a coin cyclically in that order(that is A, B, C, A, B, C, A, B......) till a head shows. Let p be the probability that the coin shows a head. Let α, β and γ be, respectively, the probabilities that A, B and C gets the first head. Prove that $\beta = (1 - p)$ α. Determine α, β and γ(in terms of P) *(1998)*

Q.5 An unbiased die, with faces numbered 1, 2, 3, 4, 5, 6, is thrown n times and the list of n numbers showing up is noted. What is the probability that among the numbers 1, 2, 3, 4, 5, 6 only three numbers appear in this list? *(2001)*

Q.6 A bag contains 12 red balls and 6 white balls. Six balls are drawn one by one without replacement of which at least 4 balls are white. Find the probability that in the next two drawn exactly one white ball is drawn. (Leave the answer in ^nC_r). *(2004)*

Q.7 A person goes to office either by car, scooter, bus or train probability of which being $\frac{1}{7}, \frac{3}{7}, \frac{2}{7}$ and $\frac{1}{7}$. respectively. Probability that he reaches offices late, if he takes car, scooter, bus or train is $\frac{2}{9}, \frac{1}{9}, \frac{4}{9}$ and $\frac{1}{9}$ respectively. Given that he reached office in time, then what is the probability that he travelled by a car? *(2005)*

Paragraph 1 (Q.8 to Q.9): Read the following Paragraph and answer the questions.

There are n urns each containing (n + 1) balls such that the ith urn contains 7 white halls and (n + 1–i) red halls. Let u, be the event of selecting ith urn, i = 1,2,3, ..., n and W denotes the event of getting a white balls. *(2006)*

Q.8 If $(u_i) \propto$ I, where i = 1, 2,3, ... , then $\lim_{n \to \infty} P(W)$ is equal to

(A) 1   (B) $\frac{2}{3}$ (C) $\frac{1}{4}$ (D) $\frac{3}{4}$

Q.9 If $P(u_i)=c$, where c is a constant, then $P(u_n/W)$ is equal to

(A) $\frac{2}{n+1}$ (B) $\frac{1}{n+1}$ (C) $\frac{n}{n+1}$ (D) $\frac{1}{2}$

Q.10 If E and F are independent events such that $0 < P(E) < 1$ and $0 < P(F) < 1$, then *(1989)*

(A) E and F are mutually exclusive

(B) E and F^c (the complement of the event F) are independent

(C) E^c and F^c are independent

(D) $P(E/F) + P(E^c/F) = 1$

Q.11 Let E and F be two independent events. The probability that both E and F happen is 1/12 and the probability that neither E nor F happen is 1/2. Then, *(1993)*

(A) P(E) = 1/3, P(F) = 1/4

(B) P(E)=1/2, P(F)= 1/6

(C) P(E)= 1/6, P(F) = 1/2

(D) P(E) = 1/4, P(F) = 1/3

Q.12 If $\bar{E}$ and $\bar{F}$ are the complementary events of E and F respectively and if $0 < P(F) < 1$, then *(1998)*

(A) $P(E/F) + P(\bar{E}/F) = 1$

(B) $P(E/F) + P(E/\bar{F}) = 1$

(C) $P(\bar{E}/F) + P(E/\bar{F}) = 1$

(D) $P(E/\bar{F}) + P(\bar{E}/\bar{F}) = 1$

Q.13 Let E and F be two independent events. The probability that exactly one of them occurs is $\frac{11}{25}$ and the probability of none of them occurring is $\frac{2}{25}$. If P(T) denotes the probability of occurrence of the event T, then *(2011)*

(A) $P(E) = \frac{4}{5}, P(F) = \frac{3}{5}$ (B) $P(E) = \frac{1}{5}, P(F) = \frac{2}{5}$

(C) $P(E) = \frac{2}{5}, P(F) = \frac{1}{5}$ (D) $P(E) = \frac{3}{5}, P(F) = \frac{4}{5}$

Q.14 One Indian and four American men and their wives are to be seated randomly around a circular table. Then the conditional probability that the Indian man is seated adjacent to his wife given that each American man is seated adjacent to his wife is *(2007)*

(A) 1/2   (B) 1/3   (C) 2/5   (D) 1/5

Q.15 Let $H_1, H_2, ...H_n$ be mutually exclusive and exhaustive event with $P(H_i) > 0$, $I = 1, 2, ..., n$ Let E be any other event with $0 < P(E) < 1$. *(2007)*

Statement-I: $P(H_i|E) > P(E | H_i). P(H_i)$ for $I = 1, 2, ...,n$

Beause

Statement-II: $\sum_{i=1}^{n} P(H_i) = 1$

(A) Statement-I is True, statement-II is true, statement-II is a correct explanation for statement-I

(B) Statement-I is True, statement-II is True, statement-II is NOT a correct explanation for statement-I

(C) Statement-I is True, statement-II is False

(D) Statement-I is False, statement-II is True.

Q.16 The letters of the word **COCHIN** are permuted and all the permutations are arranged in an alphabetical order as in an English dictionary. The number of words that appear before the work **COCHIN** is *(2007)*

(A) 360   (B) 192   (C) 96   (D) 48

Q.17 Let E^c denote the complement of an event E. Let E, F, g be pairwise independent events with $P(G) > 0$ and $P(E \cap F \cap G) = 0$. Then $P(E^c \cap F^c |G)$ equals *(2007)*

(A) $P(E^c) + P(F^c)$ (B) $P(E^c) - P(F^c)$

(C) $P(E^c) - P(F)$ (D) $P(E) - P(F^c)$

Q.18 An experiment has 10 equally likely outcomes. Let A and B be two non-empty events of the experiment. If A consists of 4 outcomes, the number of outcomes that B must have so that A and B are independent, is *(2008)*

(A) 2, 4 or 8   (B) 3, 6 or 9

(C) 4 or 8   (D) 5 or 10

Q.19 Consider all possible permutations of the letters of the word ENDEANOEL.

Match the statement/expressions in column I with the statement/expressions in column II. *(2008)*

Column I	Column II
(A) The number of permutations containing the word ENDEA is	(p) 5!
(B) The number of permutations in which the letter E occurs in the first and the last positions is	(q) 2 × 5!
(C) The number of permutations in which none of the letters D, L, N occurs in the last five position is	(r) 7 × 5!
(D) The number of permutations in which the letter A, E, O occur only in odd positions is	(s) 21 × 5!

Q.20 The number of seven digit integers, with sum of the digit equal to 10 and formed by using the digits 1, 2 and 3 only, is *(2009)*

(A) 55   (B) 66   (C) 77   (D) 88

Paragraph 2 (Q.21 to Q.23): A fair die is tossed repeatedly until a six is obtained. Let X denote the number of tosses required. *(2009)*

Q.21 The probability that X = 3 equals

(A) $\frac{25}{216}$ (B) $\frac{25}{36}$ (C) $\frac{5}{36}$ (D) $\frac{125}{216}$

Q.22 The probability that $X \geq 3$ equals

(A) $\frac{125}{216}$ (B) $\frac{25}{36}$ (C) $\frac{5}{36}$ (D) $\frac{25}{216}$

Q.23 The conditional probability that $X \geq 6$ given $X > 3$ equals

(A) $\dfrac{125}{216}$ (B) $\dfrac{25}{216}$ (C) $\dfrac{5}{36}$ (D) $\dfrac{25}{36}$

Q.24 A signal which can be green or red with probability $\dfrac{4}{5}$ and $\dfrac{1}{5}$ respectively, is received by station A and then transmitted to station B. The probability of each station receiving the signal correctly is $\dfrac{3}{4}$. If the signal received at station B is green, then the probability that the original signal was green is *(2010)*

(A) $\dfrac{3}{5}$ (B) $\dfrac{6}{7}$ (C) $\dfrac{20}{23}$ (D) $\dfrac{9}{20}$

Paragraph 3 (Q.25 to Q.26): Let U_1 and U_2 be two urns such that U_1 contains 3 white and 2 red balls, and U_2 contains only 1 white ball. A fair coin is tossed. If head appears then 1 balls is drawn at random from U_1 and put into U_2. However, if tail appears then 2 balls are drawn at random from U_1 and put into U_2. Now 1 ball is drawn at random from U_2. *(2011)*

Q.25 The probability of the drawn ball from U_2 being white is

(A) $\dfrac{13}{30}$ (B) $\dfrac{23}{30}$ (C) $\dfrac{19}{30}$ (D) $\dfrac{11}{30}$

Q.26 Given that the drawn ball from U_2 is white, the probability that head appeared on the coin is

(A) $\dfrac{17}{23}$ (B) $\dfrac{11}{23}$ (C) $\dfrac{15}{23}$ (D) $\dfrac{12}{23}$

Q.27 The total number of ways in which 5 balls of different colours can be distributed among 3 persons so that each person gets at least one ball is *(2012)*

(A) 75 (B) 150 (C) 210 (D) 243

Q.28 A ship is fitted with three engines E_1, E_2 and E_3. The engines function independently of each other with respective probabilities $\dfrac{1}{2}$, $\dfrac{1}{4}$ and $\dfrac{1}{4}$. For the ship to be operational at least two of its engines must function. Let X denote the event that ship is operational and let X_1, X_2 and X_3 denote respectively the events that the engines E_1, E_2 and E_3 are functioning. Which of the following is(are) true ? *(2012)*

(A) $P\left[X_1^c \mid X\right] = \dfrac{3}{16}$

(B) P [Exactly two engines of the ship are functioning | X] $= \dfrac{7}{8}$

(C) $P[X \mid X_2] = \dfrac{5}{16}$

(D) $P[X \mid X_1] = \dfrac{7}{16}$

Q.29 Four fair dice D_1, D_2, D_3 and D_4 each having six faces numbered 1, 2, 3, 4, 5 and 6 are rolled simultaneously. The probability that D_4 shows a number appearing on one of D_1, D_2 and D_3 is *(2012)*

(A) $\dfrac{91}{216}$ (B) $\dfrac{108}{216}$ (C) $\dfrac{125}{216}$ (D) $\dfrac{127}{216}$

Q.30 Let X and Y be two events such that $P(X \mid Y) = \dfrac{1}{2}$, $P(Y \mid X) = \dfrac{1}{3}$ and $P(X \cap Y) = \dfrac{1}{6}$. Which of the following is (are) correct ? *(2012)*

(A) $P(X \cup Y) = \dfrac{2}{3}$

(B) X and Y are independent

(C) X and Y are not independent

(D) $P(X^c \cap Y) = \dfrac{1}{3}$

Q.31 Four persons independently solve a certain problem correctly with probabilities $\dfrac{1}{2}, \dfrac{3}{4}, \dfrac{1}{4}, \dfrac{1}{8}$. Then the probability that the problem is solved correctly by the at least one of them is *(2013)*

(A) $\dfrac{235}{256}$ (B) $\dfrac{21}{256}$ (C) $\dfrac{3}{256}$ (D) $\dfrac{253}{256}$

Q.32 Of the three independent events E_1, E_2 and E_3 the probability that only E_1 occurs is α only E_2 occurs is β and only E_3 occurs is γ. Let the probability p that none of events E_1, E_2 or E_3 occurs satisfy the equations $(\alpha - 2\beta)p = \alpha\beta$ and $(\beta - 3\gamma)p = 2\beta\gamma$. All the given probabilities are assumed to lie in the interval (0, 1).

Then, $\dfrac{\text{Probability of occurence of } E_1}{\text{Probability of occurrence of } E_3} = \underline{\hspace{1cm}}$

Q.33 A pack contains n cards numbered from 1 to n. Two consecutive numbered card are removed from the pack and the sum of the numbers on the remaining cards is 1224. If the smaller of the numbers on the removed cards is k, then k − 20 = *(2013)*

Q.34 If 1 ball is drawn from each of the boxes B_1, B_2 and B_3, the probability that all 3 drawn balls are of the same colour is *(2013)*

(A) $\dfrac{82}{648}$ (B) $\dfrac{90}{648}$ (C) $\dfrac{558}{648}$ (D) $\dfrac{556}{648}$

Q.35 If 2 balls are drawn (without replacement) from a randomly selected box and one of the balls is white and the other ball is red, the probability that these 2 balls are drawn from box B_2 is *(2013)*

(A) $\dfrac{116}{181}$ (B) $\dfrac{126}{181}$ (C) $\dfrac{65}{181}$ (D) $\dfrac{55}{181}$

Q.36 Three boys and two girls stand in a queue. The probability, that the number of boys ahead of every girl is at least one more than the number of girls ahead of her, is *(2014)*

(A) $\dfrac{1}{2}$ (B) $\dfrac{1}{3}$ (C) $\dfrac{2}{3}$ (D) $\dfrac{3}{4}$

Q.37 Six cards and six envelopes are numbered 1, 2, 3, 4, 5, 6 and cards are to be placed in envelopes so that each envelope contains exactly one card and no card is placed in the envelope bearing the same number and moreover the card numbered 1 is always placed in envelope numbered 2. Then the number of ways it can be done is *(2014)*

(A) 264   (B) 265   (C) 53   (D) 67

Paragraph 4 (Q.38 to Q.39): Box 1 contains three cards bearing number 1, 2, 3 ; box 2 contains five cards bearing numbers 1, 2, 3, 4, 5; and box 3 contains seven cards bearing numbers 1, 2, 3, 4, 5, 6, 7. A card is drawn from each of the boxes. Let x_i be the number on the card drawn from the ith box, I = 1, 2, 3. *(2014)*

Q.38 The probability that $x_1 + x_2 + x_3$ is odd, is

(A) $\dfrac{29}{105}$ (B) $\dfrac{53}{105}$ (C) $\dfrac{57}{105}$ (D) $\dfrac{1}{2}$

Q.39 The probability that x_1, x_2, x_3 are in an arithmetic progression, is

(A) $\dfrac{9}{105}$ (B) $\dfrac{10}{105}$ (C) $\dfrac{11}{105}$ (D) $\dfrac{7}{105}$

Q.40 The minimum number of times a fair coin needs to be tossed, so that the probability of getting at least two heads is at least 0.96 is *(2015)*

Q.41 Let n be the number of ways in which 5 boys and 5 girls can stand in a queue in such a way that all the girls stand consecutively in the queue. Let m be the number of ways in which 5 boys and 5 girls can stand in a queue in such a way that exactly four girls stand consecutively in the queue. Then the value of $\dfrac{m}{n}$ is *(2015)*

Paragraph 5 (Q.42 to Q.45): Let n_1 and n_2 be the number of red and black balls, respectively, in box I. Let n_3 and n_4 be the number of red and black balls, respectively, in box II *(2015)*

Q.42 One of the two boxes, box I and box II, was selected at random and a ball was drawn randomly out of this box. The ball was found to be red. If the probability that this red ball was drawn from box II is $\dfrac{1}{3}$, then the correct option(s) with the possible of n_1, n_2, n_3 and n_4 is (are)

(A) $n_1 = 3, n_2 = 3, n_3 = 5, n_4 = 15$

(B) $n_1 = 3, n_2 = 6, n_3 = 10, n_4 = 50$

(C) $n_1 = 8, n_2 = 6, n_3 = 5, n_4 = 20$

(D) $n_1 = 6, n_2 = 12, n_3 = 5, n_4 = 20$

Q.43 A ball is drawn at random from box II. If the probability of drawing a red ball from box I, after this transfer, is 1/3, then the correct option(s) with the possible values of n_1 and n_2 is (are)

(A) $n_1 = 4, n_2 = 6$

(B) $n_1 = 2, n_2 = 3$

(C) $n_1 = 10, n_2 = 20$

(D) $n_1 = 3, n_2 = 6$

Q.44 A computer producing factory has only two plants T_1 and T_2. Plant T_1 produces 20% and plant T_2 produces 80% of the total computers produced. 7% of computers produced in the factory turn out to be defective. It is known that

P (computer turns out to be defective given that it is produced in plant T_1)

= 10 P (computer turns out to be defective given that it is produced in plant T_2)

Where P(E) denotes the probability of an event E. A computer produced in the factory is randomly selected and it does not turn out to be defective.

Then the probability that it is produced in plant T_2 is *(2016)*

(A) $\dfrac{36}{73}$ (B) $\dfrac{47}{79}$ (C) $\dfrac{78}{93}$ (D) $\dfrac{75}{83}$

Q.45 A debate club consists of 6 girls and 4 boys. A team of 4 members is to be selected from this club including the selection of a captain (from among these 4 members) for the team. If the team has to include at most one boy, then the number of ways of selecting the team is *(2016)*

(A) 380　　(B) 320　　(C) 260　　(D) 95

Paragraph 6 (Q.46 to Q.47): Football teams T_1 and T_2 have to play two games against each other. It is assumed that the outcomes of the two games are independent. The probabilities of T_1 winning, drawing and lo sin a game against t_2 are $\frac{1}{2}, \frac{1}{6}$ and $\frac{1}{3}$, respectively. Each team gets 3 points for a win, 1 point for a draw and 0 point for a loss in a game. Let X and Y denote the total points scored teams T_1 and T_2 respectively, after two games *(2016)*

Q.46 P (X > Y) is

(A) $\frac{1}{4}$　　(B) $\frac{5}{12}$　　(C) $\frac{1}{2}$　　(D) $\frac{7}{12}$

Q.47 P (X = Y) is

(A) $\frac{11}{36}$　　(B) $\frac{1}{3}$　　(C) $\frac{13}{36}$　　(D) $\frac{1}{2}$

Important Questions

JEE Main/Boards

Exercise 1

Q.6	Q.9	Q.14
Q.22	Q.25	Q.31
Q.36	Q.39	Q.42
Q.44		

Exercise 2

Q.1	Q.3	Q.7
Q.9		

Previous Years' Questions

Q.6	Q.7	Q.9
Q.11	Q.12	

JEE Advanced/Boards

Exercise 1

Q.5	Q.12	Q.14
Q.20	Q.22	Q.25

Exercise 2

Q.3	Q.6	Q.9
Q.15	Q.16	

Previous Years' Questions

Q.3	Q.4	Q.7
Q.11	Q.13	

JEE Main/Boards

Exercise 1

Q.1 Yes   **Q.2** Yes   **Q.3** 1/18   **Q.4** 0   **Q.5** 0.5   **Q.6** No

Q.7 1/3   **Q.8** 1.63   **Q.9** 0.86   **Q.10** 0, 1, 2, 3   **Q.11** No   **Q.12** 0.1

Q.13 1/16   **Q.14** 1.9   **Q.15** $P(r) = {}^7C_r \left(\dfrac{1}{2}\right)^7$, r = 0, 1, 2, ... 7   **Q.16** $\dfrac{25}{1296}$ **Q.17** $\dfrac{7}{12}$

Q.18 0.5   **Q.19** 1/13   **Q.20** 6/2197   **Q.21** $(0.4)^4$ **Q.22** $\dfrac{551}{15134}$

Q.23 k=0.1 (i) 0.6   (ii) 0.8   **Q.24**

X	0	1	2
P(X)	$\dfrac{144}{169}$	$\dfrac{24}{169}$	$\dfrac{1}{169}$

Q.25 $\dfrac{1}{{}^{20}C_6}$

Q.26. (i) $\dfrac{12}{25}$ (ii) $\dfrac{3}{25}$ (iii) $\dfrac{11}{25}$ (iv) $\dfrac{2}{25}$ **Q.27**

X	0	1	2	3
P(X)	$\dfrac{21}{55}$	$\dfrac{27}{55}$	$\dfrac{27}{220}$	$\dfrac{1}{220}$

Q.28 1/6

Q.29 $\dfrac{24}{29}$ **Q.30** (i) $\dfrac{7}{15}$ (ii) $\dfrac{2}{15}$ **Q.31** $\mu = \dfrac{2}{13}$, $\sigma = 0.38$

Q.32 (i) $\left(\dfrac{19}{20}\right)^5$ (ii) $\left(\dfrac{19}{20}\right)^4$ (iii) $1 - \dfrac{6}{5}\left(\dfrac{19}{20}\right)^4$ (iv) $1 - \left(\dfrac{19}{20}\right)^5$

Q.33 0.4845   **Q.34** 0.9185   **Q.35** 0.00145   **Q.36** $462(0.24)^5$ **Q.37** $\left(\dfrac{9}{10}\right)^4$

Q.38 233   **Q.39** 50   **Q.40** 0.8192   **Q.41** (i) $\left(\dfrac{2}{5}\right)^6$ (ii) $1 - \left(\dfrac{2}{5}\right)^6$ (iii) $\dfrac{1424}{3125}$

Q.42 $\dfrac{131}{243}$ **Q.43** $\left(\dfrac{2}{5} + \dfrac{3}{5}\right)^{50}$ **Q.44** 2/9   **Q.45** 1/2   **Q.46** $\dfrac{11}{1024}$

Q.47 $\dfrac{41}{112}$ **Q.48** $\dfrac{7}{15}$ **Q.49** 7/15   **Q.50** Yes independent

Exercise 2

Single Correct Choice Type

Q.1 B   **Q.2** C   **Q.3** B   **Q.4** D   **Q.5** D   **Q.6** A

Q.7 C   **Q.8** A

Previous Years' Questions

Q.1 B	**Q.2** B	**Q.3** C	**Q.4** D	**Q.5** B	**Q.6** D
Q.7 A	**Q.8** D	**Q.9** C	**Q.10** D	**Q.11** C	**Q.12.** C
Q.13 B	**Q.14** C	**Q.15** D	**Q.16** A	**Q.17** D	**Q.18** C
Q.19 D	**Q.20** A	**Q.21** D	**Q.22** B	**Q.23** B	**Q.24** C
Q.25 A	**Q.26** A	**Q.27** D	**Q.28** B	**Q. 29** D	**Q.30** A
Q.31 D	**Q.32** D	**Q.33** A	**Q.34** D	**Q.35** A	**Q.36** C
Q.37 C					

JEE Advanced/Boards

Exercise 1

Q.1 $\dfrac{13}{24}$ **Q.2** 5/9 **Q.3** 319/420 **Q.4** 120 **Q.5** 79 **Q.6** $\dfrac{\sqrt{5}-1}{2}$

Q.7 $\dfrac{1}{\sqrt{2}}$ **Q.8** 1967 **Q.9** $\dfrac{328}{625}$ **Q.10** $\dfrac{1}{9},\dfrac{1}{3}$ **Q.11** (i) 1/10 (ii) 3/10 (iii) 2/3

Q.12 0.6976 **Q.13** 19/42 **Q.14** 283 **Q.15** 407/576 **Q.16** 130

Q.17 (i) $\dfrac{125}{16^3}$ (ii) $\dfrac{63}{64}$ **Q.18** 101/1326 **Q.19** $\dfrac{189}{256}$ **Q.20** $(1-p)^{n-1}.[p_0(1-p) + np(1- p_0)]$

Q.21 (i) 4 (ii) $\dfrac{3}{35}$ (iii) 0 **Q.22** 4/35 **Q.23** (i) 0.82 (ii) 0.76

Q.24 (i) $\dfrac{1}{84}$ (ii) $\dfrac{37}{56}$ **Q.25** 20

Exercise 2

Single Correct Choice Type

Q.1 B	**Q.2** D	**Q.3** A	**Q.4** D	**Q.5** B	**Q.6** B
Q.7 D	**Q.8** B	**Q.9** D	**Q.10** C	**Q.11** D	**Q.12** B
Q.13 B	**Q.14** B	**Q.15** D	**Q.16** D		

Previous Years' Questions

Q.1 $\dfrac{1}{1260}$ **Q.2** $\dfrac{1}{3}$ or $\dfrac{1}{2}$ **Q.3** 1/5 **Q.4** $\alpha = \dfrac{p}{1-(1-p)^3}$, $\beta = \dfrac{p(1-p)}{1-(1-p)^3}$, $\gamma = \dfrac{p-2p^2+p^3}{1-(1-p)^3}$

Q.5 $\dfrac{(3^n - 3.2^n +3)\times\,{}^6C_3}{6^n}$ **Q.6** $\dfrac{{}^{12}C_2.{}^6C_4}{{}^{18}C_6}.\dfrac{{}^{10}C_1.\,{}^2C_1}{{}^{12}C_2} + \dfrac{{}^{12}C_2.{}^6C_5}{{}^{18}C_6}.\dfrac{{}^{11}C_1.{}^1C_1}{{}^{12}C_2}$ **Q.7** $\dfrac{1}{7}$

Q.8 B	Q.9 A	Q.10 B, C, D	Q.11 A, D	Q.12 A, D	Q.13 A, D
Q.14 C	Q.15 D	Q.16 C	Q.17 C	Q.18 D	
Q.19 A → p; B → s; C → q; D → q	Q.20 C	Q.21 A	Q.22 B	Q.23 D
Q.24 C	Q.25 B	Q.26 D	Q.27 B	Q.28 B, D	Q.29 A
Q.30 A, B	Q.31 A	Q.32 6	Q.33 5	Q.34 A	Q.35 D
Q.36 A	Q.37 C	Q.38 B	Q.39 C	Q.40 8	Q.41 5
Q.42 A, B	Q.43 C, D	Q.44 C	Q.45 A	Q.46 B	Q.47 C

Solutions

JEE Main/Boards

Exercise 1

Sol 1: $P(A) = \frac{1}{4}$, $P(B) = \frac{2}{3}$, $P(A \cup B) = \frac{3}{4}$

$P(A \cup B) = P(A) + P(B) - P(A \cap B)$

$\frac{3}{4} = \frac{1}{4} + \frac{2}{3} - P(A \cap B)$

$P(A \cap B) = \frac{1}{4} + \frac{2}{3} - \frac{3}{4} = \frac{3+8-9}{12} = \frac{+2}{12} = \frac{+1}{6}$

$P(A) \cdot P(B) = \frac{1}{4} \cdot \frac{2}{3} = \frac{1}{6} = P(A \cap B)$

So the events are independent.

Sol 2: $P(A) = \frac{1}{2}$, $P(B) = \frac{1}{3}$

$P(A \cap B) = \frac{1}{6}$

$P(A) \cdot P(B) = \frac{1}{2} \cdot \frac{1}{3} = \frac{1}{6} = P(A \cap B)$

So events A and B are independent.

Sol 3: A dice has 6 number on it 1, 2, 3, 4, 5, 6

In a thrown probability of getting a member 6 $P(A) = \frac{1}{6}$
Total number of possibility

4 has two possibility = 5 or 6 both are greater than 4.

So P(B) = Probability of getting 5 or 6

$\Rightarrow \frac{1}{6} + \frac{1}{6} = \frac{1}{3}$

$P(A \cap B) = P(A) P(B) = \frac{1}{6} \cdot \frac{1}{3} = \frac{1}{18}$

Sol 4: $P(A) = 0.3$ $P(B) = 0.2$

$P(B / A) = \frac{P(B \cap A)}{P(A)}$

$P(A \cap B) = 0$

Because it's given that A and B both are exclusive events

$\therefore P(B / A) = 0$

Sol 5: $P(A) = 0.4$, $P(B) = P$

$P(A \cap B) = 0$

A and B are independents so

$P(A \cap B) = P(A) \cdot P(B) = 0.4P$

$P(A \cup B) = P(A) + P(B) - P(A \cap B)$

$0.7 = 0.4 + P - 0.4P \Rightarrow 0.4 + 0.6P$

$0.7 - 0.4 = 0.3 = 0.6P$

$\Rightarrow P = \frac{0.3}{0.6} = \frac{1}{2} = 0.5$

Sol 6:

x	−2	−1	0	1	2
P(x)	0.1	0.2	−0.2	0.4	0.5

The table does not represents probability distribution as probability never can be negative.

Sol 7: $P(A) = 0.2, P(B) = 0.3$

$P(A \cap B) = 0.1$

$P(A/B) = \dfrac{P(A \cap B)}{P(B)} = \dfrac{0.1}{0.3} = \dfrac{1}{3}$

Sol 8: $n = 12, P = 1/3$

Standard deviation $= \sqrt{npq}$

$q = 1 - P = 1 - \dfrac{1}{3} = \dfrac{2}{3} = \sqrt{\dfrac{12}{1} \times \dfrac{1}{3} \times \dfrac{2}{3}} = 1.63$

Sol 9: $P(A) = 0.4, P(B) = 0.7$

$P(A/B) = \dfrac{P(B \cap A)}{P(A)} = 0.6$

$\Rightarrow P(B \cap A) = P(A) \, 0.6 = (0.4)(0.6) = 0.24$

$P(A \cup B) \, P(A) + P(B) - P(A \cap B)$

$= 0.4 + 0.7 - 0.24 = 1.1 - 0.24 = 0.86$

Sol 10: A coin is tossed 3 times

X = number of head

X = 0 (T, T, T), 1(T, H, T), 2(H, H, T), 3(H, H, H)

X can be 0, 1, 2, 3

Sol 11:

x	0	1	2
P(x)	$\frac{1}{3}$	$\frac{1}{3}$	$\frac{1}{6}$

$\Sigma \, P(x) = \dfrac{1}{3} + \dfrac{1}{3} + \dfrac{1}{6} = \dfrac{2}{3} + \dfrac{1}{6} = \dfrac{5}{6} \neq 1$

So it's not probability distribution

Sol 12:

x	0	1	2	3	4
P(x)	k	0	3k	2k	4k

$\Sigma \, P(x) = 1$

$K + 0 + 3k + 2k + 4k = 1$

$\Rightarrow 10 \, k = 1$

$k = \dfrac{1}{10} = 0.1$

Sol 13: Total spades in a deck of 52 cards = 13

In a time probability of getting one spades $= \dfrac{13}{52} = \dfrac{1}{4} = P$

P(In two times and both time gets spades)

$= P. \, P = \dfrac{1}{4} \times \dfrac{1}{4} = \dfrac{1}{16}$

Sol 14:

x	1	2	3
P(x)	0.4	0.3	0.3

Mean $= \Sigma \, P(x); \, x$

$= 1(0.4) + 2(0.3) + 3(0.3)$

$= 0.4 + 0.6 + 0.9 = 1.9$

Sol 15: A coin is tossed 7 times

P (getting heads in one time) $= \dfrac{1}{2}$

If total number of heads $= r$

There is ^7C_r way to get r heads in 7 times probability of getting r heads

$^7C_r \left(\dfrac{1}{2}\right)^r$, $r = 0 \, 1, 2, 3, 4, 5, 6, 7$

r could be any integer between 0 and 7

Sol 16: P(getting 8 in one throw of a pair dice) $= \dfrac{5}{36}$

P (in two successive throws of a pair of dice getting 8 each time)

$= \dfrac{5}{36} \times \dfrac{5}{36} = \dfrac{25}{1296}$

Sol 17: E and F are independent

$P(E) = 0.60$ and $P(E \cap F) = 0.35$

$\because \, P(E \cap F) = P(E) \cdot P(F)$

$P(F) = \dfrac{P \, |E \cap F|}{P(E)} = \dfrac{0.35}{0.60} = \dfrac{7}{12}$

Sol 18: $P(A \cup B) = 0.7 \, P(A) = 0.4$

A and B are independent 50

$P(A \cap B) = P(A) \cdot P(B)$

$\because \, P(A \cup B) = P(A) + P(B) - P(A \cap B)$

$\Rightarrow 0.7 = 0.4 + P(B) - P(A) \, P(B) P(B) \, (1 - 0.4) = 0.7 - 0.4 = 0.3$

$P(B) = \dfrac{0.3}{0.6} = \dfrac{1}{2}$

Sol 19: There is 3 possibility –

First two cards are not ace $= \dfrac{^{48}C_2}{^{52}C_2}$

One of them is ace $= \dfrac{48 \times 4}{51 \times 52}$

Both are ace = $\dfrac{4 \times 3}{51 \times 52}$

Respectively probability of getting ace in third draw

$\Rightarrow \dfrac{4}{50}, \dfrac{3}{50}, \dfrac{2}{50}$

$P = \dfrac{{}^{48}C_2}{{}^{52}C_2} \times \dfrac{4}{50} + \dfrac{48 \times 4}{52 \times 51}\dfrac{3x^2}{50} + \dfrac{4 \times 3}{51 \times 52} \times \dfrac{2}{50}$

$= \dfrac{425}{5525} = \dfrac{1}{13}$

Sol 20: There is 4 King, queen and 4 jack is a deck of 52 cards. For three drawn

P(One king, one queen, one jack) = $\dfrac{4}{52} \cdot \dfrac{4}{52} \cdot \dfrac{4}{52} \cdot 3!$

There is 3! types of way to get them = $\dfrac{6}{169 \times 13} = \dfrac{6}{2197}$

Sol 21: Total fires = 4

P = Probability of killed by one bullet = 0. 6 after 4 fire.

P(still alive) = $(1 - 0.6)^4 = (0.4)^4$

Sol 22 : Drawn randomly 5 cards

x_1, x_2, x_3, x_4, x_5 and $x_1 < x_2 < x_3 < x_4 < x_5$

$\left| \begin{array}{l} 1,2,3... \\ ...49,50 \end{array} \right|$

P = Probability of $x_3 = 30$. So, x_1 and x_2 should be less than 30 for this number of total way to get x_1 and $x_2 = {}^{29}C_2$

And $x_4, x_5 > 30 \Rightarrow$ Total way $\Rightarrow {}^{20}C_2$

Total way for $x_1 x_2 x_3 x_4 x_5 = {}^{50}C_5$

$P = \dfrac{{}^{29}C_2 \,{}^{20}C_2}{{}^{50}C_5} = \dfrac{551}{15134}$

Sol 23 :

x	-2	-1	0	1	2	3
P(x)	0.1	k	0.2	2k	0.3	k

Σ P(x); = 1(always true)

0. 1 + K + 0. 2 + 2k + 0. 3 + k = 1

4k = 1 - 0. 1 - 0. 2 - 0. 34k = 0. 4

$\Rightarrow k = \dfrac{0.4}{4} = 0.1$

(i) $P(x \le 1)$

= P(x = -2) + P(x = -1) + P (x = 0) + P(x = 1)

= 0. 1 + k + 0. 2 + 2k = 3k + 0. 3 = 0. 6

(ii) $P(x \ge 0)$

= P(x = 0) + P(x = 1) + P(x = 2) + P(x = 3)

= 0. 2 + 2k + 0. 3 + k = 0. 5 + 3k

= 0. 5 + 3(0. 1) = 0. 5 + 0. 3 = 0. 8

Sol 24: Total aces in a pack of 52 cards = 4

Two cards are drawn with replacement assume x = number of getting aces in two drawn

So (i) x_1 = 0,no aces

$P(x_1) = \left(\dfrac{52-4}{52}\right)\left(\dfrac{52-4}{52}\right) = \dfrac{48}{52} \cdot \dfrac{48}{52} = \dfrac{12^2}{13^2} = \dfrac{144}{169}$

(ii) x_2 = 1 (one ace)

$P(x_2) = 2 \times \dfrac{4}{52} \times \dfrac{48}{52} = \dfrac{24}{169}$

(iii) x_3 = 2 (both are ace)

$P(x_3) = \dfrac{4}{52} \cdot \dfrac{4}{52} = \dfrac{1}{169}$

x	0	1	2
P(x)	$\dfrac{144}{169}$	$\dfrac{24}{169}$	$\dfrac{1}{169}$

Sol 25: Number of way to get 6 numbers from 1 to 20 = ${}^{20}C_6$

Number of way to get the fixed 6 numbers = 1

$P = \dfrac{1}{{}^{20}C_6}$

Sol 26: P(A) = probability of student A passing exam

P (A) = 3/5P(B) = 4/5

$\Rightarrow$ P(A) and P(B) are independent

So, $P(A \cap B) = P(A) \cdot P(B) = \dfrac{3}{5} \cdot \dfrac{4}{5} = \dfrac{12}{25}$

(i) Both the student passing exam

$P(A \cap B) = \dfrac{12}{25}$

(ii) Only A pass the exam

= $P (A) - P(A \cap B) = \dfrac{3}{5} - \dfrac{12}{25} = \dfrac{15-12}{25} = \dfrac{3}{25}$

(iii) Only one of them passing exam

= P(A) + P(B) - 2P(A \cap B)

$= \dfrac{3}{5} + \dfrac{4}{5} - \dfrac{2.12}{25} = \dfrac{15+20-24}{25} = \dfrac{11}{25}$

(iv) None of them passing exam

$= 1 - P(A \cup B) = 1 + P(A \cap B) - P(A) - P(B)$

$= 1 + \dfrac{12}{25} - \dfrac{3}{5} - \dfrac{4}{5} = \dfrac{25 + 12 - 15 - 20}{25} = \dfrac{2}{25}$

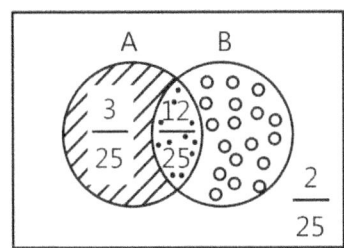

Sol 27: 3 are defective out of 12 items

A sample of 3 items is selected from the box

Where x = number of defective items

(i) $x_0 = 0$, all are in will condition

$\Rightarrow P(x_0) = \dfrac{{}^3C_1 {}^9C_3}{{}^{12}C_3} = \dfrac{21}{55}$ $\boxed{(12) = (9) + \left(\begin{array}{c}3\\ \text{def}\end{array}\right)}$

(ii) $x_1 = 1$, one defective

$P(x_1) = \dfrac{{}^3C_1 {}^9C_2}{{}^{12}C_3} = \dfrac{27}{55}$

(iii) $x_2 = 2$, two defective

$P(x_2) = \dfrac{{}^3C_2 {}^9C_2}{{}^{12}C_3} = \dfrac{27}{220}$

(iv) $x_3 = 3$, all are defective

$P(x_3) = \dfrac{{}^3C_3}{{}^{12}C_3} = \dfrac{1}{220}$

x	0	1	2	3
P(x)	$\dfrac{21}{55}$	$\dfrac{27}{55}$	$\dfrac{27}{220}$	$\dfrac{1}{220}$

Sol 28: Two dice are thrown, second dice always exhibits.

S = All possibility: (1, 4) (2, 4) (3, 4) (5, 4) (6, 4) (4, 4)

Sum of 8 → (4, 4)

$\Rightarrow P = \dfrac{x(4,4)}{n(S)} = \dfrac{1}{6}$

Sol 29 : C → Copies ans.

K → Known ans.

G → Guess ans.

R → Ans. is right

$P(C) = 1/6 \quad P(G) = 1/3$

$\Rightarrow P(K) = 1 - \dfrac{1}{6} - \dfrac{1}{3} = \dfrac{1}{2}$

$P(R / C) = \dfrac{P(R \cap C)}{P(C)} = \dfrac{1}{8}$

P(R) = P(CR) + P(KR) + P(GR) there is 4 choices for each question

So, $P(GR) = \dfrac{1}{4} \cdot P(G) = \dfrac{1}{3} \cdot \dfrac{1}{4} = \dfrac{1}{12}$

$\dfrac{P(KR)}{P(R)} = \dfrac{\dfrac{1}{2}}{\dfrac{1}{8} \times \dfrac{1}{6} + \dfrac{1}{12} + \dfrac{1}{2}} = \dfrac{24}{1 + 4 + 24} = \dfrac{24}{29}$

Sol 30 :

2 white	4 white	3 white
3 black	1 black	7 black
I	II	III

B = Ball is black

P(B) = P(BI) + P(BII) + P(BIII)

$= \dfrac{1}{3} \times \dfrac{3}{5} + \dfrac{1}{3} \times \dfrac{1}{5} + \dfrac{1}{3} \times \dfrac{7}{10}$

$= \dfrac{3}{15} + \dfrac{1}{15} + \dfrac{7}{30} = \dfrac{6 + 2 + 7}{30} = \dfrac{15}{30} = \dfrac{1}{2}$

(i) Bag II has max. number of black balls

$P\left(\dfrac{III}{B}\right) = \dfrac{P(IIIB)}{P(B)} = \dfrac{\dfrac{7}{30}}{\dfrac{1}{2}} = \dfrac{7}{15}$

(ii) Bag II has max. white balls

So $P\left(\dfrac{II}{B}\right) = \dfrac{P(IIB)}{P(B)} = \dfrac{\dfrac{1}{15}}{\dfrac{1}{2}} = \dfrac{2}{15}$

Sol 31: Two cards are drawn with replacement

X = number of getting aces from question number 29

x	0	1	2
P(x)	$\dfrac{144}{169}$	$\dfrac{24}{169}$	$\dfrac{1}{169}$

$\text{mean} = \Sigma P(x)x_1 = 0\left(\dfrac{144}{169}\right) + 1\left(\dfrac{24}{169}\right) + 2\left(\dfrac{1}{169}\right)$

$= \dfrac{24 + 2}{169} = \dfrac{2}{13}$

$\sigma = SD = \sqrt{\Sigma P(x)(x_1)^2 - (\Sigma P(x) \cdot (x_1))^2}$

$$= \sqrt{(0)^2\left(\frac{144}{169}\right) + 1^2 \cdot \left(\frac{24}{169}\right) + 2^2 \cdot \left(\frac{1}{169}\right) - \left(\frac{2}{13}\right)^2}$$

$$= \frac{1}{\sqrt{169}}\sqrt{24 + 2^2 - 2^2}$$

$$= \frac{\sqrt{24}}{13} = 0.377 = 0.38$$

$SD = 0.38$, $\mu = \frac{2}{13}$

Sol 32 : Total choose bulb = 5

P = P (bulb will fuse after 150 days) = 0.05

$q = 1 - P = 0.95 = \frac{19}{20}$, n = 5,

x = number of fuse

(i) P(none of 5 will fuse) = $(q)^5 = \left(\frac{19}{20}\right)^5$

(ii) P(not more than ore) = P(x ≤ 1)

$= P(x = 0) + P(x = 1) = (q)^5 + {}^5C_1 q^4 P^1$

$$= \left(\frac{19}{20}\right)^5 + \left(\frac{19}{20}\right)^4\left(\frac{1}{20}\right) = \left(\frac{19}{20}\right)^5\left[1 + \frac{20}{19} \cdot \frac{1}{20}\right]$$

$$= \frac{20}{19} \times \left(\frac{19}{20}\right)^5 = \left(\frac{19}{20}\right)^4$$

(iii) P(x > 1) = 1 - P(x = 0) - P(x = 1) = $1 - \frac{6}{5}\left(\frac{19}{20}\right)^4$

(iv) P(x ≥ 1) = 1 - P(x = 0) = $1 - \left(\frac{19}{20}\right)^5$

Sol 33 : x = number of hurdles; he down

Total hurdles = 10 = n

$P = 5/6 \Rightarrow q = 1 - P = 1/6$

P(x < 2) = P(x = 0) + P(x = 1)

$$= \left(\frac{5}{6}\right)^{10} + 10 \times \left(\frac{5}{6}\right)^9 \times \frac{1}{6} = 0.4845$$

Sol 34 : Total ship = 5

X = n (arrive safe ships)

$q = \frac{1}{10} \Rightarrow P = 1 - q = \frac{9}{10}$

P(x ≥ 4) = P(x = 4) + P(x = 5)

$$= {}^5C_1\left(\frac{9}{10}\right)^4 \frac{1}{10} + \left(\frac{9}{10}\right)^5 = 0.9185$$

Sol 35 : n = 10, P = 0.7

x = number of seed germinate

$P(x = 2) = {}^{10}C_2 (0.7)^2 (1 - 0.7)^8$

$$= \frac{10 \times 9}{2} \times (0.49)(0.3)^8 = 0.00145$$

Sol 36 : P(F) = P(step forward) = 0.4

P(B) = P(step backward) = 0.6

N = 11 (number of step forward)

$x_F + x_B = 11 | x_F - x_B| = 1$

$\Rightarrow |11 - 2x_B| = 1$

$\Rightarrow x_B = \frac{11 - 1}{2} = 5$ or $2x_B - 11 = 1$

$\Rightarrow 2x_B = 1 + 11 = 12$

$x_B = 6$ $x_F = 11 - 5$

or 11 - 6 = 6 or 5

$P(x_F = 6, 5) = P(x = 6) + P(x = 5)$

$= {}^{11}C_6 (0.4)^6 (0.6)^5 + {}^{11}C_5 (0.4)^5 (0.6)^6$

$= {}^{11}C_5 [(0.4 + 0.6)(0.24)^5]$

$= {}^{11}C_5 (0.24)^5 = 462 (0.24)^5$

Sol 37 : 4 balls are drawn with replacement from the bag. Assume x = number ball marked with 6

$$P(x = 0) = \left(\frac{10 - 1}{10}\right)^4 = \left(\frac{9}{10}\right)^4$$

(∵ There is only one ball which is marked 6 out of 10)

Sol 38 : 6 dice are thrown assume x_5 = number of dice which show 5 or 6

$P = \frac{2}{6} = \frac{1}{3}$, $q = \frac{4}{6} = \frac{2}{3}$

$P(x_5 \geq 3) = P(x = 3) + P(x = 4) + P(x = 5) + P(x = 6)$

$$= {}^6C_3\left(\frac{2}{6}\right)^3\left(\frac{4}{6}\right)^3 + {}^6C_4\left(\frac{2}{6}\right)^4\left(\frac{4}{6}\right)^2$$

$$+ {}^6C_5\left(\frac{2}{6}\right)^5\left(\frac{4}{6}\right) + {}^6C_6\left(\frac{2}{6}\right)^6 = \frac{233}{729}$$

Expected number $\rightarrow 729 \times \dfrac{233}{729} = 233$

$= 1 - \dfrac{112}{3^5} = \dfrac{243 - 112}{243} = \dfrac{131}{243}$

Sol 39 : 200 families, $P(B) = \dfrac{1}{2} = P(G)$ Total children = 4 for each family

$P(3B, 1G) = {}^4C_1 \times \left(\dfrac{1}{2}\right)^4 = \dfrac{4}{4 \times 4} = \dfrac{1}{4}$

expectation $\rightarrow 200 \times \dfrac{1}{4} = 50$ families

Sol 40 : P = probability of success of a operation

P = 0. 8, n = 4x = number of successful operation

$P(x \geq 3) = P(x = 3) + P(x = 4)$

$= [{}^4C_3 (0.8)^3 (1-0.8) + {}^4C_4 (0.8)^4]$

$= [4 \times 0.8^3 \cdot 0.2 + (0.8)^4]$

$= 0.8^4 \times 2 = 0.8192$

Sol 41 : P = Probability of graduate student

P = 0. 6

$\Rightarrow q = 1 - 0.6 = 0.4$

x = number of students will graduate

(i) $P(x = 0) = {}^6C_0 (0.4)^6 = \left(\dfrac{2}{5}\right)^6$

(ii) $P(x \geq 1) = 1 - P(x = 0) = 1 - \left(\dfrac{2}{5}\right)^6$

(iii) $P(x \leq 3) = P(x = 0) + P(x = 1) + P(x = 2) + P(x = 3)$

$= \left(\dfrac{2}{5}\right)^6 + {}^6C_1 \left(\dfrac{2}{5}\right)^5 \dfrac{3}{5} + {}^6C_2 \left(\dfrac{2}{5}\right)^4 \left(\dfrac{3}{5}\right)^2 + {}^6C_3 \left(\dfrac{2}{5}\right)^3 \left(\dfrac{3}{5}\right)^3$

$= \dfrac{1}{5^6} [2^6 + 6 \cdot 3 \cdot 2^5 + 15 \cdot 2^4 \cdot 3^2 + 20 \cdot 2^3 \cdot 3^3]$

$= \dfrac{7120}{5^6} = \dfrac{1424}{3125}$

Sol 42 : P = P(bomb hitting a target)

$P = \dfrac{1}{3} \Rightarrow q = 1 - \dfrac{1}{3} = \dfrac{2}{3}$

Two bombs are enough to destroy target

n = 5x = number of bombs that hit the target

so for destroy bridge

$P(x \Rightarrow 2) = 1 - P(x = 0) - P(x = 1)$

$= 1 - \left[\left(\dfrac{2}{3}\right)^5 + {}^5C_1 \left(\dfrac{2}{3}\right)^4 \times \dfrac{1}{3}\right] = 1 - \dfrac{1}{3^5} [2^5 + 5.2^4]$

Sol 43: $(p + q)^n$

variance $\sigma^2 = npq$ mean $\mu = np$

Given $\rightarrow np + npq = 42$

$\Rightarrow np (1 + q) = 42$ ...(i)

and $(np)(npq) = 360$

$\Rightarrow (np)^2 q = 360$ ...(ii)

$(1)^2 (2) \rightarrow \dfrac{(np)^2 (1 + q)^2}{(np)^2 q} = \dfrac{422}{360} = \dfrac{49}{10}$

$\Rightarrow 10(1 + q^2 + 2q) = 49 q$

$\Rightarrow 10q^2 + 202 - 49q + 10 = 0$

$\Rightarrow 10q^2 - 29q + 10 = 0$

$\Rightarrow 10q^2 - 25q - 4q + 10 = 0$

$\Rightarrow 5q (2q - 5) - 2 (2q - 5) = 0$

$\Rightarrow (2q - 5) (5q - 2) = 0$

$q = \dfrac{5}{2}$ or $\dfrac{2}{5}$, q < 1

So, $q = \dfrac{2}{5} \Rightarrow P = 1 - q = \dfrac{3}{5}$

and $(np)^2 q = 360$

$n^2 = \dfrac{360}{215 \times \left(\dfrac{3}{5}\right)^2} = \dfrac{360 \times 5^3}{2 \times 9}$

$n^2 = 10^2 \times 5^2 n = 10 \times 5 = 50 \Rightarrow \left(\dfrac{2}{5} + \dfrac{3}{5}\right)^{50}$

Sol 44 : $R_1 = I^{st}$ ball is red

$R_2 = 2^{nd}$ ball is red $\begin{vmatrix} 3 \text{ red} \\ 7 \text{ black} \end{vmatrix}$

Total 2 balls are related without replacement $P\left(\dfrac{R_1}{R_2}\right) > ?$

$\Rightarrow \dfrac{P(R_1 \cap R_2)}{P(R_2)}$

$P(R_2)_2 = \dfrac{7}{10} \times \dfrac{3}{9} + \dfrac{3}{10} \times \dfrac{2}{9} = \dfrac{21 + 6}{90} = \dfrac{22}{90} = \dfrac{3}{10}$

$P(R_1 \cap R_2) = \dfrac{3}{10} \times \dfrac{2}{9} = \dfrac{6}{90} = \dfrac{1}{15}$

$P\left(\dfrac{R_1}{R_2}\right) = \dfrac{1/15}{3/10} = \dfrac{10}{3 \times 15} = \dfrac{2}{9}$

3.47

Sol 45: Five dice are thrown

Success = odd number

There is number of odd number and even number are same which is 3 (1, 3,5 and 2, 4, 6)

So, P(success) = P(no success)

but P(success) + P (no success) = 1

∵ There is either odd or even number

$P \text{ (success)} = \dfrac{1}{2}$

Sol 46: n = 10x = number of getting even number

$P = \dfrac{1}{2} \Rightarrow q = 1 - \dfrac{1}{2} = \dfrac{1}{2}$

$P(x \geq 9) = P(x = 9) + P(x = 10)$

$= {}^{10}C_9 \left(\dfrac{1}{2}\right)^{10} + 1\left(\dfrac{1}{2}\right)^{10} = \left(\dfrac{1}{2}\right)^{10}(1 + 10) = \dfrac{11}{2^{10}} = \dfrac{11}{1024}$

4 red	3 red	5 red
4 green	5 green	2 green

Sol 47:   A   B   C

↓   ↓   ↓

one ball   one ball   one ball

= 2 green balls and one is red

$P(2G, 1R) = P[(GRG) + (GGR) + (RGG)]$

$P(2G, 1R) = \dfrac{4}{8}\left[\dfrac{5}{8} \cdot \dfrac{5}{7} + \dfrac{3}{8} \cdot \dfrac{2}{7}\right] + \dfrac{4}{8} \times \dfrac{5}{8} \cdot \dfrac{2}{7}$

$= \dfrac{1}{2}\left[\dfrac{25}{56} + \dfrac{6}{56}\right] + \dfrac{5}{56} = \dfrac{1}{56} \times \dfrac{1}{2}[25 + 6 + 10] = \dfrac{41}{112}$

4 red
3 black
3 white

Sol 48: Two balls are drawn from

$P\text{(none of the ball is white)} = \dfrac{{}^7C_2}{{}^{10}C_2} = \dfrac{7}{15}$

Sol 49: P(A) = probability of A selection $= \dfrac{1}{3}$

$P(B) = \dfrac{2}{5}$

Total post = 2

P(only one of them will selected) = P(A) + P(B) − 2P (A ∩ B)

Since both events are independent

So, $P(A \cap B) = P(A)\, P(B) = \dfrac{1}{3} \cdot \dfrac{2}{5} = \dfrac{2}{15}$

$\Rightarrow P = \dfrac{1}{3} + \dfrac{2}{5} - 2 \cdot \dfrac{2}{15} = \dfrac{5 + 6 - 4}{15} = \dfrac{7}{15}$

Sol 50 : E = 1st throw is head

F = Last throw is tail

$P(E) = \dfrac{1}{2} \times \dfrac{2}{2} \cdot \dfrac{2}{2} = \dfrac{1}{2}$ $P(F) = \dfrac{2}{2} \cdot \dfrac{2}{2} \cdot \dfrac{1}{2} = \dfrac{1}{2}$

P(E∩F) = P(1st is head and last is tail)

$= \dfrac{1}{2} \times \dfrac{2}{2} \times \dfrac{1}{2} = \dfrac{1}{4}$

$P(E) \cdot P(F) = \dfrac{1}{4} = P(E \cap F)$

So, E and F are independent.

Exercise 2

Single Correct Choice Type

Sol 1: (B) P(special card at n^{th} drawn)

$\Rightarrow \dfrac{51}{52} \cdot \dfrac{1}{51} = \dfrac{1}{52}$

Sol 2: (C) n = 4 (r person)

$P = \dfrac{1}{6}$ (correct ans. by one)

x = number of correct ones

$P(x \geq 1) = 1 - P(x = 0) = 1 - \left(1 - \dfrac{1}{6}\right)^4 = 1 - \left(\dfrac{5}{6}\right)^4$

Sol 3: (B) n = 2

x = get spade

Total spades in 52 cards = 13

P = P(he fails exactly first two times)

$P = \left(\dfrac{52 - 13}{52}\right)\left(\dfrac{52 - 13}{52}\right) \cdot \dfrac{13}{52} = \dfrac{39}{52} \cdot \dfrac{39}{52} \cdot \dfrac{13}{52} = \dfrac{9}{64}$

Sol 4: (D) 5 is to be chosen from 9 people there is a couple in group of 9

P = P(couple chosen)

q = P(couple don't chose)

$$P + q = \frac{1}{^9C_5}\left[{}^7C_5 + {}^7C_3 \cdot \frac{2}{2}\right]$$

$$= \frac{1.2.3.4}{9.8.7.6}\left[\frac{7.6}{1.2} + \frac{7.6.5}{2.3}\right] = \frac{7.8.4}{9.8.7} = \frac{4}{9}$$

Sol 5: (D) $P(n)$ = probability of shown in fall

$$P\left(\frac{1}{1 \cup 2}\right) = \frac{P(1 \cap (1 \cup 2))}{P(1 \cup 2)} = \frac{P(1)}{P(1) + P(2)}$$

$$\therefore P(1 \cap 2) = 0$$

$$= \frac{0.1}{0.1 + 0.32} = \frac{0.10}{0.42} = \frac{5}{21}$$

Sol 6: (A) $P(A + B) = P(A) + P(B) - P(AB)$

and $P(A), P(B) > P(AB)$

$\therefore P(A + B) > P(A), (AB)$

and $P(A), P(B), P(AB), P(A + B) \geq 0$

$\therefore$ option (A) is correct

$P(AB) \leq P(A) \leq P(A + B) \leq P(A) + P(B)$

Sol 7: (C) $x \in \{1, 2,......,50\}$

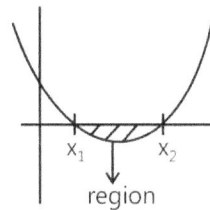

region

$$x + \frac{100}{x} > 50; \quad \Rightarrow x^2 - 50x + 100 > 0$$

$$\Rightarrow x = \frac{50 \pm \sqrt{50^2 - 4(100)}}{2}$$

$$x = 25 \pm \frac{1}{2}\sqrt{2500 - 400}; \quad x = 25 \pm \frac{10\sqrt{21}}{2}$$

$$x = 25 \pm 5\sqrt{2}$$

$x_1 = 2.0871$ and $x_2 = 47.912$

In region integer $y \in \{3, 4,......,47\}$

So only for $x = 1, 2, 48, 49, 50$;

y is greater than 0 or $x + \frac{100}{x} > 50$

$$\Rightarrow P = \frac{n(1, 2, 48, 49, 50)}{50} = \frac{5}{50} = \frac{1}{10}$$

Sol 8: (A) $P = \frac{1}{2}$ events

P = probability of win test match

L = Lose match; W = win match

$$P(L) = 1 - \frac{1}{2} = \frac{1}{2}$$

There is 5 match series.

Total possibility = 2^5

P(India's second win occurs at the 3rd test)

$P(LWW) + P(WLW)$

$$= \frac{1}{2} \times \frac{1}{2} \times \frac{1}{2} + \frac{1}{2} \times \frac{1}{2} \times \frac{1}{2} = \frac{1}{2} \times \frac{1}{2} \times \frac{1}{2} \times 2 = \frac{1}{4}$$

Previous Years' Questions

Sol 1: (B) Given that, $P(A) = 0.4$, $P(\bar{A}) = 0.6$

P(the event A happens at least one

= 1 − P(none of the event happens)

= 1 − (0.6) (0.6) (0.6) = 1 − 0.216 = 0.784

Sol 2: (B) Since, $P(A/\bar{B}) + P(\bar{A}/\bar{B}) = 1$

$\therefore P(\bar{A}/\bar{B}) = 1 - P(A/\bar{B})$

Sol 3: (C) Given, $P(A \cup B) = 0.6$, $P(A \cap B) = 0.2$

$\therefore P(\bar{A}) + P(\bar{B}) = [1 - P(A)] + [1 - P(B)]$

$= 2 - [P(A) + P(B)] = 2 - [P(A \cup B) + P(A \cap B)]$

$= 2 - [0.6 + 0.2] = 1.2$

Sol 4: (D) Let X be the number of coins shoeing heads. Let X be a binomial variate with parameter n = 100 and p.

Since, $p(X = 50) = P(X = 51)$

$$\Rightarrow {}^{100}C_{50}P^{50}(1-p)^{50} = {}^{100}C_{51}(p)^{51}(1-p)^{49}$$

$$\Rightarrow \frac{(100)!}{(50!)(50!)} \frac{(51!) \times (49!)}{100!} = \frac{p}{1-p}$$

$$\Rightarrow \frac{p}{1-p} = \frac{51}{50} \quad \Rightarrow p = \frac{51}{101}$$

Sol 5: (B) The number of ways of placing 3 black balls without any restriction is $^{10}C_3$. Since, we have total 10 places of putting 10 balls in a row. Now the number of ways in which no two black balls put together is equal to the number of ways of choosing 3 places marked '..........' Out of eight places

–W–W–W–W–W–W–W–

The can be done in $^8C_3$ ways

$\therefore$ Required probability $= \dfrac{^8C_3}{^{10}C_3} = \dfrac{8 \times 7 \times 6}{10 \times 9 \times 8} = \dfrac{7}{15}$

Sol 6: (D) It is given that $P(E) \leq P(F) \Rightarrow E \subseteq F$ ..... (i)

and $P(E \cap F) > 0 \Rightarrow E \subset F$ ..... (ii)

(a) Occurrence of $E \Rightarrow$ occurrence of F [from Eq.; (i)]

(b) Occurrence of $F \Rightarrow$ occurrence of E [from Eq. (ii)]

(c) Non-occurrence of $E \Rightarrow$ non-occurrence of F [from Eq. (i)]

Sol 7: (A) Given, $P(B) = \dfrac{3}{4}$ $(A \cap B \cap \bar{C}) = \dfrac{1}{3}$ and

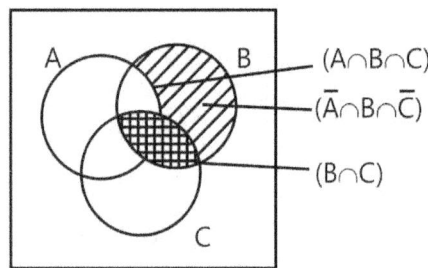

$P = (\bar{A} \cap \bar{B} \cap \bar{C}) = \dfrac{1}{3}$ Which can be shown in Venn diagram.

$\therefore P(B \cap C) = P(B) - \{P(A \cap B \cap \bar{C}) + P(\bar{A} \cap B \cap \bar{C})$

$= \dfrac{3}{4} - \left(\dfrac{1}{3} + \dfrac{1}{3}\right) = \dfrac{3}{4} - \dfrac{2}{3} = \dfrac{1}{12}$

Sol 8: (D) Since, three distinct number are to be selected from first 100 natural numbers.

$\Rightarrow n(S) = {}^{100}C_3$

$E_{\text{(favourable events)}}$ = All three of them are divisible by both 2 and 3.

$\Rightarrow$ divisible by 6 i.e., $\{6, 12, 18, ... , 96\}$

Thus, out of 16 we have to select 3.

$\therefore n(E) = {}^{16}C_3$

$\therefore$ Required probability $= \dfrac{^{16}C_3}{^{100}C_3} = \dfrac{4}{1155}$

Sol 9: (C) Let E = event when each American man is seated adjacent to his wife and

A = Event when Indian man is seated adjacent to his wife.

Now, $n(A \cap E) = (4!); \times (2!)^5$

Even when each American man is seated adjacent to his wife

Again, $n(E) = (5!) \times (2!)^4$

$\therefore P(A / E) = \dfrac{n(A \cap E)}{n(E)} = \dfrac{(4!) \times (2!)^5}{(5!) \times (2!)^4} = \dfrac{2}{5}$

Alternate Solution: Fixing four American couples and one Indian man in between any two couples; we have 5 different ways in which his wife can be seated, of which 2 cases are favourable.

$\therefore$ Required probability $= \dfrac{2}{5}$

Sol 10: (D) Since, $P(A) = \dfrac{2}{5}$

For independent events,

$P(A \cap B) = P(A)P(B) \Rightarrow P(A \cap B) \leq \dfrac{2}{5}$

$\Rightarrow P(A \cap B) = \dfrac{1}{10}, \dfrac{2}{10}, \dfrac{3}{10}, \dfrac{4}{10}$

(Maximum 4 outcomes may be in $A \cap B$)

1. Now, $P(A \cap B) = \dfrac{1}{10}$

$\Rightarrow P(A).P(B) = \dfrac{1}{10}$

$\Rightarrow P(B) = \dfrac{1}{10} \times \dfrac{5}{2} = \dfrac{1}{4}$, not possible

2. Now, $P(A \cap B) = \dfrac{2}{10}$

$\Rightarrow \dfrac{2}{5} \times P(B) = \dfrac{2}{10}$

$\Rightarrow P(B) = \dfrac{5}{10}$,

Outcomes of B = 5

3. Now, $P(A \cap B) = \dfrac{3}{10}$

$\Rightarrow P(A) P(B) = \dfrac{3}{10} \Rightarrow \dfrac{2}{5} \times P(B) = \dfrac{3}{10}$

$P(B) = \dfrac{3}{4}$, not possible

4. Now, $P(A \cap B) = \dfrac{4}{10}$

$\Rightarrow$ P(A).P(B) = $\dfrac{4}{10}$

$\Rightarrow$ P(B) = 1, outcomes of B = 10

Sol 11 : (C) Sample space A dice is thrown thrice, n(s) = 6 × 6 × 6.

Favourable events $\omega^{r_1} + \omega^{r_2} + \omega^{r_3} = 0$

i.e. (r_1, r_2, r_3) are ordered 3-triple which can take values.

(1,2,3), (1,5,3), (4,2,3), (4,5,3)
(1,2,6), (1,5,6), (4,2,6), (4,5,6)

i.e. 8 orc and each can be arranged in 3! Ways = 6

$\therefore$ n(E) = 8 × 6

$\Rightarrow$ P(E) = $\dfrac{8 \times 6}{6 \times 6 \times 6} = \dfrac{2}{9}$

Sol 12: (C) From the tree-diagram it follows that

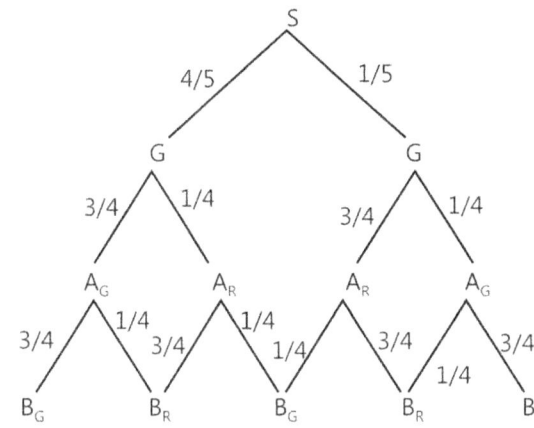

$P(B_G) = \dfrac{46}{80}$

$P(B_G|G) = \dfrac{10}{16} = \dfrac{5}{8}$

$\therefore P(B_G \cap G) = \dfrac{5}{8} \times \dfrac{4}{5} = \dfrac{1}{2}$

$P(G/B_G) = \dfrac{1/2}{P(B_G)} = \dfrac{1}{2} \times \dfrac{80}{46} = \dfrac{20}{23}$.

Sol 13: (B) $\dfrac{P(A \cap B)}{P(B)} = \dfrac{1}{2}$, $\dfrac{P(A \cap B)}{P(A)} = \dfrac{2}{3}$

Hence $\dfrac{P(A)}{P(B)} = \dfrac{3}{4}$. (But P(A) = 1/4)

$\Rightarrow P(B) = \dfrac{1}{3}$

Sol 14: (C) A = {4, 5, 6}, B = {1, 2, 3, 4}

Obviously P(A $\cup$ B) = 1

Sol 15: (D) Mean of a, b, 8, 5, 10 is 6

$\Rightarrow \dfrac{a + b + 8 + 5 + 10}{5} = 5$

$\Rightarrow a + b = 7$ ...(i)

$\therefore$ Variance = $\dfrac{\Sigma(x_1 - A)^2}{n}$

$= \dfrac{(a-6)^2 + (b-6)^2 + 4 + 1 + 16}{5} = 6.8$

$\Rightarrow a^2 + b^2 = 25$

$a^2 + (7 - a)^2 = 25$ (Using (i))

$\Rightarrow a^2 - 7a + 12 = 0$

$\therefore$ a = 3, 3 and b = 3, 4.

Sol 16: (A) $x_1 + x_2 + x_3 + x_5 = 6$

$^{5+6-1}C_{5-1} = ^{10}C_4$

Sol 17: (D) Other than S, seven letters M, I, I, I, P, P, I can be arranged in $\dfrac{7!}{2!4!} = 7 \cdot 5 \cdot 3$.

Now four S can be placed in 8 spaces in $^8C_4$ ways.

Desired number of ways = 7 . 5 3 . $^8C_4$ 7 . $^6C_4$. $^8C_4$

Sol 18: (C) Mean $(\bar{x})$ = $\dfrac{\text{sum of quantities}}{n} = \dfrac{\frac{n}{2}(a+1)}{n}$

$= \dfrac{1}{2}[1 + 1 + 100d] = 1 + 50d$

M.D. $= \dfrac{1}{n}\Sigma |x_1 - \bar{x}| \Rightarrow 255$

$= \dfrac{1}{101}[50d + 49d + 48d + ... + d + 0 + d + ... + 50d]$

$= \dfrac{2d}{101}\left[\dfrac{50 \times 51}{2}\right]$

$\Rightarrow d = \dfrac{255 \times 101}{50 \times 51} = 10.1$

Sol 19: (D) 4 novels can be selected from 6 novels in $^6C_4$ ways. 1 dictionary can be selected and 3 dictionaries in $^3C_1$ ways. As the dictionary selected is fixed in the middle, the remaining 4 novels can be arranged in 4! Ways.

∴ The required number of ways of arrangement $= {}^6C_4 \times {}^3C_1 \times 4! = 1080$

Sol 20: (A) $S = \{00, 01, 02, ..., 49\}$

Let A be the even that sum of the digits on the selected ticket is 8 then

A = {08, 17, 26, 35, 44}

Let B be the event that the product of the digits is zero

B = {00, 01, 02, 03, ..., 09, 10, 20, 30, 40}

A ∩ B = {8}

Required probability $P(A/B) = \dfrac{P(A \cap B)}{P(B)} = \dfrac{\frac{1}{50}}{\frac{14}{50}} = \dfrac{1}{14}$

Sol 21: (D) Statement-II is true

Statement-I: Sum of n even natural numbers = n (n + 1)

Mean $(\bar{x}) = \dfrac{n(n+1)}{n} = n+1$

Variance

$= \left[\dfrac{1}{n}\Sigma(x_1)^2\right] - (\bar{x}) = \dfrac{1}{2}[2^2 + 4^2 + ... + (2n)^2] - (4+1)^2$

$= \dfrac{1}{2}2^2[1^2 + 2^2 + ... + n^2] - (n+1)^2$

$= \dfrac{4}{n}\dfrac{n(n+1)(2n+1)}{6} - (n+1)^2$

$= \dfrac{(n+1)[2(2n+1) - 3(n+1)]}{3}$

$= \dfrac{(n+1)[4n+2 - 3n - 3]}{3}$

$= \dfrac{(n+1)(n-1)}{3} = \dfrac{n^2-1}{3}$

∴ Statement-I is false.

Sol 22: (B) $N(S) = {}^{20}C_4$

Statement-I:

Common difference is 1 ; total number of cases = 17

Common difference is 2 ; total number of cases = 14

Common difference is 3 ; total number of cases = 11

Common difference is 4 ; total number of cases = 8

Common difference is 5 ; total number of cases = 5

Common difference is 6 ; total number of cases = 2

Prob. $= \dfrac{17 + 14 + 11 + 8 + 5 + 2}{{}^{20}C_4} = \dfrac{1}{85}$

Sol 23: (B) $S_1 = \displaystyle\sum_{j=1}^{10} j(j-1)\dfrac{10!}{j(j-1)(j-2)!(10-j)!}$

$= 90\displaystyle\sum_{j=2}^{10} \dfrac{8!}{(j-2)!(8-(j-2))!} = 90 \cdot 2^8$

$S_2 = \displaystyle\sum_{j=1}^{10} \dfrac{10!}{j(j-1)!(9-(j-1))!}$

$= 10\displaystyle\sum_{j=1}^{10} \dfrac{9!}{(j-1)!(9-(j-1))!} = 10 \cdot 2^9$

$S_3 = \displaystyle\sum_{j=1}^{10} [j(j-1) + j]\dfrac{10!}{j!(10-j)!}$

$= \displaystyle\sum_{j=1}^{10} j(j-1)\,{}^{10}C_j = \displaystyle\sum_{j=1}^{10} j\,{}^{10}C_j = 90 \cdot 2^8 + 10 \cdot 2^9$

$= 90 \cdot 2^8 + 20 \cdot 2^8 = 110 \cdot 2^8 = 55 \cdot 2^9$

Sol 24: (C) The number of ways $= {}^3C_2 \times {}^9C_2$

$= 3 \times \dfrac{9 \times 9}{2} = 3 \times 36 = 108$

Sol 25: (A) $n(S) = {}^9C_3$

$n(E) = {}^3C_1 \times {}^4C_1 \times {}^2C_1$

$= \dfrac{3 \times 4 \times 2}{{}^9C_3} = \dfrac{24 \times !}{9!} \times 6! = \dfrac{24 \times 6}{9 \times 8 \times 7} = \dfrac{2}{7}$

Sol 26: (A) $\sigma_x^2 = 4$, $\sigma_y^2 = 5$, $\bar{x} = 2$, $\bar{y} = 4$

$\dfrac{\Sigma x_i}{5} = 2 \qquad \Sigma x_i = 10; y_i = 20$

$\sigma_x^2 = \left(\dfrac{1}{2}\Sigma x_i^2\right) - (\bar{x})^2 = \dfrac{1}{5}(\Sigma y_i^2) - 16$

$\Sigma x_i^2 = 40$, $\Sigma y_i^2 = 105$

$\sigma_2^2 = \dfrac{1}{10}\left(\Sigma x_i^2 + \Sigma y_i^2\right) - \left(\dfrac{\bar{x}+\bar{y}}{2}\right)^2$

$= \dfrac{1}{10}(40 + 105) - 9 = \dfrac{145 - 90}{10} = \dfrac{55}{10} = \dfrac{11}{2}$

Sol 27: (D) Number of ways of selecting one or more balls from 10 white, 9 green, and 7 black balls

$$= (10+1)(9+1)(7+1) - 1$$
$$= 11 \times 10 \times 8 - 1 = 879.$$

Sol 28: (B) Let A be the event that maximum is 6.

B be event that minimum is 3

$P(A) = \dfrac{^5C_2}{^8C_3}$ (the numbers < 6 are 5)

$P(B) = \dfrac{^5C_2}{^8C_3}$ (the numbers > 3 are 5)

$P(A \cap B) = \dfrac{^2C_1}{^8C_3}$

Required probability is $P\left(\dfrac{B}{A}\right) = \dfrac{P(A \cap B)}{P(A)} = \dfrac{^2C_1}{^5C_2} = \dfrac{2}{10} = \dfrac{1}{5}$

Sol 29: (D) If initially all marks were x_1 then

$$\sigma_1^2 = \frac{\sum(x_1 - \bar{x})^2}{N}$$

Now each is increased by 10

$$\sigma_2^2 = \frac{\sum[(x_1 + 10) - (\bar{x} + 10)]^2}{N} = \sigma_1^2$$

So, variance will not change whereas mean, median and mode will increase by 10.

Sol 30: (A) $P(\overline{A \cup B}) = \dfrac{1}{2} \Rightarrow P(A \cup B) = 1 - \dfrac{1}{6} = \dfrac{5}{6}$

$P(\bar{A}) = \dfrac{1}{4} \Rightarrow P(A) = 1 - \dfrac{1}{4} = \dfrac{3}{4}$

$\because \quad P(A \cup B) = P(A) + P(B) - P(A \cap B)$

$\dfrac{5}{6} = \dfrac{3}{4} + P(B) - \dfrac{1}{4}$

$P(B) = \dfrac{1}{3}$

$\because \quad P(A) \neq P(B)$ so they are not equally likely.

Also $P(A) \times P(B) = \dfrac{3}{4} \times \dfrac{1}{3} = \dfrac{1}{4} = P(A \cap B)$

$\because \quad P(A \cap B) = P(A) \cdot P(B)$ so A and B are independent.

Sol 31: (D) Variance $= \dfrac{\sum x_i^2}{N} - (\bar{x})^2$

$\Rightarrow \quad \sigma^2 = \dfrac{2^2 + 4^2 + \dots + 100^2}{50} - \left(\dfrac{2 + 4 + \dots + 100}{50}\right)^2$

$= \dfrac{4(1^2 + 2^2 + 3^2 + \dots + 50^2)}{50} - (51)^2$

$= 4\left(\dfrac{50 \times 51 \times 101}{50 \times 6}\right) - (51)^2 = 3434 - 2601$

$\Rightarrow \quad \sigma^2 = 833$

Sol 32: (D) ▯▯▯▯ Number of integer greater than

6000 may be 4 digit or 5 digit

C-1 when number is of 4 digit

C-2 when number is of 5 digit = 5! = 120

total = 120 + 72 = 192 digit

{6, 7, 8}

▯▯▯▯

3 4 3 2 = 72

Sol 33: (A) There seems to be ambiguity in the question. It should be mentaned that boxes are different and one particular box has 3 balls:

Then,

Number of ways $= \dfrac{^{12}C_3 \times 2^9}{3^{12}} = \dfrac{55}{3}\left(\dfrac{2}{3}\right)^{11}$

Sol 34: (D) $\dfrac{x_1 + x_2 \dots x_{16}}{16} = 16$

If $x_1 = 16$

$\dfrac{x_1 + x_2 \dots x_{10} - 16 + 3 + 4 + 5}{18}$

$= \dfrac{16 \times 10 - 16 + 12}{18} = \dfrac{240 + 12}{18} = \dfrac{252}{29} = 14$

Sol 35: (A) Standard deviation of numbers 2, 3, a and 11 is 3.5

$\therefore \quad (3.5)^2 = \dfrac{\sum x_1^2}{4} - (\bar{x})^2$

$\Rightarrow \quad (3.5)^2 = \dfrac{4 + 9 + a^2 + 121}{4} - \left(\dfrac{2 + 3 + a + 11}{4}\right)^2$

On solving, we get $3a^2 - 32a + 84 = 0$

Sol 36: (C) E_1 : {(4, 1), ...(4, 6)} 6 cases

E_2 : {(1, 2),........ (6, 2)} 6 cases

E_3 : 18 cases (sum of both are odd)}

$P(E_1) = \dfrac{6}{36} = \dfrac{1}{6} = P(E_2)$

$P(E_3) = \dfrac{18}{36} = \dfrac{1}{2}$

$P(E_1 \cap E_2) = \dfrac{1}{36}$

$P(E_2 \cap E_3) = \dfrac{1}{12}$

$P(E_3 \cap E_1) = \dfrac{1}{12}$

$P(E_1 \cap E_2 \cap E_3) = 0$

$\therefore$ E_1, E_2, E_3 are not independent

Sol 37: (C) SMALL

$A\_\_\_\_ \# \dfrac{4!}{2!} = 12$

$L\_\_\_\_ \# 4! = 24$

$M\_\_\_\_ \# \dfrac{4!}{2!} = 12$

$SA\_\_\_\_ \# \dfrac{3!}{2!} = 3$

$SL\_\_\_ \# 3! = 6$

$\underline{S}\,\underline{M}\,\underline{A}\,\underline{L}\,L \# 1$

58th position

JEE Advanced/Boards

Exercise 1

Sol 1:

5 science		3 science
3 engg.		5 engg.
I		II

E = engg. subject select

if, 3 or 5 came after thrown a dice then, select a subject from I otherwise from II

$\Rightarrow P(E) = \left(\dfrac{2}{6}\right) \times \dfrac{3}{8} + \left(\dfrac{4}{6}\right) \cdot \dfrac{5}{8} = \dfrac{1}{8} + \dfrac{5}{12} = \dfrac{3+10}{24} = \dfrac{13}{24}$

Sol 2: two dice throw possibility 10

$\Rightarrow$ (4, 5) (1, 5) (2, 5) (3, 5) (6, 5)......

= At least one should be greater than 4

$\Rightarrow P = \dfrac{6 \times 1 + 1 \times 6 - 1}{36} + \dfrac{6 \times 1 + 1 \times 6 - 1}{36} - \dfrac{2}{36}$

$P = \dfrac{20}{36} = \dfrac{5}{9}$

Sol 3: Odds for A, B, C, D

$\Rightarrow$ 1: 3, 1: 4, 1: 5, 1: 6

$P(A) = \dfrac{1}{1+3} = \dfrac{1}{4}$, $P(B) = \dfrac{1}{5}$,

$P(C) = \dfrac{1}{6}$, $P(D) = \dfrac{1}{7}$

P(one of them will wins)

= P(A) + P(B) + P(C) + P(D)

$= \dfrac{1}{4} + \dfrac{1}{5} + \dfrac{1}{6} + \dfrac{1}{7} = \dfrac{319}{420}$

Sol 4: Assume number of roses = x

P(sweety win when she start first) = P(S)

$= \dfrac{x}{(60+x)} + \dfrac{60^2 x}{(60+x)^3} + \dots$

$= \dfrac{x}{60+x}\left[1 + \left(\dfrac{60}{60+x}\right)^2 + \dots\right]$

$= \dfrac{x}{60+x}\left[\dfrac{1}{1 - \left(\dfrac{60}{60+x}\right)^2}\right] = \dfrac{60+x}{720+x}$

So P(sweety wins) = P(w) = $1 - \dfrac{60+x}{120+x}$

Its given that(s) = 3P (w)

$\dfrac{60+x}{120+x} = \dfrac{3.60}{120+x}$

x = 180 − 160 = 120

Sol 5: Total roll = 9(3, 3, 3)

$\dfrac{m}{n} = \dfrac{{}^3C_1 {}^3C_1 {}^3C_1}{{}^9C_3} \cdot \dfrac{{}^2C_1 {}^2C_1 {}^2C_1}{{}^6C_3} \cdot \dfrac{{}^1C_1 {}^1C_1 {}^1C_1}{{}^3C_3} = \dfrac{9}{70} = \dfrac{m}{n}$

m + n = 9 + 70 = 79

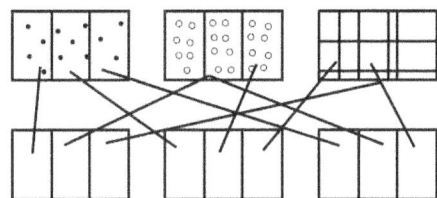

Sol 6: Probability to head shown = P'

P'(A wins, when first A start)

$= P' + (1 - P') P'^2 + (1 - P') P(1 - P') P'^2 + \ldots$

$= P[1 + (1 - P')P'((1 - P')P')2 + ((1 - P')P')^3 + \ldots]$

$= P'\left[\dfrac{1}{1 - P'(1 - P')}\right] = \dfrac{P'}{1 - P' + P^2}$

$P(B\text{ win}) = 1 - \dfrac{P}{1 - P' + P'^2}$

$= \dfrac{1 - P' + P'^2 - P'}{1 - P' + P'^2} = \dfrac{1 - 2P' + P'^2}{1 - P' + P'^2}$

For first to both

P(A) = P(B)

$\Rightarrow \dfrac{1 - 2P' + P^2}{1 - P' + P'^2} = \dfrac{P'}{1 - P' + P'^2}$

$\Rightarrow P'^2 - 3P' + 1 = 0$

$\Rightarrow P' + P = 1 \Rightarrow P'(1 - P)$

$\Rightarrow (1 - P)^2 - 3(1 - P) + 1 = 0$

$\Rightarrow P^2 - 2P + 1 - 3 + 3P + 1 = 0$

$\Rightarrow P^2 + P - 1 = 0$

$P = \dfrac{-1 \pm \sqrt{1 + 4}}{2} = \dfrac{\sqrt{5} - 1}{2} = P$

$\therefore 0 < P < 1$

Sol 7: a, b, c, d → integer $\begin{vmatrix} a & b \\ \times \\ c & d \end{vmatrix}$

| D | = ad − bc

if (D) is even than

ad and bc are even ⟨ a or d is even and b or c is even

ad and bc are odd ↘

a, d are odd   and b, c are odd

$P(|D|\text{ is even}) = (P)^4 + (1 - P^2)(1 - P^2) = \dfrac{1}{2}$

$\Rightarrow P^4 + P^4 - 2P_{+1}^2 = \dfrac{1}{2}$

$4P^4 + 4P^2 + 1 = 0$

$(2P^2 - 1)^2 = 0$

$\Rightarrow 2P^2 - 1 = \Rightarrow P^2 = 1/2 \text{ or}$

$P = \dfrac{1}{\sqrt{2}} \quad \because 0 \le P \le 1$

Sol 8: P = P(out cons of the 5th throw was already thrown)

P = 1 − P (out com of 5th throw was first line throw)

$P = 1 - 6\left[\dfrac{5}{6} \times \dfrac{5}{6} \times \dfrac{5}{6} \times \dfrac{5}{6} \times \dfrac{1}{6}\right]$

$P = 1 - \dfrac{625}{1296} = \dfrac{671}{1296} = \dfrac{a}{b}$

a + b = 671 + 1296 = 1967

Sol 9: x = number of bomb hitting on target

n = 4

P = 0.4(probability of hitting the target for and bomb)

x ≥ 2 for destroy bridge

P(x ≥ 2) = P(x = 2) + P(x = 3) + P(x = 4)

$= {}^4C_2 (0.4)^2 (0.6)^2 + {}^4C_3 (0.4)^3 (0.6)^1 + {}^4C_4 (0.4)^4$

$= \dfrac{4 \times 3}{2} \times (0.24)^2 + 4(0.064)(0.6) + (0.0256) = \dfrac{328}{625}$

Sol 10: assume P = Ist event's probability

q = IInd event's probability

$P = q^2$

Odds against the first $= \dfrac{1 - P}{P}$

odds against the first $= \dfrac{1 - q}{q}$

$\dfrac{1 - P}{q} = \left(\dfrac{1 - q}{q}\right)^3 = \dfrac{1 - q^2}{q^2}$

$\Rightarrow \dfrac{(1 - q)(1 + q)}{q^2} = \dfrac{(1 - q)(1 - q)^2}{q^3}$

$q + q^2 = 1 + q^2 - 2q$

$3q = 1 \Rightarrow q = 1/3$

So $P = (q)^2 = \left(\dfrac{1}{3}\right)^2 = \dfrac{1}{9}; \quad (P, q) \to \left(\dfrac{1}{9}, \dfrac{1}{3}\right)$

Sol 11: Total tubes = 5(3G, 2D)

defective = 2(D)

(i) Test stopped on the 2nd test

$$P = \frac{{}^2C_2}{{}^5C_2} = \frac{1 \times 2}{5 \times 4} = \frac{1}{10}$$

(ii) Test stopped on 3rd test

(DGD) + (GDD) + (GGG)

$$\Rightarrow \frac{{}^2C_1 \times {}^3C_1 \times 1}{{}^5C_3} + \frac{{}^3C_1 \times 2 \times 3}{5 \times 4 \times 3} + \frac{3 \times 2 \times 1}{5 \times 4 \times 3}$$

$$= \frac{1}{10} + \frac{1}{10} + \frac{1}{10} = \frac{3}{10}$$

(iii) $P\left(\dfrac{\text{first tube is non-defective}}{\text{test stopped on 3}^{\text{rd}}\text{ test}}\right) = \dfrac{\frac{1}{10} + \frac{1}{10}}{\frac{3}{10}} = \dfrac{2}{3}$

Sol 12: Hitting the plane at first = 0. 4

Hitting the plane at 2nd = 0. 3

Hitting the plane at 3rd = 0. 2

Hitting the plane at 4rd = 0. 1

P = P(gun hits the plane)

= 1 − (1 − 0. 4) (1 − 0. 3) (1 − 0. 2) (1 − 0. 1)

= 0. 6976

Sol 13: Total articles = 10

Defective (D) = 4

Non-defective (R) = 6

Number of chosen articles = 6

$\Rightarrow {}^{10}C_6$, x = number of defective articles

P (batch will be rejected)

$$= 1 - \frac{P(x = 0) + P(x = 1) + P(x = 2)}{1}$$

$$= 1 - \left(\frac{{}^6C_0}{{}^{10}C_6} + \frac{{}^6C_5\,{}^4C_1}{{}^{10}C_6} + \frac{{}^6C_4\,{}^4C_2}{{}^{10}C_6} \right)$$

$$= \left[\frac{\frac{1 + 6 \times 4 + 15 \times 6}{10 \times 9 \times 8 \times 7}}{1 \cdot 2 \cdot 3 \cdot 4} \right]$$

$$= 1 - \frac{115}{210} = \frac{210 - 115}{210} = \frac{95}{210} = \frac{19}{42}$$

Sol 14: Total side = 6

B ⇒ one red , 2 blue, 3 green

P(second blue result occurs on or before the tenth)

= 1 − P(second blue result occurs after 10th)

$$= 1 - \left(\frac{4}{6}\right)^{10} - \left(\frac{4}{6}\right)^{9}\left(\frac{2}{6}\right){}^{10}C_1 = \frac{6^{10} - 4^{10} - 4^{10} \times 5}{6^{10}}$$

$$= \frac{6^{10} - 6.4^{10}}{6^{10}} = \frac{6^9 - 4^{10}}{6^9} = \frac{3^9 - 2^{11}}{3^9}$$

$$= \frac{3^P - 2^q}{3^r}, \; P = r, \; q = 11, \; r = 9$$

P² + q² + r² = 9² + 9² + 11² = 283

Sol 15: Probability of good book (G) = $\dfrac{1}{2}$

So P(bod book) (B) = $\dfrac{1}{2}$

$P\left(\dfrac{P}{G}\right) = 2/3$ and $P\left(\dfrac{P}{B}\right) = \dfrac{1}{4}$

n = 2

P(at lead one book published)

= 1 − P(no-book published)

= 1 − P(GP' BP') − P(GP' GP') − P(B'P BP')

$$= 1 - 2 \times \frac{1}{2} \times \frac{1}{2} \times \frac{1}{3} \times \frac{3}{4} - \frac{1}{2} \times \frac{1}{2} \times \left(\frac{1}{3}\right)^2 - \frac{1}{2} \times \frac{1}{2}\left(\frac{3}{4}\right)^2$$

$$= 1 - \frac{169}{576} = \frac{407}{576}$$

Sol 16: $P(A) = \dfrac{1}{3}$

$P\left(\dfrac{B}{A_2}\right) = 1, P\left(\dfrac{B}{A_1}\right) = \dfrac{1}{2}, P\left(\dfrac{B}{A_0}\right) = 0$

Assume x = number of trial when A occur

n = 4

P(B) = 1. P(x ≥ 2) + 0: P(x = 0) + 1/2 P(x = 1)

$$= P(x = 2) + P(x = 3) + P(x = 4) + 0 + \frac{P(x = 1)}{2}$$

$$= {}^4C_2 \times \left(\frac{1}{3}\right)^2 \times \left(\frac{2}{3}\right)^2 + {}^4C_3\left(\frac{1}{3}\right)^3\frac{2}{3}$$

$$\quad + {}^4C_4\left(\frac{1}{3}\right)^4 + {}^4C_1\left(\frac{1}{3}\right)\left(\frac{2}{3}\right)^3 \times \frac{1}{2}$$

$$= \frac{1}{81}\left[6 \times 4 + 4 \times 2 + 1 + 4 \times \frac{4}{2}\right] = \frac{49}{81} = \frac{m}{n}$$

$$\Rightarrow m + n = 49 + 81 = 130$$

Sol 17: faces $\rightarrow$ 1, 2, 3, 4

When two dice thrown together

(i) Exactly 6 on each of successive throws

$6 \left\{ \begin{array}{l} \rightarrow (4,2) \dfrac{1}{4} \times \dfrac{2}{4} = \dfrac{2}{10} \\[2mm] \rightarrow (2,4) \dfrac{2}{4} \times \dfrac{1}{4} = \dfrac{2}{16} \\[2mm] \rightarrow (3,3) \dfrac{1}{4} \times \dfrac{1}{4} = \dfrac{1}{16} \end{array} \right.$

$\Rightarrow$ For 3 throws

$$\Rightarrow \left(\frac{2}{16} + \frac{2}{16} + \frac{1}{16}\right)^3 \Rightarrow \left(\frac{5}{16}\right)^3 = \frac{125}{16^3}$$

(ii) More than 4 on at least are of the three throws.

$\Rightarrow$ 1 – P(less than 4 or 4 on all thrown)

$\Rightarrow$ Less than 4 or 4 $\rightarrow$ (2, 2)

$$\Rightarrow \frac{2}{4} \times \frac{2}{4} = \frac{1}{4} \Rightarrow 1 - \left(\frac{1}{4}\right)^3 = \frac{64-1}{64} = \frac{63}{64}$$

Sol 18: Total red card out of 52 = 26 (R)

Total green = 4 (Q)

q card are red and green both (RQ)

Two cards drawn

$\Rightarrow$ P(one is red & one is green)

$\Rightarrow$ [26 × 2 + 25 × 1 + 25 × 1 – (1) (1)]

$$\Rightarrow \frac{101}{1326}$$

Sol 19: Total coin = 4

Discard those which turn up tails

P = (at least 3 coins discard after 2nd flip)

= P(3) + P(4) = 1 – P(0) – P(1) – P(1)

= 1 – [P(HHHH, HHHH)] – P[(HHHT, HHH) + (HHHH, HHHT)]

    – P [(HHHH, HHTT) + (HHTT, HH) + (HHHT, HHT)]

$$= 1 - \left(\frac{1}{2}\right)^8 - \frac{1}{2^8}\left[\,^4C_1 \times 2 \times 1 + \,^4C_1\right]$$

$$-\frac{1}{2^8}\left(\frac{4!}{2!2!} + \frac{2^2.4!}{2!2!} + 4 \times 2 \times 3\right)$$

$$= 1 - \frac{1}{28}[1 + 12 + 30 + 24] = \frac{256 - 67}{256} = \frac{189}{256}$$

Sol 20: Total passengers before stop = n

P(get down) = P

P(boarding the bus at next stop) = 1 – P$_0$

P(n passenger are in bus after stop)

= $(1 - P)\,^nP_0 + \,^nC_1\, P(-P)^{n-1}\, (1 - P_0)$

= $(1 - P)^{n-1}\, [P_0(1 - P) + nP(1 - P_0)]$

Sol 21: Total balls = 2n (n while, n black)

n person each draw 2 balls

(i) P(each of n person drawn both balls)

$$= \frac{2^n (n \times n)(n-1)^2 (n-2)^2}{2n!} - 3^2 . 2^2$$

$$= \frac{n \lfloor n \rfloor 2^n}{2n} = \frac{2^n}{\,^{2n}C_n} = \frac{8}{35}$$

$$n = 4 \rightarrow \frac{2^4}{\,^8C_4} = \frac{16 \times 4 \times 3 \times 2}{1 \times 7 \times 1 \times 5} = \frac{8}{35}$$

So n = 4

(ii) n = 4

Each of 4 draw the balls of same colour

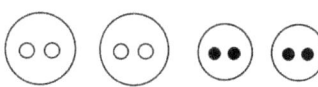

$$P = \frac{\,^4C_1 \times^3 C_1 \,^2C_1 \,^1C_1 \times 4!}{8!} \frac{}{2!2!} \times^4 C_2 . 2^2$$

$$P = \frac{4!\ 4 \times 3 \times 2 \times 1}{4 \times 8 \times 7 \times 6 \times 5 \times 4}\ 4 \times 6 = \frac{3}{35}$$

(iii) n = 7 there is 7 white balls and 7 black ball there is no way to each of the person draw balls of same colour.

$\because$ Number of same color ball is 7 (odd)

7 = 2 + 2 + 2 + 1 – ? (1)$\Rightarrow$not possible

P = 0

Sol 22: Between P$_i$ and P$_j$

If i < j $\rightarrow$ P$_i$ win

P(player P$_4$, reaches the Final)

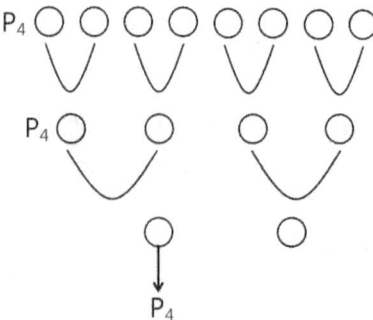

to get in final for P_4 exactly 3 match will be there and P_4 all win.

So other 3 are P_5, P_6, P_7 or P_8

$$P = \frac{{}^4C_1\,{}^3C_1\,{}^2C_1\,.4!\,2^3}{8!}$$

$$P = \frac{4\times3\times2\times4!\,2^3}{8\times7\times6\times5\;4!} = \frac{4}{35}$$

Sol 23: $P(A) = 0.4 \to P(\overline{A}) = 1 - 0.4 = 0.6$

$P(B) = 0.8 \Rightarrow P(\overline{B}) = 1 - 0.8 = 0-2$

$$P(\overline{A}/\overline{B}) = \frac{P(\overline{A}\cap\overline{B})}{P(\overline{B})} = 0.1$$

(i) $P(\overline{A}\cup B) = ?$, $P(\overline{A}\cap\overline{B}) = P(\overline{B})\,10.1$

$= 0.2 \times 0.1 = 0.02$

$1 - P(\overline{A}\cap\overline{B}) = P(A\cup B) = 1 - 0.02 = 0.98$

$P(\overline{A}\cup B) = P(\overline{A}) + P(B) - P(\overline{A}\cap B)$

$P(\overline{A}\cap B) = P(\text{only } B)$

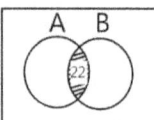

$= P(B) - P(A\cap B)$

$P(A\cap B) = P(A) + P(B) - P(A\cup B)$

$= 0.4 + 0.8 - 0.98 = 0.22$

$P(\overline{A}\cap B) = 0.8 - 0.22 = 0.58$

$P(\overline{A}\cup B) = 0.6 + 0.8 - (0.58)$

$= 1.4 - 0.58 = 0.82$

(ii) $P[(\overline{A}\cap B)\cup(A\cap\overline{B})]$

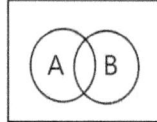

$P(\overline{A}\cap B) = 0.58$

$P(A\cap\overline{B}) = P(\text{only } A)$

$= P(A) - P(A\cap B) = 0.4 - 0.22 = 0.18$

$P[(\overline{A}\cap B)\cap(A\cap\overline{B})] = 0$ from dia.

So $P[(\overline{A}\cap B)\cup(A\cap\overline{B})]$

$= P(\overline{A}\cap B) + P(A\cap\overline{B}) = 0.18 + 0.58 = 0.76$

Sol 24: (i) Number of ways = ${}^8C_3$

choose 3 digits from set B.

Total ways = ${}^8C_3 \times {}^9C_3$

$P(\text{A and B have same 3-digit number}) = \dfrac{{}^8C_3}{{}^8C_3 \times {}^9C_3} = \dfrac{1}{84}$

(ii) Case-I: Mr A's number contains 9

$$P_1 = \frac{{}^8C_2 \times {}^8C_3}{{}^9C_3 \times {}^8C_3} = \frac{{}^8C_2}{{}^9C_3}$$

Case-II: Mr A's number do not contain 9

$$P_2 = \frac{(1 - P(\text{A \& B have same number}))}{2}$$

$\times\; P(\text{Mr A's number don't contains 9})$

$$= \left(1 - \frac{1}{{}^8C_3}\right) \times \frac{1}{2} \times \frac{{}^8C_3}{{}^9C_3}$$

$P(\text{Mr A's number} > \text{Mr. B's number})$

$= P_1 + P_2$

$$= \frac{{}^8C_2}{{}^9C_3} + \frac{1/2({}^8C_3 - 1)}{{}^9C_3} = \frac{111}{168} = \frac{37}{56}$$

Sol 25: One pair is selected

$P(\text{one is mole and one female}) = \dfrac{10}{19}$

assume total student is $= 2m$

$= n$ boy $+ n$ girl

$$P = \frac{10}{19} = \frac{{}^nC_1.{}^nC_1}{{}^{2n}C_2} = \frac{10}{19}$$

$\Rightarrow 19\,n = 10(2n - 1) = 20\,n - 10$

$\Rightarrow 20n - 19\,n = 10$

$\Rightarrow n = 10$

Total student is $2n = 20$

Exercise 2

Single Correct Choice Type

Sol 1: (B) $P(A) = P$(person A lives above 35 years)

So, $P(A') = \dfrac{9}{9+7} = \dfrac{9}{16}$

And $P(B') = \dfrac{3}{3+2} = \dfrac{3}{5}$

So, P(at least one lives)

$= 1 - P(\text{no lives}) = 1 - P(A') \cdot P(B')$

$= 1 - \dfrac{9}{16} \cdot \dfrac{3}{5} = \dfrac{80-27}{80} = \dfrac{53}{80}$

Sol 2: (D) $\Sigma P = 1$ for a close event

$0 \le P \le 1$

$P_1 + P_2 + P_3 + P_4 = 0.2 + 0.3 + 0.4 + 0.1 = 1$

$0 \le P_1, P_2, P_3, P_4 \le 1$

Sol 3: (A) P = Probability in a group of 4 person

all are born on different days of the week.

$P = \dfrac{7 \times 6 \times 5 \times 4}{7 \times 7 \times 7 \times 73} = \dfrac{120}{343}$

$\dfrac{1}{3} < P < \dfrac{1}{2}$

Sol 4: (D) x is a integer

$P(x^4$ ends in the digits 6)

$1^4 = 1,\ 2^4 = 16,\ 3^4 = 81,\ 4^4 = 256,$

$5^4 = 625,\ 6^4 = \ldots\ldots6,\ 7^4 = \ldots\ldots1$

$8^4 = \ldots\ldots6,\ 9^4 = \ldots\ldots1,\ 0^4 = 0$

There are 4 out of 6 digits which has 6 in the last for 4th power

So, $P = \dfrac{4}{10} = 40\%$

Sol 5: (B) 2 match each with two other teams.

x = 0, 1, 2(Points)

For one match

$P(x = 0) = 0.45$

$P(x = 1) = 0.05$

$P(x = 2) = 0.50$

For all matches

max point = 2 × 4 = 8

$P(x \ge 7) = P(x = 7) + P(x = 8)$

$= {}^4C_1\, P(2, 2, 2, 1) + P(2, 2, 2, 2)$

$= 4 \times (0.5)^3 (0.05) + (0.5)^4$

$= (0.5)^3 [0.2 + 0.5] = (0.5)^3\, 0.7 = 0.0875$

Sol 6: (B) Total key = n

There is only one key to open door

P(last key is the right key)

$= \dfrac{n-1}{n} \cdot \dfrac{(n-1)}{(n-1)} \ldots\ldots \dfrac{2}{3} \cdot \dfrac{1}{2} \cdot \dfrac{1}{2} = \dfrac{1}{n}$

Sol 7: (D) A and B are independent

$P(A) = P;\ P(B) = P \Rightarrow P(A \cap B) = P(A) \cdot P(B) = P^2$

$\therefore P\left(\dfrac{A}{A \cup B}\right) = \dfrac{A \cap (A \cup B)}{(A \cup B)} = \dfrac{P(A \cap A \cup (A \cap B))}{P(A \cup B)}$

$= \dfrac{P(A \cup (A \cap B))}{P(A \cup B)} = \dfrac{P(A)}{P(A \cup B)}$

$P(A \cup B) = P(A) + P(B) - P(A \cap B)$

$P(A \cup B) = P + P - P^2 = 2P - P^2$

$\dfrac{P(A)}{P(A \cup B)} = \dfrac{P}{P(2-P)} = \dfrac{1}{2-P}$

Sol 8: (B) x = number of getting tails

$n = bp = q = \dfrac{1}{2}$

More tail than heads so $x > 3 \Rightarrow P(x > 3)$

$P(x = 4) + 0\ (x = 5) + P(x = 6)$

$\Rightarrow {}^6C_4 \left(\dfrac{1}{2}\right)^6 + {}^6C_5 \left(\dfrac{1}{2}\right)^6 + {}^6C_6$

$= \dfrac{1}{2^6}\left[\dfrac{6 \times 5}{1.2} + 6 + 1\right] = \dfrac{1}{2^6}[15 + 7] = \dfrac{22}{2^6} = \dfrac{22}{64}$

Sol 9: (D) Success = head = H

Success = one rupee win = 0

Lose = Tail → lose one rupee = T

P (he loses)

$= P(T) + P(HT) + P(HHT) + P(HHHT) + P(HHHHT) + P(HHHHHT) + P(HHHHHH)$

$= \dfrac{1}{2} + \dfrac{1}{2^6}(1 + 1 + 1 + 1 + 1 + 1) = \dfrac{16+6}{32} = \dfrac{22}{32}$

Sol 10: (C) Let $\{1, 2, 3,......,50\}$

x^x should be perfect square

So $x \to$ seven

Or $x \to$ perfect

x = all even $\to 1, 2^2, 3^2, 4^2, 5^2, 6^2, 7^2$

$8^2 > 50$

So $x \neq 8^2$

Total number of $x = 25 + 7 - (3)$

because $2^2, 4^2, 6^2$ all are even and perfect square both.

$P = \dfrac{25+7-3}{50} = \dfrac{29}{50}$

Sol 11: (D) $P(A) = P$(a solve problem)

$P(A) = P$

$P(B) = 1/2$

P(they will make same mistakes)$= \dfrac{1}{100}$

$P\left(\dfrac{\text{their ans. is correct}}{\text{they get same ans.}}\right) = \dfrac{300}{301}$

$\Rightarrow P$(they get same ans.)

$= P \times \dfrac{1}{2} + \dfrac{1}{100} = \dfrac{1}{100} + \dfrac{P}{2}$

P(their ans. is correct) $= P \times \dfrac{1}{2} = \dfrac{P}{2}$

$\Rightarrow \dfrac{300}{301} = \dfrac{P/2}{\dfrac{1}{100}+\dfrac{P}{2}} = \dfrac{100 \times P}{2+100P}$

$\Rightarrow 3(2 + 100P) = 301P$

$\Rightarrow 6 + 300P = 301P$

$\Rightarrow P = 6, P \neq 6 \because = P < 1$

Sol 12: (B) Two dice are thrown until a 6 appears

P(For the Ist time, 46 appear in second throw)

$= \dfrac{5}{6} \times \dfrac{5}{6} \times \dfrac{11}{36} = \dfrac{275}{1296}$

Therefore there is 1 case to show at least one 6

(6, 1) (1, 6) (2, 6) (6, 2) (3, 6) (6, 3) (4, 6) (6, 4) (5, 6) (6, 5) (6, 6)

3 white	2 white
2 red	4 red

Sol 13: (B) A   B

$\downarrow$ $\downarrow$

2 balls   2 balls

P(all balls are white) $= \dfrac{{}^3C_2 \cdot {}^2C_2}{{}^5C_2 \cdot {}^6C_2} = \dfrac{3 \times 1}{\dfrac{5 \times 4}{2} \cdot \dfrac{6 \times 5}{2}}$

$= \dfrac{1}{50} = 2\%$

Sol 14: (B) $P(\bar{A}) = 0.7$

$\Rightarrow P(A) = 1 - P(\bar{A}) = 1 - 0.7 = 0.3$

$P(\bar{B}) = a$

$\Rightarrow P(B) = 1 - a$

$P(A \cup B) = P(A) + P(B) - P(A \cap B)$

$P(A \cap B) = P(A) \cdot P(B)$

$\because$ A and B are independent

$\Rightarrow 0.8 = 0.3 + 1 - a + (0.3)(1-a)(-1)$

$\Rightarrow a(1-0.3) = 1 + 0.3 - 0.3 - 0.8 = 0.2$

$\Rightarrow a = \dfrac{0.2}{0.7} = \dfrac{2}{7}$

Sol 15: (D) P(son will post the letter) $= \dfrac{1}{2} = P(5_p)$

P(letter react its destination) $= \dfrac{5}{6} = P(D)$

$P(R)$ = probability of letter was received

$P\left(\dfrac{S_p^1}{R'}\right) = \dfrac{P(S_p^1 \cap R')}{P(R')}$

$P(R') = P(S_p \cap D') + P(S_p^1) = \dfrac{1}{2} + \dfrac{1}{2} + \dfrac{1}{6} = \dfrac{6+1}{12} = \dfrac{7}{12}$

$P\left(\dfrac{S_p^1}{R'}\right) = \dfrac{1/2}{7/12} = \dfrac{6}{7}$

Sol 16: (D) Set $= \{1, 2, 3...,20\}$

$x_1 x_2 \in$ set

$x_1 x_2$ is even

when $x_1 + x_2$ = odd

possibility for this condition = (one odd, one even)

(not of possibility /or this condition)

$= {}^{20}C_1 \cdot {}^{10}C_1 = 200$

Total possibility $= 20 \times 20 = 400$

$P = \dfrac{200}{400} = \dfrac{1}{2}$

1, 2, 4, 5, 7, 8, 10, 11, 12, 14, 15, 16, 20, 21, 24, 25

3, 6, 7, 13, 17, 18, 19, 22, 23.

Previous Years' Questions

Sol 1: Since, the drawn balls are in the sequence black, black, white, white, white, red, red and red.

Let the corresponding probabilities be

$P_1, P_2, \ldots\ldots . P_9$

Then, $P_1 = \dfrac{2}{9}$; $P_2 = \dfrac{1}{8}$, $P_3 = \dfrac{4}{7}$;

$P_4 = \dfrac{3}{6}$, $P_5 = \dfrac{2}{5}$

$P_6 = \dfrac{1}{4}$, $P_7 = \dfrac{3}{3}$, $P_8 = \dfrac{2}{2}$, $P_6 = 1$

$\therefore$ Required probabilities

$P_1.P_2.P_3........P_9$

$= \left(\dfrac{2}{9}\right)\left(\dfrac{1}{8}\right)\left(\dfrac{4}{7}\right)\left(\dfrac{3}{6}\right)\left(\dfrac{2}{5}\right)\left(\dfrac{1}{4}\right)\left(\dfrac{3}{3}\right)\left(\dfrac{2}{2}\right)(1) = \dfrac{1}{1260}$

Sol 2: Given, $P(A).P(B) = \dfrac{1}{6}$, $P(\bar{A}).P(\bar{B}) = \dfrac{1}{3}$

$\therefore [1 - P(A)]\,[1 - P(B)] = \dfrac{1}{3}$

Let $P(A) = x$ and $P(B) = y$

$\Rightarrow (1 - x)(1 - y) = \dfrac{1}{3}$ and $xy = \dfrac{1}{6}$

$\Rightarrow 1 - x - y + xy = \dfrac{1}{3}$ and $xy = \dfrac{1}{6}$

$\Rightarrow x + y = \dfrac{5}{6}$ and $xy = \dfrac{1}{6}$

$\Rightarrow x\left(\dfrac{5}{6} - x\right) = \dfrac{1}{6}$

$\Rightarrow 6x^2 - 5x + 1 = 0 \Rightarrow (3x - 1)(2x - 1) = 0$

$\Rightarrow x = \dfrac{1}{3}$ and $\dfrac{1}{2}$

$\therefore P(A) = \dfrac{1}{3}$, $P(B) = \dfrac{1}{2}$

Sol 3: The total number of ways to answer the question

$= {}^4C_1 + {}^4C_2 + {}^4C_3 + {}^4C_4 = 2^4 - 1 = 15$

P(getting marks) = P(correct answer in I chance) + P(correct answer in II change) + P(Correct answer in III chance)

$= \dfrac{1}{15} + \left(\dfrac{14}{15}.\dfrac{1}{14}\right) + \left(\dfrac{14}{15}.\dfrac{13}{14}.\dfrac{1}{13}\right) = \dfrac{3}{15} = \dfrac{1}{5}$

Sol 4: Let $Q = 1 - P =$ probability of getting the tail. We have $\alpha =$ probability of A getting the head on tossing firstly

$= P\,(H_1 \text{ or } T_1T_2T_3H_4 \text{ or } T_1T_2T_3T_4T_5T_6H_7 \text{ or } \ldots..$

$= P(H)P(T) + P(H)P(T)^3 + P(H)P(T)^6 \ldots..$

$= \dfrac{P(H)}{1 - P(T)^3} = \dfrac{P}{1 - Q^3}$

Also $\beta =$ probability of B getting the heat on tossing secondly

$= P(T_1H_2 \text{ or } T_1T_2T_3T_4H_5 \text{ or } T_1T_2T_3T_4T_5T_6T_7H_8 \text{ or } \ldots..)$

$= P(H)\,P(T) + P(H)\,P(T)^4 + P(H)\,P(T)^7 + \ldots$

$= P(T)\,[P(H) + P(H)\,P(T)^3 + P(H)P(T)^6 + \ldots]$

$= Q\alpha = (1 - P)\alpha = \dfrac{P(1 - P)}{1 - Q^3}$

Again, we have $\alpha + \beta + \gamma = 1$

$\Rightarrow \gamma = 1 - (\alpha + \beta) = 1 - \dfrac{P + P(1 - P)}{1 - Q^3} = 1 - \dfrac{P + P(1 - P)}{1 - (1 - P)^3}$

$\gamma = \dfrac{1 - (1 - P)^3 - 2P + P^2}{1 - (1 - P)^3} = \dfrac{P - P^2 + P^3}{1 - (1 - P)^3}$

Also, $\alpha = \dfrac{P}{1 - (1 - P)^3}$, $\beta = \dfrac{P(1 - P)}{1 - (1 - P)^3}$

Sol 5: The total no. of outcomes $= 6^n$

We can choose three numbers out of 6 in ${}^6C_3$ ways. By using three numbers out of 6 we can get 3^n sequences of length n. But these sequences of length n which use exactly two umbers and exactly one number.

The number of n – sequences which use exactly two numbers

$= {}^3C_2\left[2^n - 1^n - 1^n\right] = 3\left(2^n - 2\right)$ and the number of n sequence which are exactly one number.

$= \left({}^3C_1\right)\left(1^n\right) = 3$

Thus, the number of sequences, which use exactly three numbers

$= {}^6C_3\left[3^n - 3\left(2^n - 2\right) - 3\right] = {}^6C_3\left[3^n - 3\left(2^n\right) + 3\right]$

$\therefore$ Probability of the required event.

$= {}^6C_3\left[3^n - 3\left(2^n\right) + 3\right]/6^n$

3.61

Sol 6: Let A_1 be the event exactly 4 white balls have been drawn. A_2. Be the event exactly 5 white balls have been drawn. A_3 be the event exactly 6 white balls have been drawn. B be the event exactly 1 white ball is drawn from two draws.

Then,

$$P(B) = P\left(\frac{B}{A_1}\right)P(A_1) + P\left(\frac{B}{A_2}\right)P(A_2) + P\left(\frac{B}{A_3}\right)P(A_3)$$

But $P\left(\dfrac{B}{A_3}\right) = 0$ ($\because$ there are only 6 white balls in the bag)

$$\therefore P(B) = P\left(\frac{B}{A_1}\right)P(A_1) + P\left(\frac{B}{A_2}\right)P(A_2)$$

$$= \frac{^{12}C_2 \cdot {}^{6}C_4}{^{18}C_6} \cdot \frac{^{10}C_1 \cdot {}^{2}C_1}{^{10}C_2} + \frac{^{12}C_1 \cdot {}^{6}C_5}{^{18}C_6} \cdot \frac{^{11}C_1 \cdot {}^{1}C_1}{^{12}C_2}$$

Sol 7: As, the statement shows problem is to be related to Baye's law.

Law C, S, B, T be the events when when person is going by car, scooter, bus or train respectively.

$$\therefore P(C) = \frac{1}{7}, \ P(S) = \frac{3}{7}, \ P(B) = \frac{2}{7}, \ P(T) = \frac{1}{7}$$

Again, L be the event of the person reaching office late.

$\therefore \bar{L}$ be the event of the person reaching office in time.

Then, $P\left(\dfrac{\bar{L}}{C}\right) = \dfrac{7}{9}$, $P\left(\dfrac{\bar{L}}{S}\right) = \dfrac{8}{9}$, $P\left(\dfrac{\bar{L}}{B}\right) = \dfrac{5}{9}$

And $P\left(\dfrac{\bar{L}}{T}\right) = \dfrac{8}{9}$

$$\therefore P\left(\frac{C}{L}\right) = \frac{P\left(\dfrac{\bar{L}}{C}\right).P(C)}{P\left(\dfrac{\bar{L}}{C}\right).P(C) + P\left(\dfrac{\bar{L}}{S}\right).P(S) + P\left(\dfrac{\bar{L}}{B}\right).P(B) + P\left(\dfrac{\bar{L}}{T}\right).P(T)}$$

$$= \frac{\dfrac{7}{9} \times \dfrac{1}{7}}{\dfrac{7}{9} \times \dfrac{1}{7} + \dfrac{8}{9} \times \dfrac{3}{7} + \dfrac{5}{9} \times \dfrac{2}{7} + \dfrac{8}{9} \times \dfrac{1}{7}} = \frac{1}{7}$$

Sol 8 : (B) Here, $P(u_i) = K_i$, $\Sigma P(u_i) = 1$

$$\Rightarrow k = \frac{2}{n(n+1)}$$

$$\therefore \lim_{n\to\infty} P(W) = \lim_{n\to\infty} \sum_{i=t}^{n} \frac{2i^2}{n(n+1)^2}$$

$$= \lim_{n\to\infty} \frac{2n(n+1)(2n+1)}{6n(n+1)^2} = \frac{2}{3}$$

Sol 9: (A) $P\left(\dfrac{u_n}{W}\right) = \dfrac{\dfrac{n}{n+1}}{\dfrac{\Sigma i}{n+1}} = \dfrac{2}{n+1}$

Sol 10: (B, C, D) Since, E and F are independent events. Therefore $P(E \cap F) = P(E).P(F) \neq 0$, so E and F are not mutually exclusive events.

Now, $P(E \cap \bar{F}) = P(E) - P(E \cap F)$

$= P(E) - P(E).P(F) = P(E)[1 - P(F)] = P(E).P(\bar{F})$

and $P(\bar{E} \cap \bar{F}) = P(\overline{E \cup F}) = 1 - P(E \cup F)$

$= 1 - [1 - P(\bar{E}).P(\bar{F})]$

($\because$ E and F are independent)

$= P(\bar{E}).P(\bar{F})$

So, E and $\bar{F}$ as well and $\bar{E}$ and $\bar{F}$ are independent events.

Now,

$$P(E/F) + P(\bar{E}/F) = \frac{P(E \cap F) + P(\bar{E} \cap F)}{P(F)} = \frac{P(F)}{P(F)} = 1$$

Sol 11: (A, D) Both E and F happen $\Rightarrow P(E \cap F) = \dfrac{1}{12}$ and neither E nor F happens

$$\Rightarrow P(\bar{E} \cap \bar{F}) = \frac{1}{2}$$

But for independent events, we have

$$P(E \cap F) = P(E) \, P(F) = \frac{1}{12} \qquad \qquad \text{..... (i)}$$

and $P(\bar{E} \cap \bar{F}) = P(\bar{E})P(\bar{F})$

$= \{1 - P(E)\} \{(1 - P(F)\}$

$= 1 - P(E) - P(F) + P(E)P(F)$

$$\Rightarrow \frac{1}{2} = 1 - \{P(E) + P(F)\} + \frac{1}{12}$$

$$P(E) + P(F) = 1 - \frac{1}{2} + \frac{1}{12} = \frac{7}{12} \qquad \text{..... (ii)}$$

On solving Equation (i) and (ii), we get

either $P(E) = \dfrac{1}{3}$ and $P(F) = \dfrac{1}{4}$ or $P(E) = \dfrac{1}{4}$ and $P(F) = \dfrac{1}{3}$

Sol 12: (A, D) $P(E/F) + P(\bar{E}/F) = \dfrac{P(E \cap F)}{P(F)} + \dfrac{P(\bar{E} \cap F)}{P(F)}$

$= \dfrac{P(E \cap F) + P(\bar{E} \cap F)}{P(F)} = \dfrac{P(F)}{P(F)} = 1$

(b) $P(E/F) + P(E/\bar{F})$

$= \dfrac{P(E \cap F)}{P(F)} + \dfrac{P(E \cap \bar{F})}{P(F)}$

$= \dfrac{P(E \cap F)}{P(F)} + \dfrac{P(E \cap \bar{F})}{1 - P(F)} \neq 1$

(c) $P(\bar{E}/F) + P(E/\bar{F}) = \dfrac{P(\bar{E} \cup F)}{P(F)} + \dfrac{P(E \cap \bar{F})}{P(\bar{F})}$

$= \dfrac{P(\bar{E} \cap F)}{P(F)} + \dfrac{P(E \cap \bar{F})}{1 - P(F)} \neq 1$

(d) $P(E/\bar{F}) + P(\bar{E}/\bar{F}) = \dfrac{P(E \cap \bar{F})}{P(\bar{F})} + \dfrac{P(\bar{E} \cap \bar{F})}{P(\bar{F})}$

$= \dfrac{P(E \cap \bar{F}) + P(\bar{E} \cap \bar{F})}{P(\bar{F})} = \dfrac{P(\bar{F})}{P(\bar{F})} = 1$

Sol 13: (A, D) $P(E \cup F) - P(E \cap F) = \dfrac{11}{25}$ ..... (i)

(i.e. only E or only F) Neither of them occurs $= \dfrac{2}{25}$

$\Rightarrow P(\bar{E} \cap \bar{F}) = \dfrac{2}{25}$ ..... (ii)

From Eq. (i) $P(E) + P(F) - 2P(E \cap F) = \dfrac{11}{25}$ ..... (iii)

From eq. (ii), $(1 - P(E))(1 - P(F)) = \dfrac{2}{25}$

$\Rightarrow 1 - P(E) - P(F) + P(E).P(F) = \dfrac{2}{25}$ ..... (iv)

From Eq. (iii) and (iv), we get

$P(E) + P(F) = \dfrac{7}{5}$ and $P(E).P(F) = \dfrac{12}{25}$

$\therefore P(E).\left\{\dfrac{7}{5} - P(E)\right\} = \dfrac{12}{25}$

$\Rightarrow (P(E))^2 - \dfrac{7}{5}P(E) + \dfrac{12}{25} = 0$

$\Rightarrow \left(P(E) - \dfrac{3}{5}\right)\left(P(E) - \dfrac{4}{5}\right) = 0$

$\therefore P(E) = \dfrac{3}{5}$ or $\dfrac{4}{5} \Rightarrow P(F) = \dfrac{4}{5}$ or $\dfrac{3}{5}$

Sol 14: (C) Let E = event when each American man is seated adjacent to his wife

A = event when Indian man is seated adjacent to his wife

Now $n(A \cap E) = (4!) \times (2!)^5$

Even when each American man is seated adjacent to his wife

Again $n(E) = (5!) \times (2!)^4$

$\Rightarrow P\left(\dfrac{A}{E}\right) = \dfrac{n(A \cap E)}{n(E)} = \dfrac{(4!) \times (2!)^5}{(5!) \times (2!)^4} = \dfrac{2}{5}$

Sol 15 : (D) Statement-I:

If $P(H_i \cap E) = 0$ for some I, then

$P\left(\dfrac{H_i}{E}\right) = P\left(\dfrac{E}{H_i}\right) = 0$

If $P(H_i \cap E) \neq 0$ for $\forall\ i = 1, 2.....n$ then

$P\left(\dfrac{H_i}{E}\right) = \dfrac{P(H_i \cap E)}{P(H_i)} \times \dfrac{P(H_i)}{P(E)}$

$= \dfrac{P\left(\dfrac{E}{H_i}\right) \times P(H_i)}{P(E)} > P\left(\dfrac{E}{H_i}\right) \cdot P(H_i)$ [as $0 < P(E) < 1$]

Hence, Statement-I may not always be true.

Statement-II: Clearly $H_1 \cup H_2 ..... \cup H_n = S$ (sample space)

$\Rightarrow P(H_1) + P(H_2) + ... + P(H_n) = 1$

Sol 16 : (C) COCHIN

The second place can be filled in $^4C_1$ ways and the remaining four alphabets can be arranged in 4! Ways in four different places. The next 97th word will be COCHIN

Hence, there are 96 word before COCHIN.

Sol 17: (C) $P\left(\dfrac{E^c \cap F^c}{G}\right) = \dfrac{P(E^c \cap F^c \cap G)}{P(G)}$

$$= \frac{P(G) - P(E \cap G) - P(G \cap F)}{P(G)}$$

$$= \frac{P(G)(1 - P(E) - P(F))}{P(G)} \qquad [\because P(G) \neq 0]$$

$$= 1 - P(E) - P(F)$$

$$= P(E^c) - P(F)$$

Sol 18: (D) $P(A \cap B) = \frac{4}{10} \times \frac{p}{10} = \frac{2p/5}{10}$

$\Rightarrow \quad \frac{2p}{5}$ is an integer

$\Rightarrow \quad = 5$ or 10

Sol 19: $A \rightarrow p;\ B \rightarrow s;\ C \rightarrow q;\ D \rightarrow q$

(A) ENDEA, N, O, E, L are five different letter, then permutation $= 5!$

(B) If E is in the first and last position then

$\frac{(9-2)!}{2!} = 7 \times 3 \times 5! = 2! \times 5!$

(C) For first four letters $= \frac{4!}{2!}$

For last five letters $= 5!/3!$

Hence $\frac{4!}{2!} \times \frac{5!}{3!} = 2 \times 5!$

(D) For A, E and O $5!/3!$ And for others $4!/2!$

Hence $\frac{5!}{3!} \times \frac{4!}{2!} = 2 \times 5!$

Sol 20 : (C) Coefficient of x^{10} in $(x + x^2 + x^3)^7$

Coefficient of x^3 in $(1 + x + x^2)^7$

Coefficient of x^3 in $(1 - x^3)^7 (1 - x)^{-7}$

$= {}^{7+3-1}C_3 - 7$

$= {}^9C_3 - 7$

$= \frac{9 \times 8 \times 7}{6} - 7 = 77$

Sol 21 : (A) $P(X = 3) = \left(\frac{5}{6}\right)\left(\frac{5}{6}\right)\frac{1}{6} = \frac{25}{216}$

Sol 22: (B) $\frac{25}{216}$

Required probability $= 1 - \frac{11}{36} = \frac{25}{36}$

Sol 23 : (D) For $X \geq 6$, the probability is

$$\frac{5^5}{6^6} + \frac{5^6}{6^7} + \dots \infty = \frac{5^5}{6^6}\left(\frac{1}{1 - 5/6}\right) = \left(\frac{5}{6}\right)^5$$

For $X \geq 3$

$$\frac{5^3}{6^4} + \frac{5^4}{6^5} + \frac{5^5}{6^6} + \dots \infty = \left(\frac{5}{6}\right)^3$$

Hence the conditional probability $\dfrac{\left(\frac{5}{6}\right)^6}{\left(\frac{5}{6}\right)^3} = \dfrac{25}{36}$.

Sol 24: (C) Event G = original signal is green

E_1 = A receives the signal correct

E_2 = B receives the signal correct

E = signal received by B is green

P(signal received by B is green)

$= P(GE_1E_2) + P(GE_1\bar{E}_2) + P(G\bar{E}_1E_2) + P(\bar{G}E_1\bar{E}_2) + P(\bar{G}\bar{E}_1E_2)$

$P(E) = \dfrac{46}{5 \times 16}$

$P(G/E) = \dfrac{40/5 \times 16}{46/5 \times 16} = \dfrac{20}{23}$

Sol 25: (B) $H \rightarrow$ ball from U_1 to U_2

$T \rightarrow$ 2 ball from U_1 to U_2

E : 1 ball drawn from U_2

P/W from $U_2 = \dfrac{1}{2} \times \left(\dfrac{3}{5} \times 1\right) + \dfrac{1}{2} \times \left(\dfrac{2}{5} \times \dfrac{1}{2}\right) + \dfrac{1}{2}$

$\times \left(\dfrac{{}^3C_2}{{}^5C_2} \times 1\right) + \dfrac{1}{2} \times \left(\dfrac{{}^2C_2}{{}^5C_2} \times \dfrac{1}{3}\right) + \dfrac{1}{2}$

$\times \left(\dfrac{{}^3C_1 \cdot {}^2C_1}{{}^5C_2} \times \dfrac{2}{3}\right) = \dfrac{23}{30}$

Sol 26: (D) $P\left(\dfrac{H}{W}\right) = \dfrac{P(W/H) \times P(H)}{P(W/T) \cdot P(T) + (W/H) \cdot P(H)}$

$= \dfrac{\dfrac{1}{2}\left(\dfrac{3}{5} \times 1 + \dfrac{2}{5} \times \dfrac{1}{2}\right)}{23/30} = \dfrac{12}{23}$

Sol 27: (B) Number of ways

$$= 3^5 - {}^3C_1 \cdot 2^5 + {}^3C_2 1^5$$

$$= 243 - 96 + 3 = 150$$

Sol 28: (B, D) $P(X_1) = \dfrac{1}{2}, P(X_2) = \dfrac{1}{4}, P(X_3) = \dfrac{1}{4}$

$P(X) = P(X_1 \cap X_2 \cap X_3^C) + P(X_1 \cap X_2^C \cap X_3)$

$+P(X_1^C \cap X_2 \cap X_3) + P(X_1 \cap X_2 X_3) = \dfrac{1}{4}$

(A) $P(X_1^C / X) = \dfrac{P(X \cap X_1^C)}{P(X)} = \dfrac{\frac{1}{32}}{\frac{1}{4}} = \dfrac{1}{8}$

(B) P [exactly two engines of the ship are

functioning $\mid$ X] $= \dfrac{\frac{7}{32}}{\frac{1}{4}} = \dfrac{7}{8}$

(C) $P\left(\dfrac{X}{X_2}\right) = \dfrac{\frac{5}{32}}{\frac{1}{4}} = \dfrac{5}{8}$

(D) $P\left(\dfrac{X}{X_1}\right) = \dfrac{\frac{7}{32}}{\frac{1}{2}} = \dfrac{7}{16}$

Sol 29: (A) Favourable : D_4 shows a number and only 1 of $D_1 \, D_2 \, D_3$ shows same number

Or only 2 of $D_1 \, D_2 \, D_3$ shows same number

Or all 3 of $D_1 \, D_2 \, D_3$ shows same number

Required probability

$$= \dfrac{{}^6C_1({}^3C_1 \times 5 \times 5 + {}^3C_2 \times 5 + {}^3C_3)}{216 \times 6}$$

$$= \dfrac{6 \times (75 + 15 + 1)}{216 \times 6} = \dfrac{6 \times 91}{216 \times 6} = \dfrac{91}{216}$$

Sol 30: (A, B) $P(X / Y) = \dfrac{1}{2}$

$\dfrac{P(X \cap Y)}{P(Y)} = \dfrac{1}{2} \Rightarrow P(Y) = \dfrac{1}{3}$

$P(Y / X) = \dfrac{1}{3}$

$\dfrac{P(X \cap Y)}{P(X)} = \dfrac{1}{3} \Rightarrow P(X) = \dfrac{1}{2}$

$P(X \cup Y) = P(X) + P(Y) - P(X \cap Y) = \dfrac{2}{3}$ \qquad (A is correct)

$P(X \cap Y) = P(X) \cdot P(X) \Rightarrow X$ and Y are

independent (B is correct)

$P(X^c \cap Y) = P(Y) - P(X \cap Y)$

$= \dfrac{1}{3} - \dfrac{1}{6} = \dfrac{1}{6}$ (D is not correct)

Sol 31: (A) P (at least one of them solves correctly) = 1 – P (none of them solves correctly)

$$= 1 - \left(\dfrac{1}{2} \times \dfrac{1}{4} \times \dfrac{3}{4} \times \dfrac{7}{8}\right) = \dfrac{235}{256}$$

Sol 32: (6) Let $P(E_1) = x$, $P(E_2) = y$ and $P(E_3) = z$

Then $(1 - x)(1 - y)(1 - z) = p$

$(1 - x)y(1 - z) = \alpha$

$(1 - x)y(1 - z) = \beta$

$(1 - x)(1 - y)(1 - z) = \gamma$

so $\dfrac{1 - x}{x} = \dfrac{p}{\alpha}$ \qquad $x = \dfrac{\alpha}{\alpha + p}$

Similarly $z = \dfrac{\gamma}{\gamma + p}$

So, $\dfrac{P(E_1)}{P(E_3)} = \dfrac{\frac{\alpha}{\alpha + p}}{\frac{\gamma}{\gamma + p}} = \dfrac{\frac{\gamma + p}{\gamma}}{\frac{\alpha + p}{\alpha}} = \dfrac{1 + \frac{p}{\gamma}}{1 + \frac{p}{\alpha}}$

Also given $\dfrac{\alpha\beta}{\alpha - 2\beta} = p = \dfrac{2\beta\gamma}{\beta - 3\gamma} \Rightarrow \beta = \dfrac{5\alpha\gamma}{\alpha + 4\gamma}$

Substituting back $\left(\alpha - 2\left(\dfrac{5\alpha\gamma}{\alpha + 4\gamma}\right)\right)p = \dfrac{\alpha \cdot 5\alpha\gamma}{\alpha + 4\gamma}$

$\Rightarrow \quad \alpha p - 6p\gamma = 5\alpha\gamma$

$\Rightarrow \quad \left(\dfrac{p}{\gamma} + 1\right) = 6\left(\dfrac{p}{\alpha} + 1\right) \Rightarrow \dfrac{\frac{p}{\gamma} + 1}{\frac{p}{\alpha} + 1} = 6$

Sol 33: (5)

Clearly, $1 + 2 + 3 + \dots + n - 2 \le 1224 \le 3 + 4 + \dots n$

$\Rightarrow \quad \dfrac{(n-2)(n-1)}{2} \le 1224 \le \dfrac{(n-2)}{2}(3+n)$

$\Rightarrow \quad n^2 - 3n - 2446 \le 0$ and $n^2 + n - 2454 \ge 0$

$\Rightarrow 49 < n < 51 \Rightarrow n = 50$

$\therefore \dfrac{n(n+1)}{2} - (2k+1) = 1224 \Rightarrow k = 25 \Rightarrow k - 20 = 5$

Sol 34: (A) P (required) = P (all are white) + P (all are red) + P (all are black)

$= \dfrac{1}{6} \times \dfrac{2}{9} + \dfrac{3}{12} + \dfrac{3}{6} \times \dfrac{3}{9} \times \dfrac{4}{12} + \dfrac{2}{6} \times \dfrac{4}{9} \times \dfrac{5}{12}$

$= \dfrac{6}{648} + \dfrac{36}{648} + \dfrac{40}{648} = \dfrac{82}{648}$

Sol 35: (D) Let A : one ball is white and other is red

E_1 : both balls are from box B_1

E_2 : both balls are from box B_2

E_3 : both balls are from box B_3

Here, P (required) $= P\left(\dfrac{E_2}{A}\right)$

$= \dfrac{P\left(\dfrac{A}{E_2}\right) \cdot P(E_2)}{P\left(\dfrac{A}{E_1}\right) \cdot P(E_1) + P\left(\dfrac{A}{E_2}\right) \cdot P(E_2) + P\left(\dfrac{A}{E_3}\right) \cdot P(E_3)}$

$= \dfrac{\dfrac{{}^2C_1 \times {}^3C_1}{{}^9C_2} \times \dfrac{1}{3}}{\dfrac{{}^1C_1 \times {}^3C_1}{{}^6C_2} \times \dfrac{1}{3} + \dfrac{{}^2C_1 \times {}^3C_1}{{}^9C_2} \times \dfrac{1}{3} + \dfrac{{}^3C_1 \times {}^4C_1}{{}^{12}C_2} \times \dfrac{1}{3}}$

$= \dfrac{\dfrac{1}{6}}{\dfrac{1}{5} + \dfrac{1}{6} + \dfrac{2}{11}} = \dfrac{55}{181}$

Sol 36: (A) Either a girl will start the sequence or will be at second position and will not acquire the last position as well.

Required probability $= \dfrac{({}^3C_1 + {}^3C_1)}{{}^3C_2} = \dfrac{1}{2}$

Sol 37: (C) Number of required ways

$= 5! - \{4 \cdot 4! - {}^4C_2 \cdot 3! + {}^4C_3 \cdot 2! - 1\} = 53$

Sol 38: (B) Case-I : One odd, 2 even

Total number of ways $= 2 \times 2 \times 3 + 1 \times 3 \times 3 + 1 \times 2 \times 4 = 29$

Case-II: All 3 odd

Number of ways $= 2 \times 3 \times 4 = 24$

Favourable ways = 53

Required probability $= \dfrac{53}{3 \times 5 \times 7} = \dfrac{53}{105}$

Sol 39: (C) Here $2x_2 = x_1 + x_3$

$\Rightarrow \quad x_1 + x_3 = $ even

Hence number of favorable ways

$= {}^2C_1 \cdot {}^4C_2 + {}^1C_1 \cdot {}^3C_1 = 11$

Sol 40: (8) Let coin was tossed 'n' times

Probability of getting atleast two heads $= 1 - \left[\dfrac{1}{2^n} + \dfrac{n}{2^n}\right]$

$\Rightarrow \quad 1 - \left[\dfrac{n+1}{2^n}\right] \geq 0.96$

$\Rightarrow \quad \dfrac{2^n}{n+1} \geq 25 \quad \Rightarrow \quad n \geq 8$

Sol 41: (5) $n = 6! \cdot 5!$ (5 girls together arranged along with 5 boys)

$m = {}^5C_4 \cdot (7! - 2 \cdot 6!) \cdot 4!$

(4 out of 5 girls together arranged with others – number of cases all 5 girls are together)

$\dfrac{m}{n} = \dfrac{5 \cdot 5 \cdot 6! \cdot 4!}{6! \cdot 5!} = 5$

Sol 42: (A, B) P (Red ball) $= P(I) \cdot P(R \mid I) + P(II) \cdot P(R \mid II)$

$P(II \mid R) = \dfrac{1}{3} = \dfrac{P(II) \cdot P(R \mid II)}{P(I) \cdot P(R \mid I) + P(II) \cdot P(R \mid II)}$

$\dfrac{1}{3} = \dfrac{\dfrac{n_3}{n_3 + n_4}}{\dfrac{n_1}{n_1 + n_2} + \dfrac{n_3}{n_3 + n_4}}$

Of the given options, A and B satisfy above condition

Sol 43: (C, D) P (Red after Transfer) = P(Red Transfer) . P(Red Transfer in II Case) + P (Black Transfer) . P(Red Transfer in II Case)

$P(R) = \dfrac{n_1}{n_1 + n_2} \cdot \dfrac{(n_1 - 1)}{(n_1 + n_2 - 1)} + \dfrac{n_2}{n_1 + n_2} \cdot \dfrac{n_1}{n_1 + n_2 - 1} = \dfrac{1}{3}$

Of the given options, option C and D satisfy above condition.

Sol 44: (C) $P(T_1) = \dfrac{1}{5}$, $P(T_2) = \dfrac{4}{5}$, $P(D) = \dfrac{7}{100}$

$$P\left(\dfrac{D}{T_1}\right) = 10 \cdot P\left(\dfrac{D}{T_2}\right). \quad \text{Let } P\left(\dfrac{D}{T_2}\right) = x$$

Now, $P(T_1) \times P\left(\dfrac{D}{T_1}\right) + P(T_2) \cdot P\left(\dfrac{D}{T_2}\right) = \dfrac{7}{100}$

$$= \dfrac{1}{5} \times 10x + \dfrac{4}{5} \times x = \dfrac{7}{100} \Rightarrow x = \dfrac{1}{40}$$

$$\therefore \quad P\left(\dfrac{T_2}{D}\right) = \dfrac{\dfrac{4}{5} \times \dfrac{39}{40}}{\dfrac{93}{100}} = \dfrac{78}{93}$$

Sol 45: (A) $= {}^6C_3 \times {}^4C_1 \times 4 + {}^6C_4 \times = 380$

Sol 46: (B) $P(X > Y) = \left(\dfrac{1}{2} \times \dfrac{1}{2}\right) + \left(\dfrac{1}{2} \times \dfrac{1}{6}\right) + \left(\dfrac{1}{6} \times \dfrac{1}{2}\right) = \dfrac{5}{12}$

Sol 47: (C) $P(X = Y) = \left(\dfrac{1}{2} \times \dfrac{1}{3} \times 2\right) + \left(\dfrac{1}{6} \times \dfrac{1}{6}\right) = \dfrac{13}{36}$